Y. Imanishi (Ed.)

Progress in
Pacific Polymer Science 2

Proceedings of the Second Pacific
Polymer Conference
Otsu, Japan, November 26–29, 1991

Springer-Verlag
Berlin Heidelberg NewYork
London Paris Tokyo
Hong Kong Barcelona Budapest

Prof. Dr. Y. Imanishi
Department of Polymer Chemistry
Kyoto University
Kyoto 606
Japan

ISBN-13: 978-3-642-77638-0   e-ISBN-13: 978-3-642-77636-6
DOI: 10.1007/978-3-642-77636-6

02/3020-543210 – Printed on acid-free paper

Preface

This book is a collection of addresses of the keynote speakers and invited lecturers as well as manuscripts of papers which
were delivered in "Polymer Science and the Arts" session at the
Second Pacific Polymer Conference organized by the Pacific Polymer Federation at the Otsu Prince Hotel in the Shiga Prefecture,
Japan, 26-29 November, 1991.

The Second Pacific Polymer Conference was a multi-topic
conference that covered a wide range of the research fields of
macromolecular science and technology. Five keynote and thirty-
five invited lectures were delivered at the conference by eminent
polymer scientists both from academic and industrial fields.
The conference was attended by 378 scientists including 130
from overseas. The conference provided an important opportunity
for the interaction between the academic and industrial polymer
scientists, and promoted polymer science and technology in all
the Pacific Basin Countries.

This book, therefore, provides a wide-angle snapshot of the
polymer research in the early 1990s. In particular, the papers
presented in the "Polymer Science and the Arts" session undoub-
tedly give perspective to the future aspects of polymer science
and technology in the 21st century. It is a useful book for all
scientists interested in polymers and the progress of the science
in the countries of the Pacific Basin. The editor hopes that many
attendees were stimulated by the meeting and that new ideas and
new collaborations will result in a further enrichment of research, and lead to new useful polymers for all countries.

The editor would like to express his most sincere appreciation to all the speakers for the excellent talks that they presented at the Second Pacific Polymer Conference and the high-qua-
lity papers that they submitted for publication.

June 1992   Y. Imanishi

Contents

VIII

Microcomposite Formation of p-Aramid with Inorganic Glass and Conductive Polymers

M. Takayanagi

Kyushu Sangyo University, Faculty of Engineering
Matsukadai, Higashi-Ku, Fukuoka, 813 Japan

Abstract: Recent trends in research of polymer alloys are classified in (1) orthodox approach to polymer alloys and (2) singular (out-of-orthodox) approach to polymer alloys. The development in orthodox approach is reviewed on the preparative methods, the prediction of alloy morphology and the relationships between morphology and properties. As singular approach to polymer alloys, the development of molecular composites and the microcomposites are reviewed. The field of microcomposites are mentioned more in detail. The microcomposites of p-aramid film with inorganic glass or electroconductive polymers are taken up as a special topic. The rigid rod-like polymers form a film with controlled void content, having a very low critical diffusion volume at the percolation threshold, which enables the infiltration of functional monomers. Technological applications are suggested for the future based on these knowledges.

INTRODUCTION

The Second Pacific Polymer Conference is held this time as its objective to promote polymer science and technology throughout all the Pacific Basin Countries. To exchange the informations on the new facts and truth found in the field of polymer science among these countries will be effective in accelerating the creative activities of the coming generation to realize the prosperous societies around the Basin.

The title of this paper does not cover an extensive area of polymer science, but only a limited field of it. However, there is an extensive background in the naming of the title: "microcomposite" was born from the concepts of polymer alloy and the molecular composites, "p-aramid" symbolizes the rigid rod-like molecule which is an epoch-making topic in the history of polymers in its behavior quite different from the flexible coil-like molecule, the "inorganic glass" is recently recognized as an attractive partner of organic polymers, and the "electroconductive polymers" have already been manufactured in Japan as an electrode of plastic battery, representing various kinds of functional polymers.

Y. Imanishi (Ed.)
Progress in Pacific Polymer Science 2
© Springer-Verlag Berlin Heidelberg 1992

RECENT DEVELOPMENT IN ORTHODOX APPROACH TO POLYMER ALLOY

Polymer alloy is attracting the interests of both the polymer scientists and engineers to its versatility in combinations of different types of polymers and improved properties of base polymers. The methods of preparation of polymer alloy are grossly divided in the miscible systems and the immiscible systems. The main subject of research in the miscible system is the prediction of miscibility of molecularly designed homo- and copolymers, based on the thermodynamical theories such as the lattice model by Flory [1] and the lattice fluid model by Sanchez and Lacombe [2]. The research subjects in the immiscible systems are concentrated to the molecular design of compatibilizers. The finding of appropriate compatibilizer realizes high performance blends never seen before. Another important factor in the success of blending is the introduction of reactive processing with twin screw extruder, with which continuous, systematic and most efficient blending is realized, even capable of control of chemical reaction and morphology of the blend. In addition to it, the block and graft copolymerizations including anion copolymerization have become industrial techniques. Thus, it may be said that we are now in the revolutional age of free use of existing polymers, which is comparable to the finding of oil colors after the ages of mosaic and tempera.

Development of Preparative Methods

Miscible systems: Recently theoretical evaluation of the interaction parameter has been extended to the systems in which the classical consideration does not allow the miscibility. The homopolymer composed of monomer A is denoted by the component 1, and the random copolymer composed of monomers C and D is denoted by the component 2. The interaction parameter between the components of 1 and 2, is evaluated by the plus contributions from the interaction parameters of the combinations of A-C and A-D, and the minus contribution from the volume-cross term for the interaction parameter of C and D. When the repulsive force between the monomeric units of C and D is large enough, the minus term is increased and the blends of 1 and 2 become miscible in the limited composition of the copolymer of C-co-D. Brinke and Karasz [3] proposed the mean field approximation to explain the miscibility for the systems including the above case. Ikawa et al. [4] further extended this approximation to predict the miscibility of the terpoymer/copolymer and the terpolymer/homopolymer and proved its propriety from the experiment.

The kinds of molecular species of copolymer units in the above examples are confined only to three. To extend the kinds more than three is an interesting field. The completion of the data base of the interaction parameters for various combinations of polymer units is useful for a more sophisticated molecular design of copolymers. Practically, for example, the improvement in heat-endurance of the matrix homopolymer is realized by adding to it the well-designed miscible copolymer, still keeping the dispersed state of rubber particles as it is. The decrease in impact strength by adding the copolymer must be avoided.

Immiscible Systems: Recently, the molecular design of compatibilizer to remove the distinct interface and to provide gradient interphase is widely employed in the polymer technologies. The SEM observation of the fractured surface of the blend using a compatibilizer shows the smeared interfacial region, being incapable of detecting the interface [5]. As a technologically successful case, the blend using nylon 6 or 66 as one component and poly(α-olefin) as another component is cited. Maleic anhydride-grafted polypropylene reacts with amine-terminated nylon 6 through the reaction of acid-amide formation. With increasing fraction of maleic anhydride the impact strength is increased [6]. The development of the blend of polyphenylene ether with nylon by using a compatibilizer is one of the expected technologies of polymer alloy.

Reactive processing: In the relation to the above example, the recent development of reactive processing must be cited. Monsanto company has developed "Santoprene", thermoplastic elastomer, which is manufactured by preparing the maleic anhydride-modified polypropylene and reacting with amine-terminated nitrile butadiene rubber to form the compatibilizing block copolymers in-situ during mixing in the screw extruder [7]. The advantages of reactive processing is said to be (1) continuous and small volume, (2) no solvent, being environmental and energy saving, (3) flexible in combination with other processes, and (4) controllable in chemical reaction and morphology for manufacturing high performance polymeric materials.

Prediction of Alloy Morphology

Neumann's Triangle: Nakamura and Inoue [8] applied the concept of Neumann's triangle to discuss the multi-phase blend morphology developed by melt-processing of a ternary blend consisting of polymers (A, B and C) which are immiscible with each other in the forms of homopolymers. For a C polymer rich system (the matrix is C polymer), four morphologies were predicted:

(a) the isolated particles of A and B, (b) the particles composed of A and B stuck together, (c) the hybrid particle of A encapsulated by B, and (d) the hybrid particle of B encapsulated by A on the basis of the balance between interfacial tensions. They further confirmed their predictions by the droplet-sandwich experiment. The parameters of Neumann's triangle could discriminate among four morphologies, while the spreading coefficient proposed by Hobbs et al. in 1988 [8] could not discriminate between the hybrid stuck particle and the separated particles of A and B. To accomplish a full list of interfacial tension for various pairs of polymers will be useful to predict the morphologies of polymer alloys.

Percolation theory applied to the two-phase system: It is a well known fact that the transition of modulus takes place in the two-phase system from a rubbery state to a glassy state at the critical region (not a point) as a function of the composition. Morphologically the transition is understood as the reverse of phases. Recently by introducing the concept of percolation concept, the transition composition is defined by the elastic percolation threshold. The scaling rule proposed by de Gennes [9] is applied to such a variation of modulus, using the critical composition at the elastic percolation threshold.

Morphology and Properties Relationship in Alloys

Brittle-to-tough transition in rubber-toughened plastics: Wu[10] found a generalized criterion for the brittle-to-tough transition for the particulate rubber toughening of pseudoductile polymer matrix such as nylon, according to which the critical matrix ligament thickness is a single parameter for brittle-to-tough transition, being independent of rubber fraction and particle size. The ligament distance is defined as the interspherical surface-to-surface distance of rubber particles, which is equal to the center-to-center interparticle distance minus the diameter of rubber particle. When Izod impact strength is plotted versus the matrix ligament distance, the transition is found to occur at a single critical value, which is independent of rubber volume fraction and particle size. He assumes that the thin ligaments form a network with sufficient connectivity for the yielding process of the matrix to propagates over the entire system at the transition.

Formulation of the brittle-to-tough transition as a percolation problem:

Margolina and Wu [11] introduced the concept of percolation to formulate the above phenomenon and analyzed the impact values in the tough region as a function of excess stress volume fraction over the critical volume fraction with the scaling rule (de Gennes [9]). Stress volume is defined by the spherical domain with the diameter of ligament distance plus rubber particle diameter, which was used to calculate the percolation threshold of the stress spheres. The percolation threshold refers to the onset of the first-path connectivity through the entire system.

Both logarithmic plots of Izod impact strength versus excess volume fraction over the critical stress volume fraction determined by the critical ligament distance gives the slope of 0.45, which is comparable with the critical exponent of 0.44 calculated for monodisperse particles by Bug and coworkers [12].

Digital image analysis applied to morphologies of polymer alloys: Tanaka and coworkers [13] presented new techniques of digital image analysis (DIA) which are suitable for analyzing particle-distributing patterns. According to them, P parameter is the index for representing the degree of homogeneity of the distribution of the particle as a point in the matrix. They further proceeded to analyze the pattern of droplets whose volume is not negligible and provided the parameter to express the distribution of the dispersed particle size, which is called S parameter. For the blend system of polyamide, ethylene-propylene-rubber (EPR) and modified EPR, it was shown that the tensile strength correlates with P parameter and the Izod impact strength correlates with S parameter [14]. The effectiveness of DIA in predicting the morphology/properties relationship of polymer alloy remains for the future.

SINGULAR (OUT-OF-ORTHODOX) APPROACH TO POLYMER ALLOY

Molecular Composites

Molecular composites as an extension of fiber reinforcement: Molecular composites is designed to use rigid rodlike molecules as reinforcement for the flexible coil molecules as matrix. The patent applications on the molecular composites were made almost in the same age independently by Takayanagi in Japan in 1977 and by Helminiak in the United States in 1978. Takayanagi proposed thermoplastic nylon reinforced by poly(p-phenylene terephthalamide)(PPTA) and Helminiak proposed wet process using poly(p-phenylene benzobisthiazole)(PBT)-reinforced poly(2,5(6)benzimidazole) (ABPBI). In molecular composite (MC) [15,16,17], the fineness of reinforcement was pursued to its limit, i.e. to the molecular dimension.

However, in reality, in unmodified PPTA, only the microfibrillar dimension
of 10 nm in diameter was attained. The rod-like molecules dispersed in
molecular level are expected to reinforce the matrix flexible molecules if
the molecular interaction between both components is strong enough. The
merits of MC are expected in the large aspect ratio of rod-like molecules.
Another possible merit is in the realization of ideal valence bond strength
in the main chain, which is free from any defects associated with the
super-structure of macroscopic fiber.

The preparation method of MC is to extrude the dilute solution of rod-like
molecule into a coagulant to avoid crystal formation as far as possible.
Liquid crystal formation of the dope is unprofitable in the meaning of
strong tendency of rod molecules to crystallize in thick molecular bundles
with low aspect ratio. Strong acids as solvent are also disadvantageous to
decompose the matrix flexible molecules. To find out a new solvent dis-
solving both rigid and flexible molecules is of primary importance.
Takayanagi-Katayose [17] found that sodium hydride (NaH) and dimethyl-
sufoxide (DMSO) form sodium methylsulfinylcarbanion and this reagent diffuses into freshly polyemrized PPTA to substitute the amide proton with
sodium ion, which dissolves PPTA in DMSO to form a homogeneous solution.
DMSO is a strong solvent for various flexible polymers and an isotropic
solution of ternary system is easily prepared.

The inspection of fractured surface of MC of PPTA/nylon 6 revealed the triangular broken edges of PPTA microfibril with 30 nm in diameter pulled out
from the smooth surface of nylon matrix, indicating the microfibrillar
reinforcement in MC. In spite of these observations, the stress-strain
curve of MC showed embrittlement owing to the heterogeneity in the
morphology, although the tensile modulus and strength were improved.

Distinct improvement of MC in toughness was accomplished by employing the
block copolymers. The typical examples are the blends of PPTA-b-nylon 6,
66/nylon 6, 66 [18] and PPTA-b-butadiene rubber (BDR)/ABS resin [19]. In
the latter case, only 2.5 wt % of PPTA in MC improved the energy-to-
fracture by a factor of four in comparison with that of unmodified ABS
resin, while the blend employing homopolymer of PPTA was brittle. Thus,
the block copolymerization of PPTA with flexible matrix polymer blocks has
advantages of the removal of defects associated with heterogeneous texture
in MC and the increase in the degree of dispersion of PPTA block in MC.

Another conspicuous feature of the PPTA-MC using block copolymers is in the effectiveness of small quantity of reinforcement. The extension of the theory of fiber reinforcement to the molecular composite well predicts the fact that the strong molecular interaction between PPTA and the matrix polymer is always effective. Another approach to this problem will be to apply the percolation concept. According to Margolina-Wu [11], the critical volume fraction of stress sphere was evaluated as 0.42 and the scaling rule was proved to hold with this threshold value. In their case, the matrix ligament distance governed the limit of the stress transfer between the neighboring rubber particles. Bug and coworkers [20] calculated theoretically that the critical density for rods is proportional to the inverse of the excluded volume. The excluded volume of rigid rodlike molecule is far larger than that of sphere and the inverse of it is far smaller than sphere. In another words, the rigid rodlike molecules in highly dispersed state and capable of stress transfer through the matrix have a very small critical value. If the both ends of rod are connected with the ductile matrix chains, the stress transfer will be efficiently carried out, bringing about the effective reinforcement.

Molecular composites in terms of synthesis: The reinforcements are also formed by the in-situ reaction in the matrix polymers. Such methods are another way to pursue the limit of molecular dispersion of reinforcing materials. One example is found in the in-situ precipitation of reinforcing silica in polydimethylsiloxane networks with the sol-gel methods [21]. Another example is the direct polycondensation of p-aminobenzoic acid or p-hydroxybenzoic acid from their monomers in solutions of polyarylate [22]. Mechanical properties of the cast films indicated increase in modulus and tensile strength at elevated temperatures.

A reversed case is found, in which the matrix polymers are formed from the reinforcing rodlike molecule [23]: PPTA polyanion was used as the initiator for the anionic polymerization of acrylamide to form the nylon 3 matrix. Composite films showed greatly improved strength and modulus over unmodified nylon 3 with no loss of flexibility. The combination of various chemical reactions will bring many possibilities.

Micro-Composites

PPTA film with easily permeable texture: The rigid rodlike feature of PPTA molecule gives a PPTA film with rather coarse super-structure in a dried state if the annealing temperature is not so high. This fact was found by the diffusional behavior of low molecular weight compounds through the PPTA film. The PPTA film was prepared by electrodeposition of a solution of PPTA polyanion in DMSO, which was washed, dried and annealed [23]. Electrolysis-polymerization of pyrrole (Py) was carried out on an ITO glass plate covered with a PPTA film in a cell containing pyrrole and a supporting electrolyte in acetonitrile. Fig. 1 shows the current density measured as a function of time during polymerization of Py in the PPTA film annealed at various temperatures as indicated in the figure [24].

Fig. 1 shows that the maximum current density, I(max), is decreased with increasing annealing temperature and becomes almost zero for the films annealed above 250°C. This means that the diffusion of Py in the PPTA film annealed at 240°C corresponds to the diffusion percolation threshold. The density of the film was 1.384 g/ml. The observed density of Kevlar fiber is 1.49 g/ml, which was assigned to the microfibrils filling the space in the dried PPTA film. Then the critical fraction of diffusion volume, ϕ_c, is 0.071 for the PPTA film annealed at 240°C. The densities of the PPTA films annealed at 220, 200, 150 °C and the unannealed (dried at room temperature) film give the diffusion volumes of ϕ = 0.078, 0.081, 0.087 and 0.099, respectively. Fig. 2 shows the plot of logarithmic I(max) versus logarithmic increment of diffusion volume fraction over the critical one at the percolation threshold. The scaling rule holds with the exponent of 1.0, which supports the correctness of the application of the percolation concept to the diffusion of Py in the PPTA film. The critical value of diffusion volume fraction of 0.071 is very low compared with 0.42 of the stress-sphere percolation threshold by Margolina-Wu. The theoretical prediction by Bug et al.[20] for various systems of rods is that the percolation threshold decreases with increasing aspect ratio. The diffusion volume in the PPTA matrix will take a prolonged shape, its locating along with the axes of PPTA micro- fibrils. For infiltration of low molecular weight compounds into the diffusion space inside of the PPTA film, the texture of the film composed of rodlike molecules has an advantage over the coil molecules.

Microcomposite formation of PPTA film with silica: The electrodeposited PPTA film was dried at 100°C to 5 wt% of absorbed water, which was soaked in an ethanol solution of tetraethyl orthosilicate (TEOS) at 60 C for one day. Cross-linking polycondensation of silanol derived from TEOS and water

Table 1 Comparison of mechanical properties of PPTA filmes before and after sol-gel glass formation.

	Sample A			Sample B			
	Glass /wt%	E/GPa	σ_B/MPa	Glass /wt%	E/GPa	σ_B/MPa	Thermal shrinkage/%
PPTA	–	5.56	170	–	3.45	160	2.03
PPTA/Glass	13.0	6.00	205	8.3	5.26	220	0.17

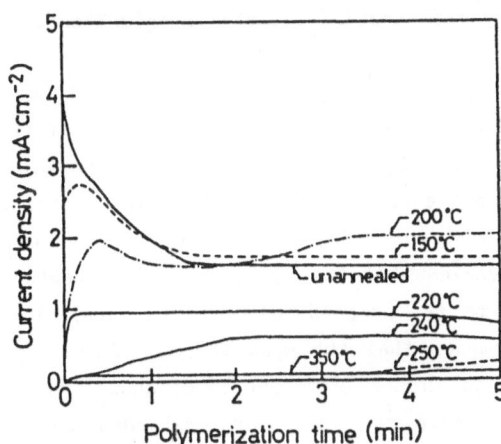

Fig. 1 Current density vs. time for polypyrrole polymerization at various annealing temperatures as indicated in the figure.

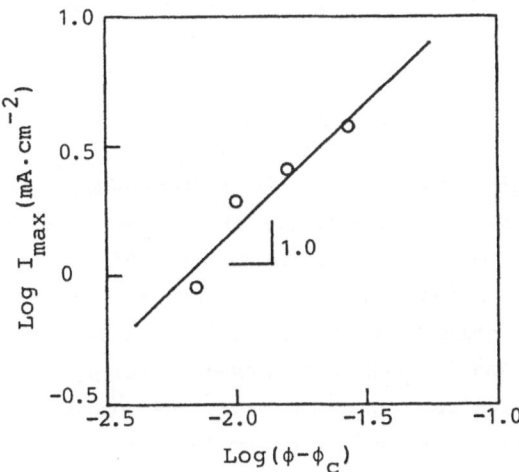

Fig. 2 Both logarithmic plots of maximum current density vs. excess diffusion volume over the critical diffusion volume evaluated from density.

in the presence of trace of nitric acid was completed by annealing at 370 °C
for 1 min.   IR absorption spectrum confirmed the formation of silica.   The
weight fraction of glass in the film was determined by the thermogravimetric analysis.   Table 1 compares the mechanical properties and thermal
shrinkage of the composite film.   Thermal shrinkage was improved to 1/10 of
that of the PPTA film by the composite formation.   Tensile strength and
modulus were raised, and their maxima were found at the glass content of 5
wt%.   The reason for reinforcement at 5 wt% glass is ascribed to the
removal of defects by infiltrating the silica into the defects in the PPTA
film.

Fig. 3 shows the SEM photograph of the composite film of PPTA/silica after
removal of PPTA with the reagent of NaH and DMSO.   The thin glass plate is
developed over the whole area and it is broken like a broken window glass.
The broken edge of glass plate forms vital defects to the tensile stress.
X-ray diffraction revealed that the hydrogen-bonded sheets show the orien-
tational variation from the perpendicular to the parallel alignment to the
film surface.   This means that the weak van der Waals force exerts along
the thickness direction in the later period of annealing and the growth of
silica tends to delaminate the interhydrogen-bonded sheets, resulting in
the development of glass sheet parallel to the film surface and the increase in thickness, for example, from 25 μm to 29 μm [26].   Lusignea et
al. [27] applied the sol-gel method for the reinforcing silica formation to
the PBT film and the compressive strength of the laminates of the
PBT/silica composite films was found largely improved.

Microcomposite formation of PPTA film with electroconductive polymers: The
gel membrane of PPTA prepared by electrodeposition of polyanion of PPTA was
dried on the ITO glass to use as a PPTA-covered electrode.   By using this
as an anode, pyrrole in acetonitrile or aniline in water was electro-
polymerized to give the microcomposite film of PPTA with electroconductive
polymers.   Table 2 shows the electroconductivities and mechanical
properties of the composite films [24, 28].   The conductivity is proportional to the fraction of conductive polymers.   Polypyrrole and polyaniline
prepared by using the perchlorate anion as the dopant did not give self-
supporting membranes while the microcomposite with PPTA gave flexible films
without any loss in their conductive functions.   Fig. 4 shows the SEM
photographs of the delaminated surface of the PPTA composites with polypyrrole prepared by using perchlorate as a dopant.   The uniformly dispersed
holes are formed after the polypyrrole bundles are pulled out from the PPTA

Table 2 Conductivity and mechanical properties of microcomposites of PPTA/conductive polymers.

Membrane	Conductivity S/cm	E/GPa	σ_B/MPa	ε_B/%	Fraction of Conductive Polymers/wt%
PPy(TsO$^-$)/PPTA	20	6.2	142	4.4	36
PPy(TsO$^-$)/PPTA	57	1.7	61	1.2	59
PPy(ClO$_4^-$)/PPTA	3	3.8	98	3.9	30
PAn(ClO$_4^-$)/PPTA	7	2.1	159	12.0	–

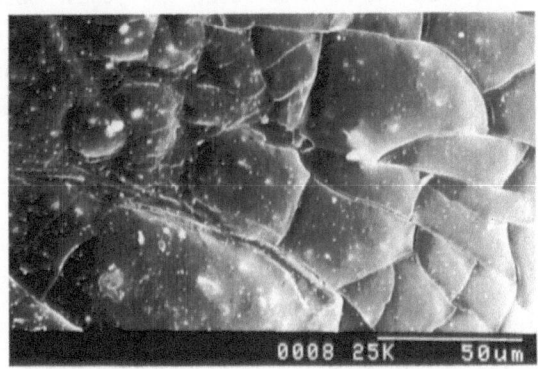

Fig. 3 Silica glass remaining after removing PPTA from the composite film viewed by SEM.

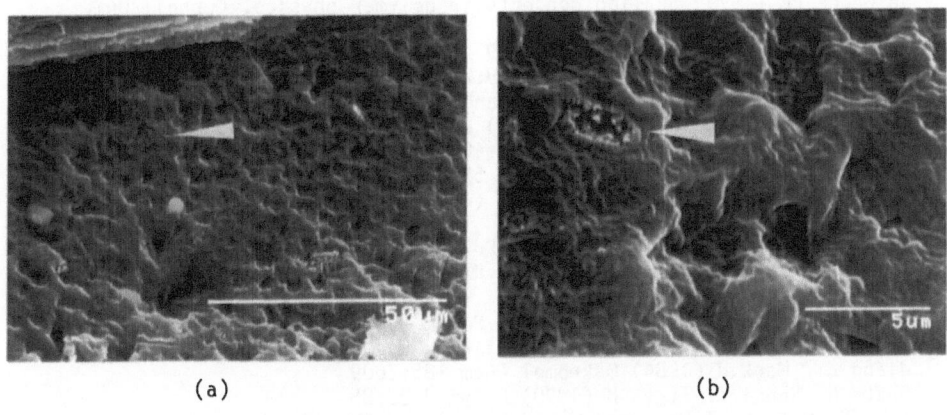

(a)   (b)

Fig. 4 The peeled surface of composite film of PPTA and PPy viewed by SEM: a) holes formed at the sites of PPy fiblils being pulled out; b) the enlarged image of a) showing the fractured bundle of PPy microfibrils in a hole of PPTA.

matrix. The enlarged image of the same photo shows the remaining polypyr-role bundles being fractured at the delaminated surface. The bundles are formed through the nucleation and growth processes and the holes do not reflect the diffusion volume as mentioned above. The SEM photograph of the etched composite film with sodium hydride and DMSO shows the direct images of polypyrrole in the composite film prepared by using p-toluenesufonate as a dopant. The diameter of polypyrrole particle is large with a pan-cake like shape near the film surface and is finely dispersed as small as 1μm in diameter inside of the film. In the composite with polyaniline, the fibrous polyanline grew from the PPTA surface. All these observations support the IPN-type texture of the composite. The microcomposite formation with PPTA provides the reinforced film without loss in the functionality of guest polymers. It is again emphasized that the coarse texture of films composed of rigid rodlike molecules are suitable for infiltration of various low molecular weight compounds capable of in-situ polymerization, whether they are organic or inorganic.

References:

1 Flory PJ, Orwoll RA, Vrij A (1964) J Am Chem Soc 86:3507
2 Sanchez IC, Lacombe RH (1976) J Phys Chem 80:2532
3 ten Brinke G, Karasz FE (1984) Macromol (1984) 17:815
4 Ikawa K, Hosoda S (1990) Polym J 22:643
5 Hosoda S, Kihara H, Kojima K, Satoh Y, Doi Y (1991) 23:277
6 Ide F, Hasegawa A (1974) J Appl Polym Sci 18:963
7 Coran AY, Patel R (1983) Rubber Chem Tech 56:1045
8 Nakamura G, Inoue T (1990) Kobunshi Ronbunshu 47:409
9 de Genne PG (1979) Scaling concepts in polymer physics. Cornell Univ Press, Ithaca London
10 Wu S (1988) J Appl Polym Sci 35:549
11 Margolina A, Wu S (1988) Polymer 29:2170
12 Bug ALR, Safran SA, Grest GS, Webman I (1985) Phys Rev Lett 55:1896
13 Tanaka H, Hayashi T, Nishi T (1989) J Appl Phys 65:4480
14 Hayashi T, Nishi T (1991) Kobunshi 40:458
15 Takayanagi M (1983) Pure Appl Chem 55:819
16 Takayanagi M (1986) In: Sedlacek M (ed) Polymer composites. Walter Gruyter, Berlin New York
17 Takayanagi M, Katayose T (1981) J Polym Sci Chem Ed 19:1133
18 Takayanagi M, Ogata T, Morikawa M, Kai T (1980) J Macromol Sci Phys 17:591
19 Takayanagi M, Goto K (1984) J Appl Polym Sci 29:2547
20 Bug ALR, Safran SA, Webman I (1985) Phys Rev Lett 54:1412
21 Jiang CY, Mark J (1984) Makromol Chem 185:2609
22 Ogata N, Sanui K, Itaya H (1990) Polym J 22:85
23 Koga K, Ueta S, Takayanagi M (1988) Polym J 20:639
24 Koga K, Iino T, Ueta S, Takayanagi M (1989) 21:303, 21:499
25 Takayanagi M unpublished
26 Takayanagi M, Koga K, Ueta S (1991) Polym Prepr Japan 40:3016
27 Haghighat RR, Kovar RF, Lusignea RW (1988) Mat Res Soc Symp Proc 121:755
28 Koga K, Yamasaki S, Narimatsu K, Takayanagi M (1989) Polym J 21:733

Recent Developments in Free Radical Polymerization

Roland P.-T. Chung and David H. Solomon

School of Chemistry, The University of Melbourne, Parkville Victoria 3052, Australia

Free radical polymerization continues to attract a significant research effort and this interest is also seen in conventional free radical chemistry. A major factor in this resurgence of interest in free radicals is the recognition that free radicals reactions including polymerizations are capable of selectivity, including stereospecificity, and control not previously associated with radical systems.

In the 1960s, free-radicals were described as neutral highly reactive species which were non-selective in attacking a substrate and these reactions were virtually uninfluenced by the solvent used. Indeed, one of the diagnostic tests for a free radical mechanism was the lack of any solvent effect.

In the 90s we have come full circle and describe radicals as nucleophilic or electrophilic with selectivity in both the point of attack and the stereochemistry. We recognise the important role of solvents and reaction conditions generally.

Polymer chemists are attracted to free-radical systems because of the relatively undemanding conditions. In marked contrast to ionic or coordination polymerizations they exhibit a tolerance for trace impurities, like oxygen and moisture. Stabilizers are often not removed in commercial practice. Indeed radical polymerizations are remarkable among chain polymerizations in that they can be conveniently conducted in aqueous media. These practical advantages and the newly developing refinements in control of polymer structure and properties contribute to the renaissance in free radical polymerization.

In reviewing this field it is convenient to focus on the major directions of the past few years. These include:

1) Structure/property relationships, particularly the importance of defect groups or structural imperfections as weak links. This is particularly important in determining the stability of the polymer chains, for example the resistance or susceptibility to the action of heat, oxidation, or radiation.

2) The synthesis of very low molecular weight functional oligomers with control of both the number and type of functional group(s).

3) Solvent effects on virtually all steps in the free radical chain process.

4) The challenge of control over the propagation step so as to avoid (or enhance) head-head addition and other side reactions and ideally to control the stereospecificity of the growth.

and

Y. Imanishi (Ed.)
Progress in Pacific Polymer Science 2
© Springer-Verlag Berlin Heidelberg 1992

5) Copolymerization and the challenge of predicting and controlling copolymer sequences.

We will illustrate the above points in the following discussion and indicate likely directions for future study.

The basic mechanism of radical polymerization was enunciated in the 1950's and 1960's, and comprises the initiation, propagation and termination steps of a chain reaction.[1]

Since molecular weights were traditionally high and propagation steps outnumbered initiation or termination events by up to 100,000 to 1, no great significance or interest was taken in the control of initiation or termination. Added to this, techniques were not readily available for the "needle in a haystack" chemistry needed to study these reactions. However, the recognition that these end groups can contribute in a significant way to properties led to the development of techniques to study these minor pathways. These developments have had a tremendous bearing on developments in free radical chemistry. The knowledge gained has contributed to our ability to control end-group functionality and to contemplate polymers with degrees of polymerization (DP) as low as 10 where the end groups represent 20% of the structure (at D.P. of 100,000 end groups represent only 0.002%).

INITIATION

We have reviewed initiation elsewhere,[2] but in brief the techniques used are:
1 Trapping of the initiating species
2 Chemical Analysis
3 Use of labelled initiators (particularly with ^{13}C NMR)

Trapping can involve either nitroxides[3] followed by separation and characterization or the use of nitroso compounds[4] and subsequent structural analysis by ESR. As an example of the former, the trapping of the radicals from the reaction of t-butoxy radicals and methyl methacrylate (MMA) by 1,1,3,3-tetramethylisoindolinyl-2-oxy (1) is shown (Scheme 1). Alkyoxyamines were isolated by conventional techniques and their pathways deduced. The methyl radical, formed by β-scission of the t-butoxy radical, is trapped as the methoxyamine, which in turn can add a further monomer unit in a thermally activated step growth addition to form (2). As an example of the latter, the radicals from the same reaction are now trapped by 2-methyl-2-nitrosopropane as the corresponding nitroxyl radicals.

Initiator derived end group in a polymer may be determined by chemical analysis, e.g. methyl methacrylate polymerization initiated by t-butoxy radicals.[5] The t-butoxy end group may be cleaved off with boron trichloride and the t-butyl chloride produced can then be quantified. Evaluation of the amounts of acetone and t-butanol formed during polymerization indicates the percentage of β-scission and abstraction from monomer respectively.

^{13}C NMR used in conjunction with appropriately labelled ^{13}C initiator (eg. azo-bis-(isobutyronitrile)-α-^{13}C) (AIBN-α-^{13}C) has proven to be enormously successful in determining the type of end group in a polymer.[6-8] Because of the low natural abundance of ^{13}C and the low proportion of initiator fragment compared to the backbone polymer, it would normally be impossible to determine the end group fragment by ^{13}C NMR. The enhanced signal of the ^{13}C labelled initiator allowed the determination of the mode of addition and the type of group adjacent to the initiator fragment.

The ^{13}C NMR technique is particularly useful with carbon centred radicals and has the added virtue that actual polymerization conditions are used. A combination of the above techniques is of great value. As an example of the way in which initiator can effect properties, we have established by nitroxide trapping and ^{13}C NMR that benzoyl peroxide initiated polystyrene (PS) was derived from two main initiating pathways: tail addition, i.e. addition to the less substituted end (93%) and head addition, i.e. addition to the more substituted end (7%) (Scheme 2).[2]

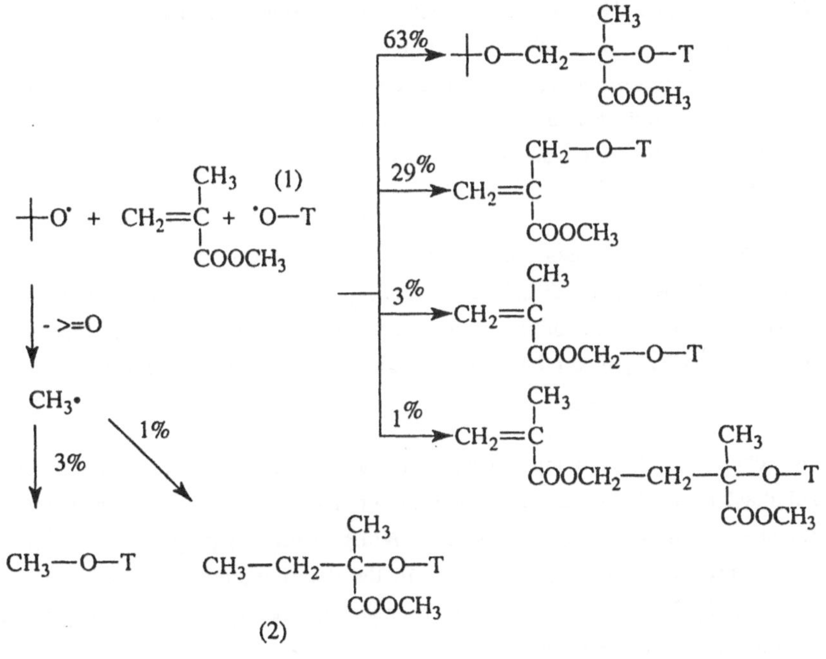

Scheme 1

The thermal stability of these end groups has been studied along with the model compounds 2,4,6-triphenyl-1-ene and 1,3,5-triphenyl-1-ene.[9] These model compounds represent the unsaturated ends formed on elimination of the benzoate end groups from (3) and (4) respectively.

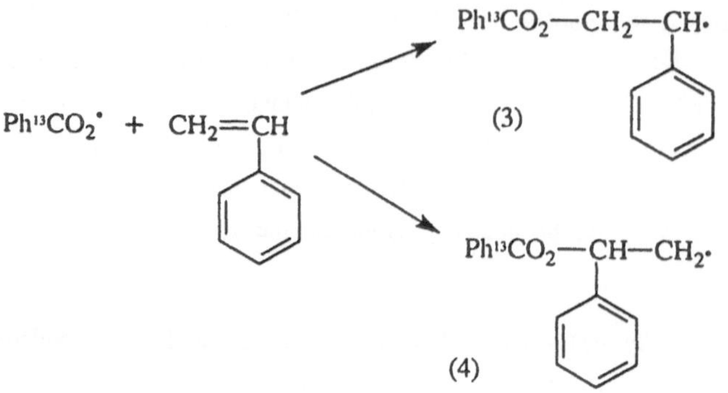

$$Ph^{13}CO_2{}^\bullet \ + \ CH_2\!\!=\!\!CH$$

$$Ph^{13}CO_2\!\!-\!\!CH_2\!\!-\!\!CH\!\bullet \qquad (3)$$

$$Ph^{13}CO_2\!\!-\!\!CH\!\!-\!\!CH_2\!\bullet$$

$$(4)$$

Scheme 2

The secondary benzoate (4) is labile below 300°C whereas the primary benzoate (3) is stable. Both the model compounds were found to be labile at 300°C. Hence it can be concluded that the stability of the polymers is directly related to the stability of the respective benzoates. This study clearly shows the effect of one weak link per chain on the polymer properties . An additional route to secondary benzoates is discussed below under termination. In related studies on styrene we have established that both the cyanoisopropyl radicals (from AIBN)[7] and t-butoxy radicals[2] are much more selective in this system and add virtually exclusively to the tail to give polymers that are thermally stable.

A wide range of monomers and initiators has now been studied and reported. An example[10] of the versatility of the nitroxide trapping technique, particularly with oxygen centred radicals, is the reaction with allyl methacrylate. In contrast with the attack on styrene, the complex pattern of initiation gave rise to a variety of products (addition, hydrogen abstraction and β-scission), which were isolated and characterized.

The ^{13}C NMR technique is also proving valuable, for example, in its application in assigning relative rates of attack in copolymerization. Moad et al.[11] has shown that labelled monomer can be very useful in studying the NMR of polymers. NMR of PMMA prepared from MMA carbonyl-^{13}C has proved to be very convenient in the determination of the tacticity of homopolymers while NMR of copolymers prepared from labelled monomers can rapidly provide information on monomer sequence distribution.

Bevington et al.[12] have found with the use of ^{13}C-labelled benzoyl peroxide that the polymerization of methyl methacrylate or styrene in the presence of stilbene resulted in a high proportion of the initiator fragment attached to the stilbene unit. The stilbene was found to react readily with oxygen-centred benzoyloxy radical, but was very unreactive to carbon-centred radicals. The stilbene is not polymerized.

In a later study, Barson *et al.*[13] had found that the acetylenic compounds (PhC≡CH, PhC≡CPh and PhC≡CC≡CPh) are significantly less effective in reacting with benzoyloxy radicals than their corresponding olefins (styrene, stilbene and 1,4-diphenylbuta-1,3-diene respectively).

We have carried out studies using azo-bis(methyl isobutyrate-α-^{13}C) (AIBMe-α-^{13}C) and AIBN-α-^{13}C with methyl methacrylate, styrene (S) and vinyl acetate (VAc). The quantity of tail and head adducts formed in VAc polymerization can be determined on the basis of ^{13}C chemical shift of the quaternary carbon. We were able to conclude that the major polymer obtained at 30°C contained predominantly the tail adduct.[14]

TERMINATION

The termination mechanism is a radical-radical reaction leading to either combination or disproportionation. In the absence of other reactants most radical polymerizations terminate by both processes and so we expect termination to lead to a mixture of:
1   Head-head linkage
2   Unsaturated chain end
3   Saturated chain end

There are various techniques that are available to study the process of termination.[15] These include:
1   Molecular weight distribution.
2   End Group Analysis
3   ^{13}C NMR
4   Gelation Technique
5   Model Studies.

The molecular weight distribution may easily be measured by GPC. A polydispersity (Mw/Mn) of 2.0 is obtained if termination occurs exclusively by disproportionation and 1.5 if termination is by combination. Since the polydispersity is conversion dependent, this method is limited to low conversion polymers.

End group analysis requires precise quantitative determination of the end groups on the polymer chains as well as the number average molecular weight. The initiator-derived end groups are usually labelled and can therefore be determined by chemical, radiochemical or spectroscopic (NMR, IR, UV) means. In the ideal case, two initiator residues per molecule is obtained if termination is by combination while one is obtained by disproportionation. The ratio of disproportionation to combination (k_d/k_c) is thus calculated from the relationship $k_d/k_c = (2-x)/2(x-1)$ where x is the number of initiator fragments per molecule.

Among the methods mentioned, NMR offers the most versatility as initiator residues incorporated into polymer chains can be distinguished from unchanged initiator, end groups formed by transfer to initiator or by copolymerization of initiator by-products. This would preclude any possibility of overestimating k_d/k_c in circumstances where transfer to monomer, solvent, etc., are significant. The other virtues of NMR have already been discussed for initiation.

The "gelation technique", developed by Bamford et al.,[16] involves the measurement of the time required for gelation when a polymeric halo compound, e.g. poly(vinyl trichloroacetate), is subjected to photolysis in the presence of $Mn_2(CO)_{10}$ and an appropriate monomer. Termination by combination and disproportionation will lead to crosslinking and graft formation respectively. The gelation time is a measure of the degree of cross-link and thus k_d/k_c. The original results were calibrated against styrene polymerization where k_d/k_c was taken to be 0.

In cases where the determination of k_d/k_c from direct analysis of a polymerization or the resultant polymer requires data on other areas of the polymerization mechanism which is not easily obtainable, it is more appropriate to investigate the self-reaction of the corresponding low molecular weight models. Studies on model systems provide a means of isolating the termination reaction and analyzing it in detail. Not only is the k_d/k_c easier to determine, competing side reactions can also be readily monitored and accounted for in any evaluation of k_d/k_c.

Most of the model systems gave comparable findings to that of the corresponding polymer systems. However a notable exception is that of methacrylonitrile (MAN). Bamford and coworkers[16] had found that % combination is only 35%, whereas Solomon et al.[17] had % combination to be greater than 90%.

The simplest model for the polymerization of MAN is that of the decomposition of AIBN.[17] The cyanoisopropyl radical mimicks the action of a propagating poly(methacrylonitrile) radical.

In the decomposition of AIBN, besides the formation of the products (tetramethyl succinonitrile (TMSN), isobutyronitrile (IBN) and MAN) from the normal Carbon-Carbon reaction of the cyanoisopropyl radicals, a ketenimine (5) from a Carbon-Nitrogen reaction also occurs. This arises because one of a pair of cyanoisopropyl radicals can tautomerize to a keteniminyl radical which subsequently reacts with the second cyanoisopropyl radical to form the ketenimine. This reaction is thermally reversible and the ultimate products would be TMSN, IBN and any polymerized MAN.

$$CH_3-\underset{\underset{CN}{|}}{\overset{\overset{CH_3}{|}}{C}}-N=C=C\overset{CH_3}{\underset{CH_3}{<}}$$

(5)

The decomposition of an unsymmetrical azonitrile (6) results in the formation of 2 ketenimines (7) and (8) in addition to the normal products (Scheme 3).

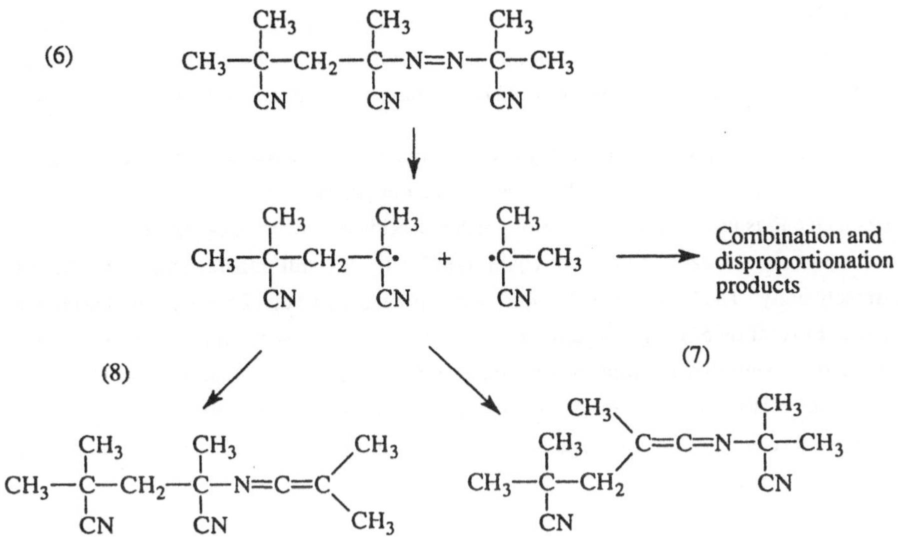

(6)

$$CH_3-\underset{\underset{CN}{|}}{\overset{\overset{CH_3}{|}}{C}}-CH_2-\underset{\underset{CN}{|}}{\overset{\overset{CH_3}{|}}{C}}-N=N-\underset{\underset{CN}{|}}{\overset{\overset{CH_3}{|}}{C}}-CH_3$$

↓

$$CH_3-\underset{\underset{CN}{|}}{\overset{\overset{CH_3}{|}}{C}}-CH_2-\underset{\underset{CN}{|}}{\overset{\overset{CH_3}{|}}{C}}\cdot \quad + \quad \cdot\underset{\underset{CN}{|}}{\overset{\overset{CH_3}{|}}{C}}-CH_3 \quad \longrightarrow$$

Combination and disproportionation products

(8)

(7)

$$CH_3-\underset{\underset{CN}{|}}{\overset{\overset{CH_3}{|}}{C}}-CH_2-\underset{\underset{CN}{|}}{\overset{\overset{CH_3}{|}}{C}}-N=C=C\underset{CH_3}{\overset{CH_3}{<}}$$

$$CH_3-\underset{\underset{CN}{|}}{\overset{\overset{CH_3}{|}}{C}}-CH_2 \quad \underset{CH_3}{\overset{CH_3}{>}}C=C=N-\underset{\underset{CN}{|}}{\overset{\overset{CH_3}{|}}{C}}-CH_3$$

Scheme 3

Recent work [18] has shown that ketenimine (7) is favoured over (8) by a ratio of 6 to 1. This shows that the larger keteniminyl radical is preferentially formed.

In current studies[19] it was found that in the decomposition of ketenimine (5), the cage reaction favours disproportionation over combination. It was further found that the cage reaction is favoured to the encounter reaction with increasing viscosity. The implication of these results is that as the viscosity of a polymerization mixture increases with conversion, the cage reaction and therefore the disproportionation mechanism would increasingly be favoured. This observation would serve to account for the discrepancies in the MAN results obtained by the Solomon and Bamford research groups. High conversion resulted in the lower % combination of the Bamford groups whereas low conversion favours the high % combination of the Solomon group.

To illustrate the importance of the different structural units, i.e. head-head linkage, saturated and unsaturated chain ends, let us consider polymethyl methacrylate which has been the subject of detailed study in recent years.

Studies with model polymers have shown that the head-head linkage and the unsaturated chain end are less stable thermally than the saturated end.[20] This work provides on explanation for previous observations that anionically polymerized methyl methacrylate is more stable thermally than that prepared by a free-radical mechanism. The anionic polymer lacks the head-head linkage and the unsaturated end. It also provides yet another example of the importance of defect groups and their major influence on the properties of the polymer.

Radical-radical reaction can also occur by primary radical-polymer reaction and chain transfer to initiator. This is a possible route to additional secondary benzoates observed by us in benzoyl peroxide initiated polystyrene. Clearly it would be desirable to avoid radical-radical termination.

Chain transfer agents[21] are thus of great interest as they not only avoid both the head-to-head linkage and the unsaturated chain end, but also give the preferred saturated chain end. The saturated chain end is preferred for reasons of greater stability of the resulting polymer.

The need is to have available a range of chain transfer agents with transfer constants (C_T) of approximately unity. The C_T is defined as the ratio of the rate constant of propagation (k_p) to the rate constant of reaction of the propagating radical with the transfer agent (k_{tr}), i.e. k_p/k_{tr}. A C_T of unity would ensure that the ratio of transfer agent to monomer is constant throughout the polymerization. This in turn would ensure a narrow molecular weight distribution. An additional interest in chain transfer agents comes from the present interest in low molecular weight functional polymers.

MONO- AND DI-FUNCTIONAL OLIGOMERS

With high molecular weight polymers, functionality can readily be introduced by the use of appropriate comonomers, for example, 2-hydroxyethyl methacrylate (HEMA) to methyl methacrylate (MMA).

The incorporation is as expected in a statistical copolymerization and yield a polymer with the following characteristics:

1   Functional groups along the chain and not at the chain end.
2   Distribution of functional groups depending on copolymerization parameters.

The need to prepare materials with much lower molecular weight, e.g. with DP of between 10 to 100, results in two major problems:

1   Some chains may have no functional groups while others may have excess; hence in a subsequent crosslinking reaction, some chains are too highly functional while others are not incorporated into the network.
2   The inability to place the functional group at chain ends for easier chemical accessibility and possible greater flexibility of the final network.

This has led to the recent intense search for a new chain transfer agent. The two major aims in this search are:

1   Monofunctional reactive oligomers for use in, for example, block copolymerization.
2   Difunctional reactive oligomers for use in subsequent thermosets by either step growth or chain growth reaction with themselves or other entities.

Previously these mono-functional oligomers have been prepared by the use of functional chain transfer agents, e.g. thioglycolic acid.[21] In the case of di-functional oligomers the technique referred to as "dead-end" polymerization has been used, whereby high concentrations of a functional initiator are used. This approach requires termination to occur exclusively by either combination or primary radical termination. A satisfactory incorporation of functionality is only obtained with monomers that satisfy this criterion. Mixtures are obtained at best, for example, with polystyrene it is claimed that the degree of functionality f=1.8 rather than the required f=2.0.

Consequently the challenge has been to prepare new chain transfer agents which meet the following criteria:

1   The reaction with a whole range of polymer radicals must have a C_T approaching unity.
2   It must be able to carry a variety of functional substituents, e.g. hydroxy, carboxylic, amide, amino groups.
3   It must be non-toxic and cheap.
4   Initiation must be 'clean' in that it must proceed by either H-abstraction or addition to form monofunctional oligomers, or by addition to form bi-functional oligomers.

Selected examples of each type of functional oligomers are shown below. In the case of mono-functional oligomers, the functional group can be derived from the chain transfer agent, e.g. by using allylic sulfides with one functional residue. Difunctional oligomers can be obtained by the addition-elimination mechanism, e.g. from allylic sulfides with 2 functional residues[22] or from styryl ethers.[23]

This is a very active field of research and further developments along the general lines suggested can be expected.

CATALYTIC CHAIN TRANSFER

Square planar cobalt complexes including porphyrin and cobaloxime derivatives rapidly and reversibly add carbon-centred radicals. An alternative pathway to the reverse reaction is an elimination to form the olefin and a cobalt hydride species, which can initiate polymerization by donating hydrogen to monomer.[24]

Thus these cobalt complexes function as extremely efficient catalytic chain transfer agents, with each polymer chain having an olefinic end group.

INHIBITORS

Careful choice of inhibitors can give functional polymers similar to those discussed under chain transfer agents. The inhibitors may donate one or preferably two hydrogens to one or two propagating radicals respectively. The end group functionality is provided by the initiator. In the case of mono-functional oligomers, the inhibitor, e.g. 1,5-dimethyl-1,4-cyclohexadiene, donates two hydrogens to two separate propagating radicals, with the aromatization of the inhibitor to the 1,3-dimethylbenzene providing the driving force.[25]

In the case of difunctional polymers, captodative olefins[26] rapidly scavenge propagating radicals to give new radicals which are unable, or slow, to reinitiate polymerization. These radicals have no β-hydrogens and therefore are unable to decay by disproportionation. Thus they react exclusively by combination. The functional groups are again provided by the initiator.

INIFERTERS

The concept of iniferters (initiator-transfer-termination) was proposed by Otsu[27] as a method for the design of polymer chain end structure. Radical polymerization in the presence of iniferters give a polymer with two initiator fragments at its chain end. With an appropriate choice of iniferter, e.g. the photoiniferter tetraethylthiuram disulphide (TD), a polymer is obtained with end groups still containing iniferter functionality. In the case where the 'polymer' is polystyrene (PS), the Et_2NCSS end groups in TD-PS is very close to 2, i.e. TD is a difunctional photoiniferter for styrene polymerization. This polymer is capable of living radical polymerization with other monomers, e.g. Vinyl acetate (VAc), to form a block copolymer.

Recently, the use of thermal iniferters in radical polymerization have also been discussed in a review by G.Clouet and C.P.R. Nair.[28]

SOLVENT EFFECTS

There is still a considerable amount of work to be done before a full understanding of the effects of solvents on the various steps in radical polymerization can be achieved. We expect these effects to be of greater significance as a further degree of sophistication is introduced into the control of the various steps.

It is convenient to discuss effects where the solvent is:

     1   chemically involved in the polymerization

and   2   the ratios of competing pathways vary by virtue of the solvent.

1   Chemical Incorporation

An example which illustrates this process is the initiation of methyl methacrylate with t-butoxy radicals in cyclohexane.   Under appropriate conditions the end groups are derived solely from the solvent .[29]

Such a system offers possibilities for functional group incorporation by appropriate choice of solvent. The absence of any initiator-derived residue in the polymer is also noted.

2   Competitive Pathways Modification

i Initiation: The t-butoxy initiation of methyl methacrylate in the absence of solvent proceeds as shown (Scheme 2).

The ratio of products is solvent dependent.   With carbon tetrachloride as solvent, the product ratio derived from addition: H-abstraction : β-scission is 56:23:18 while in acetonitrile, the ratio is 31:21:45 respectively.[30]

ii   Propagation: In a classical study of solvent effect on propagation, Hatada *et.al.*[31] found that the polymerization of vinyl acetate is subjected to significant influence by the solvent. With benzene as solvent essentially no branching was observed. However, in ethyl acetate 0.7 branch per chain (for $M_n \cong 20000$) was observed. The reason postulated is that the benzene complexes with the propagating radical to make it more selective.

The rate constant of propagation (k_p) is also lower in benzene.   Furthermore, a variation of c.100-fold in k_p (at 30°C) was observed in vinyl acetate polymerization on changing the solvent from ethyl acetate to benzonitrile.

In other systems shown to be solvent dependent alternative hypothesis[32] include:

1   Reversible complexing to effectively lower the radical concentration.   In the case of aromatic solvent like benzene, this would involve the formation of a cyclohexadienyl radical.

2   Alteration of electrophilicity and hence propensity of the radical for tail addition. The radical formed by head addition being likely to undergo alternative reaction.

Although we have made significant progress from the concept that solvents are inert, much remains to be done in order to fully understand solvent effects.

PROPAGATION CONTROL

Perhaps the outstanding challenge remaining in free radical polymerization is to be able to control each successive propagation step in a manner approaching that found in Zeigler-Natta systems. We have already discussed solvent effects which suggest that radical complexes with various solvents are a possibility for achieving control. Similarly, cobalt complexes, notably cobalt porphyrins, suggest that suitable metal complexes could function by radical complexation.

Ideally, a complexing agent should possess the following properties:

1 selectively complex with the propagating radical and not to the monomer which is present in a very much higher concentration than the radical.
2 The equilibrium and rate constants associated with complex formation must be high.
3 The reactivity of the complexed radical must be many times greater than that of the uncomplexed radical.

In a preliminary experiment[33], the copolymerization of styrene and methyl methacrylate was carried out with azo-bis-(methyl isobutyrate-α-^{13}C) (AIBMe-α-^{13}C) as initiator and in the presence of low concentration (4×10^{-3}M) of Diethyl aluminium sesquichloride (DEASC) as the complexing agent.

The effect observed is quite dramatic. The specificity of radical is greatly enhanced by DEASC. The proposed mechanism involves complex formation between the initiator and DEASC leading to a complexed radical on photolysis. This radical is more electron deficient than the corresponding uncomplexed form and shows enhanced reactivity towards styrene, relative to methyl methacrylate.

Much remains to be done but this is an exciting development which we are actively pursuing at present.

SUMMARY

Free radical polymerization is undergoing a metamorphosis. We now have at our disposal ways of systematically selecting the polymerization conditions, i.e. the solvent, initiator, monomer types and concentrations. This gives rise to exciting new chemistry which could lead to oligomers of very low molecular weight and controlled functionality. The possibility also exists to combine the attractions of free radical systems with the control of polymer structure, usually associated with more complex ionic/coordination catalyzed systems.

ACKNOWLEDGMENTS

This paper represents a summary of the contribution from the research groups which Prof. Solomon established at CSIRO Division of Chemicals and Polymers, the new group at the University of Melbourne, and the sponsored research carried out for us by Griffith University. The early work was supported by the Potter Foundation following a suggestion by Prof. Sir Geoff Allen FRS. Subsequent support by the Australian Research Grants is gratefully acknowledged.

REFERENCES

1 Bevington JC (1989) In: Eastmond GC, Ledwith A, Russo S, Sigwalt P (Eds), Comprehensive Polymer Science 3:65, Pergamon, London.
2 Solomon DH, Moad G (1987) Makromol Chem, Macromol Symp, 10/11, 109.
3 Griffiths PG, Rizzardo E and Solomon DH (1982) J Macromol Sci-Chem, A17(1):45.
4 Sato T. and Otsu T. (1977) Makromol Chem 178:1941.
5 Rizzardo E, Solomon DH (1979) J Macromol Sci-Chem 13:1005.
6 Moad G, Solomon DH, Johns SR, Willing RI (1982) Macromolecules 15:1188.
7 Moad G, Solomon DH, Johns SR, Willing RI (1984) Macromolecules 17:1094.
8 Moad G, Rizzardo E, Solomon DH, Johns SR, Willing RI (1984) Makromol Chem, Rapid Commun 5:793.
9 Krstina J, Moad G, Solomon DH (1989) Eur Polym J 25:767.
10 Thang SH (1987) PhD Thesis, Griffith University.
11 Moad G (April 1991) Chem in Aust 122.
12 Bevington JC, Huckerby TH (1985) Macromolecules 18:176.
13 Barson CA, Bevington JC, Huckerby TN (1991) Polym Bull 25(1):83.
14 Krstina J, Moad G, Solomon DH (1991) Eur Polym J in press.
15 Moad G, Solomon DH (1989) In: Eastmond GC, Ledwith A, Russo S, Sigwatt P (Eds) Comprehensive Polymer Science 3:147, Pergamon, London.
16 Ref 15, ref 69 therein.
17 Serelis AK, Solomon DH (1982) Polym Bull 7:39.
18 Danek SK, Jones SL, Kelly DP, Moad G, Solomon DH, unpublished results.
19 Chung RPT, Quach C, Solomon DH, unpublished results.
20 Meisters A, Moad G, Rizzardo E, Solomon DH (1988) Polym Bull 20:499.
21 Ref 1, ref 34 therein.
22 Meijs GF, Rizzardo E, Thang SH (1988) Macromolecules 21:3122.
23 Meijs GF, Rizzardo E (1988) Makromol Chem, Rapid Commun. 9:547.
24 Burczyk AF, O'Driscoll KF, Rempel GL (1984) J Polym Sci, Polym Chem Ed 22:3255.
25 Solomon DH, unpublished results.
26 Mignani S, Janousek R, Merenyi R, Viehe HG, Riga J, Verbist J (1984) Tetrahedron Lett 25:1571.
27 Otsu T, Matsunaga T, Kuriyama A, Yoshioka M (1989) Eur Polym J 25:643.
28 Nair CPR, Clouet G (1991) J Macromol Sci, Rev Macromol Chem Phys C31 (2-3):311.
29 Grant RD, Griffiths PG, Moad G, Rizzardo E, Solomon DH (1983) Aust J Chem 36:397.
30 Ref 29, 36:51.
31 Hatada K, Terawaki Y, Kitayama T, Kamachi M, Tamaki M (1981) Polym Bull 4:451.
32 Moad G, Solomon DH (1990) Aust J Chem 43:215.
33 Krstina J, Moad G, Solomon DH (1991) Polym Bull in press.

Changing Demands on the R & D Organisation in an International Cooperation

U. H. Felcht and J. P. Riggs

Hoechst AG and Hoechst Celanese Corporation

Frankfurt/M - Postfach 80 03 20 - Summit, NJ 07901 - 86 Morris Ave.

My purpose in this discussion today is really two-fold -- I have a dual agenda: It is to develop a dialogue on the specifics and interrelationships of the impact, critical issues and future challenges in the area of polymeric materials science and engineering and relate this to the implications of the globalization of technology and the resulting new demands on the R&D organization in an international corporation like Hoechst.

But first, a few background comments on Hoechst: The Hoechst Group is one of the major chemical and materials enterprises in the world, with sales of $28B in the areas of chemicals and dye stuffs/colorants, agriculture, fibers and films, polymers, life sciences and technical information systems, industrial gases and engineering ceramics. Hoechst is, indeed, active in virtually all the main areas covered by the chemical industry, with a very broad product range. Materials science and engineering is the foundation for well over half the product revenue, with polymeric-based materials accounting for about 35%.

Within Hoechst there is a large tradition of maintaining a strong science and technology base, and the research activities of Hoechst AG have played a vital role in the company's progress for over 125 years. We have this emphasis not only to improve the quality of products and the efficiency of processes, but also -- and above all -- to raise safety levels and to improve the general quality of life for mankind -- through the improvement of current technology and products and the development of new technologies and new products. Equally important aspects of the "Hoechst High Chem" strategy are respect for the environment and economical use of the world's finite energy and material resources. These aspects make an all-inclusive approach necessary -- one that encompasses not just research, production and marketing, but also recycling -- and this latter issue is particularly relevant to aspects of future polymer developments and will be addressed in more depth shortly.

Research is, without question, the most important kind of strategic investment, and high quality research ensures a long-term future for the company. Hoechst currently invests, world-wide, $1.7B annually in R&D, with extensive facilities in the major regions of the world (research centers in 16 countries) employing 15,000 R&D personnel. It is important to note here, also, the growing expenditure abroad on R&D -- from a mere 5% spent outside of Germany in 1970 to the 38% of today, with the main centers of foreign research being in France, the U.S. and Japan, respectively -- with an expanding commitment to Japan. It is clear that our international R&D effort has become a matter of prime importance, and Hoechst is addressing critical issues developing from the accelerating globalization of technology and the resulting changing demands on the R&D organization in an international corporation; and the materials/polymer science area serves as an excellent example of the needs and opportunities in this respect.

Y. Imanishi (Ed.)
Progress in Pacific Polymer Science 2
© Springer-Verlag Berlin Heidelberg 1992

Materials science and engineering (MS&E) has emerged as a pivotal, enabling force for industry and society as a whole (1). It has provided the basis for economic well-being and improved quality of life the world-over, and continuing advances indicate this role will only be enhanced. Indeed, the developing capabilities to address specific needs by the specific design and tailoring of materials at the molecular level and to control macroscopically synthesis, process, property and structure interactions for different end-use applications point to the still further importance of MS&E in the future -- it is entering a period of unprecedented intellectual challenge and productivity. Within this context polymeric materials are of particular significance in providing unique advantages and opportunities and present an excellent case-study framework for the international management of technology resources.

Before proceeding further, let me remind all of us (Figure 1)

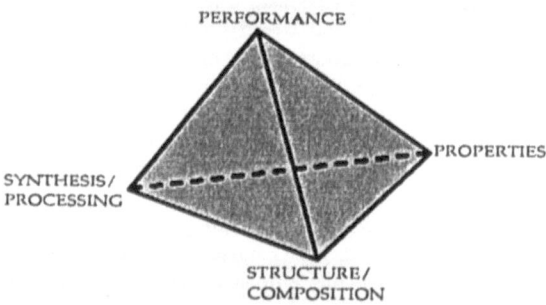

The Four Elements of Materials Science and Engineering.....

PERFORMANCE

PROPERTIES

SYNTHESIS/ PROCESSING

STRUCTURE/ COMPOSITION

.....Particularly Relevant to Polymers

Figure 1

of the four key elements of materials science and engineering and their interactions -- synthesis/processing, structure/composition, properties and performance -- because the basic understanding of these inter-relationships is the corner stone of modern materials science and because they present degrees of freedom with respect to the design, development and use of polymers that are truly unique. Within this framework, it should be noted that the integration of materials synthesis with materials processing to identify commercially viable materials options will be a major challenge of the 1990's and will heighten the need for highly interdisciplinary team approaches to problem definition and solution.

Polymeric materials today have a broad impact, world-wide, on current, established industrial technology and the economy with many, varied uses familiar to all of us -- clothing, furniture and furnishings, packaging and containers of all sorts and selected applications in the construction and building, electronics and transportation sectors, to name but a few. As important -- perhaps more so in several important respects -- polymers are also playing a critical role in new, advanced materials developments with cross-cutting impact on the growth of a range of new technologies, capabilities and emerging industries. Those include both structural polymers, where the thrust is metals replacement and the property profile sought emphasizes characteristics such as stiffness, tensile strength, impact resistance, corrosion resistance, and thermal and chemical stability, and functional polymers, where the focus is on developing materials with, e.g., unique electrical, optical, separation or barrier and biological properties -- or combinations of these functions designed in at the molecular level. Specific examples include high performance structural polymers (including composite materials) with applications dominant in the automotive/transportation and aircraft/aerospace industries; polymeric materials for electrical/electronic applications (from lithography to components to packaging); polymeric-based gas and liquid membrane separation uses; biopolymers (with important applications encompassing artificial organs and prostheses through controlled release of pharmaceuticals -- and in this same general category, the imitation of biological/living systems which represent an evolutionary strategy to develop complex, high performance materials); and, perhaps the newest area of impact, polymer materials for optical applications (such as nonlinear optics, photorefractive materials and displays, among others) with uses developing in both passive and active devices and components in telecommunications, data communications, information storage and other areas.

The current and future significance of these materials is further emphasized by examining the role of advanced polymeric materials as compared to metals and other inorganics in both structural and functional applications. Over 50% of the high performance materials application area is based on polymers; the total dollar volume associated with all of these higher performance structural materials is projected to grow to about $50B by the year 2000, and to about $35B for the functional materials. Of this, over $40B could be polymeric based, with the highest growth area being in functional materials.

It is this increasing scope of impact of polymeric materials science and technology -- and the implications this has for changing demands on the R&D organization and the need for increased cooperation among industry, government and universities -- that makes this area and this conference particularly appropriate for addressing organizational and operational needs with respect to international R&D management and the globalization of technology.

In the late 1970's in a major study on the management of technology (2), one (and in retrospect, critical) external issue -- how to manage R&D resources on a global basis -- was conspicuously absent from the list of concerns raised by industrial managers. Today, this R&D management issue is a very high priority concern and this reflects what has been called "one of the major transformations of the final two decades of the twentieth century: the globalization of technology." (3) There are a number of reasons for this (2,3), including factors directly concerned with science and technology, market factors, government related and regulatory factors and competitive factors. A key challenge in dealing with these is to put in place a global R&D and technology infrastructure that maximizes for a particular corporation the strategic advantage of technology and enables development of better technology identification and assessment capabilities, more effective routes to commercialization -- including increasing emphasis on joint, international precompetitive phase research with universities and other companies with complementary expertise -- and stronger market focus in technology and product development; and to do this in an environmentally sensitive manner.

Let us now consider specific issues and approaches to research management in an international company -- with all of the functions and disciplines that are involved -- and in a framework that for Hoechst stresses multinational cooperation in research as a central strategic element that will be consistently pursued. Hoechst is addressing the integration and coordination of its international technology operations (Figure 2)

Figure 2

through the development and implementation of a tri-regional concept for research and technology and business development comprising the US, Europe and East Asia (particularly Japan), although the broadening to other regions in the future should not be excluded. This tri-regional approach is a natural consequence of intensifying competition and a need to accelerate the process of converting research results to marketable products and to insure that correct priority is given to market development and technological progress throughout the course of the R&D activity. This means that when deciding about a particular research project, the right market and technology environment has to be selected and that this must be done with the optimum regional focus. It also recognizes the criticality of having employees on site in a given region that are familiar with specific market conditions and needs and science and technology capabilities and interests, and combine this with a knowledge of the world-wide technological resources and developments within the corporation. Central to this tri-regional operation are team management, flexible project organization and well-conceived and well-communicated product-technology - marketing goals that insure full use of technology synergies and the best applications of all resources. In summary, then, the major objectives being pursued are to, on a global scale: a) Balance the classical issues of technology - versus market-driven opportunity pursuit; b) Enhance the linkage among the key functions critical to business development; c) Establish a common strategic and operational framework that provides a shared purpose, commitment and key values; enhances promotion of technology and business development synergies; creates a supportive environment; and emphasizes the importance of being a learning organization, and d) Develop true strengths from a cultural diversity -- which means being highly sensitive to and appreciative of cultural differences and insuring that these are taken into account in the management of technology.

Controlling this international and interdisciplinary cooperative enterprise represents a major challenge, and to achieve this we have taken a number of steps: We have formed (Figure 3)

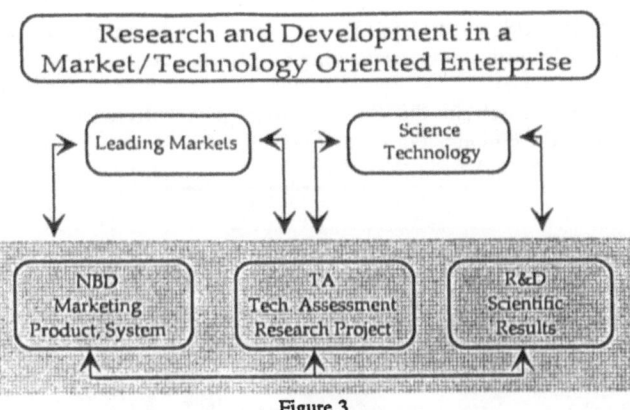

Figure 3

in one group , on a local level, capabilities in R&D, technology assessment (TA) and new business development (NBD) to permit a coordinated and integrated activity in identifying and evaluating technology and market opportunities, developing the science and technology and moving this into commercialization; this group, which encompasses a central research activity and a new business development focus also coordinates closely with business unit R&D. On a global scale, analogous organizational concepts (Figure 4)

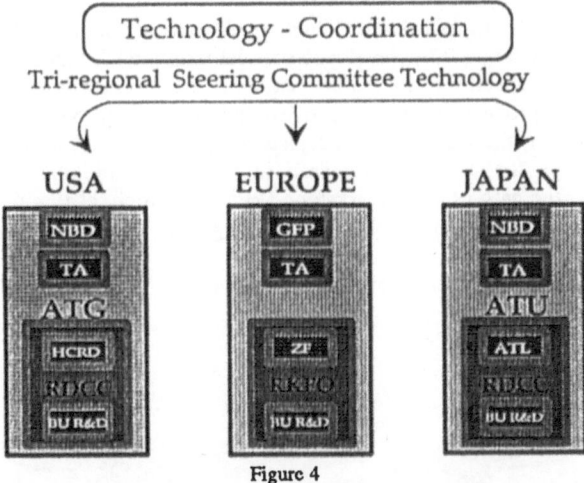

Figure 4

have been implemented in the three major regions. On this international level, we have built up on various levels a multilayered coordination and management system directed at insuring optimum use of our world-wide research potential. It is also designed to permit the greatest possible flexibility in decision making and to have this occur at the appropriate management levels. A Technology Steering Committee has been established as the most senior coordination body at the international level. In addition to this international forum, leading representatives of Hoechst Celanese and Hoechst Japan research also belong to various research bodies at Hoechst AG. Furthermore, a regular exchange of personnel at all levels between the US, Japan and Europe also effects an innovative transfer of technology, as well as fostering an understanding of different cultural perspectives.

We pursue strategic research in the three regions with differences in emphasis and responsibilities among the regions. For example (Figure 5)

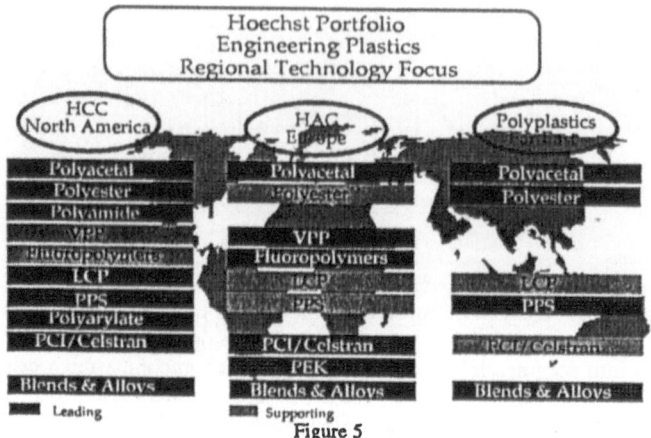

Figure 5

in the field of engineering plastics and high performance polymers, in an area such as polyacetal, which is a large, fully-developed product with a wide range of uses, each region plays a leading role, specific to its market requirements, in providing technology support. In the case of other products, such as thermotropic liquid crystal polymers (LCP), which is a developing product, the European and Japanese regions play a global supporting role to the U.S., which is the center of research in the field and where the know-how is concentrated. In the case of polyetherketone (PEK), research, at this point, is carried out centrally in Europe, but the overall objective is also global.

Research in the area of functional materials (Figure 6)

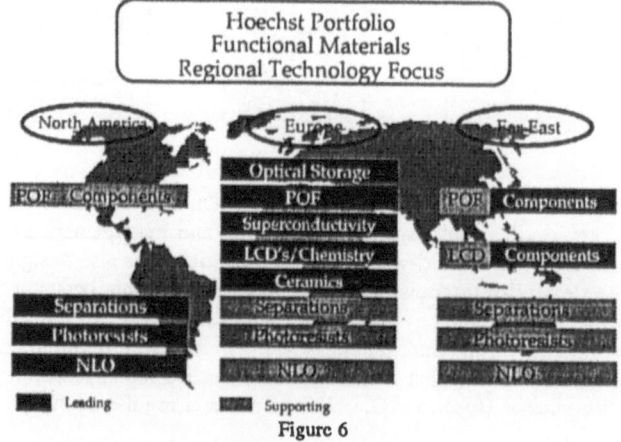

Figure 6

on the other hand, because of its generally, but not completely, earlier stage of development, is organized such that one region exclusively takes the lead in the technology development with support provided as appropriate from one or both of the other regions. Europe is the center for work in optical storage, polymer optical fiber (POF), high temperature superconductivity, ferroelectric liquid crystals (LCD's) and ceramics. The U.S. is the focal point for work in nonlinear optics (NLO), photoresists and separation membrane technology. Complementing these is the important task of our new Japanese research center to transform this research into applications in various market sectors, and to develop as quickly as possible its own focus for lead research.

A final, and in many aspects, the most important, consideration with respect to our efforts to build effective, innovative international technology management teams regards people and the organizational environment supporting their efforts. In the international arena there are a combination of demanding requirements on the R&D manager. Above all, this person has to be open-minded, but must also possess, or be willing to work at developing, a broad range of traits and skills that are essential to building a strong organization. Among these are a) Having people development as a high personal priority, insuring opportunity for individuals and continuity for the organization; b) Encouraging and supporting creativity and innovation; c) Recognizing the importance of clear, timely communication; and the sharing of information; d) Being a risk taker and creating a risk-rewarding environment; e) Being able to cross disciplinary boundaries and being able to foster multifunctional working groups; f) Having a strongly developed sense of, and appreciation for, cultural differences; g) Being service oriented and being business oriented; being able to communicate effectively with customers and understand their needs; h) Willing to relocate; recognizing the importance of experiencing the business and social environment in a foreign country -- and the value of this to the organization and to customers; i) Having a long-term commitment to strategic research and to the organization and to customers; j) Being a non-conformist; having an open mind for "different" ideas and the personal strength and conviction to open new avenues of thought.

At the beginning of this discussion I said there was a dual agenda for the presentation; namely, to examine the impact, critical issues and future challenges posed by polymeric materials science and engineering and the globalization of technology, with the implications this has for new demands on the R&D organization, and the interactive requirements for success. We have, at this point, dealt with the first two -- impact and critical issues -- and I would like to conclude with a re-emphasis of the issues and a view of future challenges.

Polymeric materials science and engineering has provided the foundation for a wealth of world-wide economic and social benefits and clearly can continue to do so in the future, although the emphasis and challenges are now quite different. An agenda for the 1990's -- and perhaps beyond -- is as follows:

o Define the property limits of polymers in both structural and functional applications: This can entail improved or new synthetic methods which lead to highly pure starting materials; improved or new synthetic methods which omit weak linkages in macromolecules and, hence, lead to a major property increase of known materials; a much better understanding of characteristics such as strength, resistance to wear and various degradation mechanisms and how these relate to processing; definition of the limits of the use of polymers in electrical/electronic and, in particular, electrooptic and all optical applications where the promise of organics and the potential of polymer technology appears very large; design and synthesis of polymers with multi-functionality to extend and define, for example, the combinations of optical and electronic characteristics that can be built into a molecule.

o Address manufacturability and processing issues with a different perspective: The application of high performance, structural polymers -- including composites -- has been characterized by very long and costly development cycles and penetration of the major structural applications -- that is, metal replacement -- resulting from well over thirty years of activity is relatively very small. Perhaps the major obstacles to a more timely introduction have been delays in development of effective fabrication technologies for end-use conversion, along with an incomplete understanding of the potential cost impact on the total system and a definition of standards to help sort out what has become a rather confusing array of polymeric-based materials options. There are many reasons for this including cost and timing constraints associated with replacing or modifying installed capital, inherent conceptual difficulties in devising new or modified fabrication concepts and inadequate early research into fabrication and manufacturing implications -- and environmental concerns related to waste management and recycling are becoming, with very good reason, more and more demanding with respect to the application of polymeric materials. Failure to more creatively address these types of issues may deny full utilization of the materials and design advances that have been , and are being, made. This would seem to be a very fruitful area for both innovative process and engineering research and development and a new approach to the integration of materials synthesis with materials processing to define more commercially viable materials options.

o Extend and make more effective use of molecular design and optimization and morphological optimization: A number of points related to these opportunities have been noted above; namely, design of multifunctional molecules and structures and the integration of materials synthesis with materials processing. Recalling the earlier discussion on the four elements of materials science and engineering -- synthesis/processing, structure/composition, properties, performance -- and their inter-relationships, I would like to re-emphasize the unique advantages that can accrue to polymer engineering and applications through a renewed effort at a fundamental understanding of these relationships.

o Develop a better understanding of the biology/materials interface and the relevance of biomimetics: As has been noted, the utility of many modern materials is limited by fabrication technology and finished part costs. In polymeric materials, there are the additional problems of optimizing the molecular chain backbone, specifying the overall part morphology and, in composite materials, controlling the geometry and interface between hard and soft segments. These issues are especially relevant for very high performance materials, i.e., continuous fiber reinforced composites, materials designated for use under conditions of environmental extremes, materials which enable accelerated switching speeds in electrical or optical devices and parts which need to function reliably for extended periods of time. Biological systems perform such highly sophisticated functions and represent a strategy for the design of materials, parts and components which need to meet complex performance criteria in a variety of environments. These evolutionary engineered realities have emerged slowly and represent a successful materials synthesis and fabrication methodology. Examples of this biological approach" include: a) Creation of broad range of functional polymers (proteins) through controlled sequence assembly (enzymatic catalysis) of a fixed raw material supply (20 amino acids); b) In situ chemical production (cells) of parts and systems; c) Repetitive use of proven strategies as exemplified by hierarchical structures; d) Extreme control of the processing environment, allowing steady state operation far from equilibrium.

The possibility of extracting the principles of "evolutionary strategy" into modern engineering and scientific practice is an emerging focus of polymer science.

o <u>Maximize the value of computer-based capabilities:</u> Computer- based methodologies are currently in wide-spread use in the development of polymer science and the applications of polymeric materials -- and for very good reason. From component design to systems analysis through molecular design and interaction simulation, the advantages in speed, cost, visualization and insight are well established. In the areas of synthesis, processing and material properties predictions, future computing technologies are already coming into the picture. The structural data bases and reaction data bases can be maintained with optical recognition of chemical structures. Object-oriented data bases will allow the symbols used in chemical nomenclature to be as readily manipulated by computer as the letters of the alphabet. Automated synthesis schemes will find their way into the laboratory, so that, for example, polymer sequencing can be accomplished in the same way as currently can be done for synthetic protein synthesis. Neural networks will be used to understand the non-linear components to materials production for advanced scale-up capabilities. And inverse design, the ability to design structures based on desired properties, will become fully automated. Processing of synthetic materials will take advantage of the intrinsic properties of synthetic materials, so, for example, anisotropic materials will be chosen for a particular application because of unique advantages, rather than just as a metal replacement. Design for composite materials will have a stronger engineering base for fabrication and application. Computers will be used to recognize and separate plastics for recycling from other materials and from each other. Expert systems will be used to help troubleshoot process improvements. Molecular modeling will have developed a stronger correlation with experimental results. Each of the polymer properties prediction methods will be based on a more basic level so that properties are estimated accurately without having to do synthesis. Only the optimum materials are then synthesized. In this area, adhesion mechanisms will be understood, good estimates of free energy of interactions and entropy factors will be possible, and the correlation with physical properties established. Liquid crystal phase transitions and copolymer properties will be accessible and coupled with process history. Current work suggests also that membrane properties, amorphous polymer properties and even tensile strength predictions are achievable.

o <u>Fully evaluate and understand the different aspects of commercialization of material specialties:</u> With increasing urgency for the chemical/materials industry in a highly competitive environment, it is necessary to re-examine positioning on the value-added chain -- specifically to re-examine where optimum value can be obtained for the materials science and engineering skills inherent to the chemical industry and invaluable to other industries. This is particularly the case with respect to functional polymeric materials where the material in small quantity is bringing a critical, enabling capability to a device, component and/or system, and high performance structural materials and composites where materials replacement and design flexibility result in component and systems advantages in physical properties, weight, corrosion resistance, wear characteristics and others. The issue is not just extracting more value for the classical materials developer and supplier; it is just as much thoroughly understanding with respect to all parties involved at which point in the chain the best application of knowledge and skills results in the overall most cost/performance effective product and how capabilities among different types of companies and industries are most effectively shared.

o Be a role-model for the responsible development of industrial ecology: The chemical industry, and in our context here specifically the plastics industry, is under a hot, bright light with respect to materials safety and waste management issues and how it is dealing with lifecycle, toxicity, and recycle of various materials. To a certain extent, at this stage, the plastics industry is not getting objective treatment by the public -- the effects of Green Power and emotions are strong and cloud the situation; the chemical and plastics industries world-wide have an excellent record of safe and healthy practices. Plastics are at present only ca. 7% by weight of the composition of municipal and solid waste in the U.S. and Europe, for example (this is dominated by paper/paper board -- 40 to 25% -- and yard wastes -- 20 to 30% -- respectively (4,5). However, there is clearly a growing problem since the production and use of plastics is projected to increase at a rapid rate during the coming decade. In the U.S. alone, plastics use is expected to increase from an estimated 55B pounds in 1987 to 72B pounds in 2000, with post-consumer waste going from 36B pounds in 1990 to 49B pounds in 2000. Plastics packaging is expected to account for about 46% of all post-consumer plastics during the coming decade and remain by far the largest single contributor to the waste stream. Plastics recycling is still in its infancy but economics, including the costs in both money and quality of life to society of disposing of its wastes, will eventually determine the course. Integrated waste management will recycle the most valuable resins, incinerate the rest for their high fuel value and buy the ash. Recovery of most resins will have a negligible effect on the total market for virgin resins (4,5).

The fact is that we are dealing with finite resources on a crowded planet and there is a clear responsibility of industry to address the issues. The plastics industry has done this, and continuous to do so, in an effective manner. Indeed, this is an area where world-wide cooperation in sharing ideas, approaches and relevant technology can clearly and immediately benefit all.

Finally, resolving the issues and meeting the challenges of the technology that have been outlined are, as has been discussed earlier, dependent in no insignificant manner on building a global R&D and technology infrastructure that maximizes the strategic advantage of technology. To summarize, the principal objectives of this are to: a) Enhance the linkage between all the key functions critical to business development; b) Develop a basis for working with customers to provide innovative options that are regionally specific, as required; c) Accelerate technology transfer; d) Develop multi-disciplinary, multi-functional capabilities that can be efficiently shared; e) Develop the required new skills and traits in the R&D manager and in the acquisition of knowledge and the processing of information -- become a new kind of learning organization;

f) Proactively address problems of the environment; g) Develop true strengths from a cultural diversity.

I hope, in these few minutes, I have been able to convey to you some sense of our conviction of the central role polymers have in many new materials technologies; of the impact advances in polymers have had, and will continue to have, on progress and new developments in other technologies across a range of important industries; of the quite extraordinary range of applications now being, and potentially, served by polymers and the social and economic value of this service; and of the critical issues and future challenges facing polymeric materials science and engineering.

I also hope that I have been able to demonstrate to you that we are proactively addressing the demands and challenge of the globalization of technology and that we regard the tri-regional research concept as an investment for the future which will secure and strengthen our company at home and abroad. With this international network of research, Hoechst will keep pace with the intensifying global competition and provide leadership in science, technology and product development in its chosen areas of focus. The significance of international cooperation in the world-wide implementation of research plans incorporating leading technologies will grow, and Hoechst is committed to being a valued partner in this cooperation.

Factors on the path to success which deserve particular mention are the overcoming of functional, cultural and geographical barriers by emphasizing tolerance, motivation, encouragement and the targeted international control of resources. We are convinced that with our research network we have laid the foundation for numerous developments within the framework of our international research family, and that these developments will help us to continue to secure a leading position in research and business development and that our "Hoechst High Chem" message will quickly be supplemented by the word "international".

Thank you very much for your attention. It has been a privilege to participate in this conference.

1   Some introductory comments taken from an address by Professor Peter Eisenberger (Director, Princeton Materials Institute, Princeton University), Complex Materials: A Research and Strategy Challenge, on the occasion of the celebration of the 25th Anniversary of Hoechst Japan, Ltd., Tokyo, Japan; October 22, 1991

2   Booz, Allen & Hamilton, Inc., Outlook Magazine, 1987: "Global Management of R&D Resources"

3   Westney, D. Eleanor, Massachusetts Institute of Technology MIT JP 91-05, 1991: "The Globalization of Technology and the Internationalization of R&D"

4   Kirkman, Angela and Kline, Charles H., Chemtech, October 1991, 606: "Recycling Plastics Today"

5   Curlee, T. Randall and Das, Sujit, Materials and Society, 1991, Vol. 15, No. 1, 41: "Plastics Recycling: Quantity Projections and Cost Estimates"

Key Issues for the Japan's Chemical Industries Toward the 21st Century

Yotaro Nishida, Senior Managing Director, Ube Industries, Ltd., 12-32 Akasaka 1-chome, Minato-ku, Tokyo 107, Japan

We are, indeed, in the midst of very drastic changes on a global scale. In the politics, we have just observed the landsliding development of the Soviet Union, which was triggered by the German re-unification coincided with the revolutionary developments in the Eastern Block countries. Situation in the Middle East also continues to develop.

In the economic scene, there are also a number of vital movements around the world, which are too numerous to mention. To cite a few instances, we see such developments as the Unified EC toward the end of 1992 which accompanies an agreement between EC and EFTA on EEA, the formation of a free trade block in North America, current negotiations of the GATT Urguai Round, and so forth.

What all of us are greatly concerned is how a next new order will be re-established in the post cold-war world. All the industries will strongly be affected whether the world heads for a borderless free-trade system or a block economy system under protectionism.

Secondly, as another important issue, we have to think of global environmental problems. I am of opinion that the environmental protection should have the highest priority over all our activities as our business norm of utmost importance. This means that we have to prepare ourselves for incessantly increasing expenditures for R&D and facility investments in improving environmental performances.

Particularly, our chemical industry is in position to play the leading role in solving the problems, and the world also expects our positive posture in this respect. It is needless to say that, without coping with this environmental issue, there will be no future for the chemical industries. It should also be noted that we may find new business opportunities in the course of solving environmental problems.

Thirdly, I have to point out an issue relating to energy and materials. Since the resources of energy and materials for the chemical industries are limited, these prices will inevitably go up in the long run. So, our efforts should be paid not only to the above mentioned environmental issues but also to the resource conservation and energy saving. It is our important task to continue to develop alternative energy and alternative materials, where process innovation should be emphasized.

Taking account of these backgrounds, I would like to give my consideration to the key issues for the Japan's chemical industries toward the 21st century.

Y. Imanishi (Ed.)
Progress in Pacific Polymer Science 2
© Springer-Verlag Berlin Heidelberg 1992

1.   OUTLINE OF THE JAPANESE CHEMICAL INDUSTRIES

The outline of the Japanese chemical industries in 1990 is
summarized in contrast to the previous year as seen in the
Table I.

Table I   Chemical Industries in Japan

	1989	1990	Changes
Production Index	128.7	134.4	+ 4.4%
Manufacturer's Shipment Index	127.2	134.0	+ 5.3%
Total Shipment Valve (MM$)	170,362	181,538	+ 6.6%
Ethylene Production (1,000 MT)	5,603	5,810	+ 3.7%
Export Value (MM $)	14,776	15,872	+ 7.4%
Import Value (MM $)	15,948	16,045	+ 0.6%
Facility Investment (MM$)	15,452	17,892	+15.8%
Workforce Index	103.2	103.7	+ 0.5%
Domestic Wholesale Price Index	91.4	92.5	+ 1.2%

Source: Japan Chemical Industry Association
Notes: 1985 Indices = 100, Exchange Rate = Yen130/$

The production of the chemical industry in Japan has been very
positive, showing a consecutive growth over the last 9 years
since 1982.   The chemical industry represented 7.4% of the total
shipment value of the whole Japanese manufacturing industries in
1989.   This was ranked at the 4th position after the three big
sectors, namely, electric/electronic industries (17.1%),
transportation industries (14.1%), and machinery industries
(10.0%).

The export value of the chemical industries was 5.4% of the
Japan's total export while the import value was 7.6% of the total
amount.   In the chemical sector, the import has exceeded the
export since 1982.

Investments have actively been made for the chemical industries,
which have been largest among the basic material manufacturing
industries during the last four years from 1987 to 1990.

The shipment value of the world chemical industries in 1989 was
as follows:

	Value (100 MM$)
U.S.A.	2,555
Japan	1,611
Former West Germany	850
France	546
United Kingdom	442
Italy	442

Source: CEFIC

As seen in the foregoing presentations, our Japanese chemical
industries as a whole had been in very good shape from the
macroscopic point of view.   However, we observed the downtrend
started around the spring of 1991.   By now, many chemical
companies are obliged to revise their sales and profit plans in a
downward direction.   Investment cutbacks of manufacturing
facilities are reported, while production curtailment is
considered in certain sectors.

Furthermore, when we think of the future of our industries, we
will be faced with the problems to be discussed in the following
parts.

2.   PROBLEMS BY WHICH THE JAPAN'S CHEMICAL INDUSTRIES ARE
      CONFRONTED

(1)   Size of Individual Company

Table II shows a 1989 ranking of the world chemical companies in
order of sales volume.

Table II   World Chemical Industries (1989)

Rank	Company	Sales MM$	Profit MM$	Profit % Sales
1	DuPont	35,209	2,480	7
2	BASF	25,317	1,071	4
3	Hoechst	24,403	1,026	4
4	Bayer	23,021	1,107	5
5	ICI	21,889	1,733	8
6	Dow	17,730	2,487	14
7	Ciba-Geigy	12,598	952	8
8	Rhone-Poulenc	11,463	642	6
9	Enimont	11,191	522	5
10	Norsk Hydro	9,599	389	4
11	Akzo	8,838	450	5
12	Union Carbide	8,744	573	7
13	Monsant	8,681	679	8
14	Mitsubishi Kasei	8,649	157	2
15	Asahi Glass	8,088	427	5
16	Sumitomo Chemical	6,938	253	4

Source: Fortune June. 1990

Even the largest-scale Japanese companies, that is, Mitsubishi
Kasei and Sumitomo Chemical were respectively ranked at the 14th
and the 16th position among the world chemical companies.   The
profit level of the Japanese companies was fairly lower than that
of the large-scale U.S. and European Companies.

Japanese chemical industries, when viewed on a world-scale, are
very fragmented, namely, there are a number of medium- and small-
sized companies which are severely competing with each other in
one place, and are also overlapping in many R&D subjects.

A comparison was made in the Table III in terms of ethylene
production capacity of the top 10 ethylene producing companies,
respectively in the world and in Japan.

Table III   Ethylene Capacity (1990)

(Thousand Tons)

WORLD		JAPAN	
1. Shell	4,300	1. Ukishima PC	947
2. Dow	3,500	2. Idemitsu PC	751
3. Exxon	2,700	3. Showa Denko	709
4. NOVA	2,000	4. Mitsubishi PC	672
5. Enimont	1,700	5. Maruzen PC	480
6. Union Carbide	1,600	6. Tohnen PC	463
7. Lyondell Petchem.	1,500	7. Sanyo Ethyl.	440
8. Oxy Petchem.	1,500	8. Mizushima Ethyl.	400
9. Pemex	1,400	9. Tohso	377
10. Atochem	1,400	10. Sumitomo Chem.	367

Source: Takahashi, S. Kagaku Keizai, Feb. 1991

Such a wide gap in the production size will eventually lead to a gap in overall business competitiveness, especially, for petrochemical industries where a scale factor is one of the key economic elements.

We have to admit that our Japanese chemical companies lack so-called 'bodily powers' on an individual company basis, when we think of further increasing investments in the R&D and the environmental protections as well as in the globalizing activities.

(2) Globalization

Table IV indicates how the leading U.S. and European chemical companies have already expanded their activities in the world market.

Table IV Globalization (Sales by Destination %)

Company	Year	Destination and Sales Percentage				
Du Pont		U.S.A.	Europe	Others		
	1989	60	29	11		
	1982	67	21	12		
Dow		U.S.A.	Europe	Others		
	1989	46	31	23		
	1980	43	31	21		
ICI		U.K.	Conti. Europe	America	Asia Pacific	Others
	1989	22	25	29	18	6
	1980	42	19	16	16	8
Akzo		Nether land	Germany	Rest of Europe	North America	Others
	1989	9	15	37	21	18
	1980	12	18	38	19	13
Hoechst		Europe	North America	Latin America	Africa, Asia Oceania	
	1989	58	22	6	14	
	1980	67	11	6	16	
Bayer		Europe	North America	Asia Pacific	Latin America	
	1989	66	20	9	5	
	1980	68	15	8	9	
BASF		Europe	North America	Asia Pacific	Latin America	
	1989	64	22	6	8	
	1980	76	11	5	8	

Source: Annual Reports, and Higashi, K.'Kagaku Keizai'Jan. 1991

In contrast to the above situation, only around 10% of the total sales is made outside Japan in the case of the Japanese companies. Foreign investments by the Japanese chemical companies are gradually increasing as were 910 million dollars in 1987, 1,292 million dollars in 1988 and 2,109 million dollars in 1989. However, we should say that we are still in the very early stage in this direction. An immediate and important task is how to globalize our business in the three major industrialized regions, namely, North America, Europe and Asia.

A recent significant tendency is an increasing interest from the U.S. and European companies toward Japan and the Asian countries as a strategic target area. The major reason for this movement is that a higher growth potential is expected in this region than in any other part of the world. Namely, substantial market is to be added in the future.

Shown in the following tables is a long-term forecast made by the British Chemical Industry Association:

Table V Regional Share of Chemical Production

Region	1989 (%)	2000 (%)	Change (%)
W. Europe	27.5	24.6	-10.5
N. America	23.2	21.8	- 6.0
Japan	16.3	17.2	+ 5.5
E. Europe	15.3	15.3	0
Far East	5.0	6.4	+28.0
Others	12.7	14.0	+15.7

Source : CWI April 25, 1990

Table VI 1989 - 2000 Average Annual Growth Rate (% per Year)
- GDP and Demands for Chemicals -

Region	GDP		Demand for Chemicals	
	'89 - '95	'95 - 2000	'89 - '95	'95 - 2000
W. Europe	2.8	3.0	3.0	3.2
N. America	2.8	3.2	3.4	3.8
Japan	4.3	4.5	4.6	5.0
Asia Pacific	6.2	6.5	7.0	7.5
E. Europe	1.5	3.5	3.0	5.0
M. and S. America	4.0	5.5	5.0	7.0
Indian Sub-Cont.	4.7	5.0	6.0	6.5
Africa	2.5	2.5	3.5	3.5
[OECD]	2.9	3.3	3.6	4.0
World	2.9	3.5	3.8	4.6

Source : ECN Jan. 22, 1990

In this respect, many U.S. and European chemical companies are actively investing in Japan and other Asian countries. In particular, such big companies like Hoechst, Bayer, BASF, du Pont, ICI and the like have already established substantial business in Japan, amounting to, in terms of group turnover, one hundred and several ten billion Yen, or more than two hundred billion Yen, which may corresponds to the business scale of a medium-sized Japanese chemical company.

The following are the reasons why Japan attracts attention of these global industries:

a) As the economic center of Far East, Japan has already a very big existing market. Also, it is regarded as one of the areas which have the biggest future potential.

b)   Japan is expected to continue to have the leading edge on
     the world in many high-tech business, namely, in the field
     of automotive, electronics, information devices, electric
     appliances, opto-electronics, bio-industries and so forth.

c)   Because of the stringent requirements for the highest
     quality, any product, if successful in the Japanese market,
     will be well accepted in the rest of the world.

d)   Japan's geographical position will increase its importance
     as the relation develops in both economics and technology
     between Japan and other Asian countries.

Historically, foreign companies started activities in Japan by
way of products importation or establishing joint ventures.
Today's approach is by way of establishing research labs and
production facilities on their own so as to lay a solid
foundation in preparation for the future development.

Japanese chemical companies will inevitably be exposed to severe
competition with those global companies not only in Asian
countries but also in Japan.

(3)   Environmental Issues

Top priority should be given to environmental issue as a norm of
utmost importance not only in the chemical industries but in all
business activities. What is required for the sound development
of our business is a harmony with the surrounding environments
and good social acceptance.

As a matter of fact, we have to admit that the chemical
industries have so far a rather poor public image in respect of
environment and safety. Of course, efforts are being made by the
top management members themselves to propagate how the world
chemical companies are devoted to improve the environments at
respective companies. In fact, expenditures for the
environmental protection increase year by year.

Many of the environmental problems, whether global or local, can
be best solved by means of chemical technology. Therefore, it is
a task of our chemical industries to make a contribution to the
world in this respect. This will again provide us with new
business opportunities.

Now, as an example of the environmental issues in Japan, I would
like to take a look at the today's situation and the future
outlook of plastic waste management.

A special committee for the plastic waste management at Japan
Petrochemical Industry Association made a survey of plastics
volumes in 1988 in terms of the produced, wasted and recycled
amounts with respect to the certain plastics, as seen in the
Table VII.

Table VII Produced, Wasted and Recycled Plastics in 1988

[Volume: Thousand Tons, Ratio: %]

Category	LDPE	HDPE	PP	PS,AS ABS	PVC	Others	Thermo-setting	Total
Recycled(A)	111	77	134	140	58	60	0	580
Produced(B)	1,422	959	1,559	1,891	1,838	1,461	1,887	11,016
Wasted (C)	1,586		802	1,060	761	312	357	4,878
Recycle ratio								
A/B	7.8	8.0	8.6	7.4	3.2	4.1	0	5.3
A/C	11.9		16.7	13.2	7.6	19.2	0	11.9

Source: Special Committee for Waste Management, Japan Petrochemical Industry Association

The total production of all plastics materials was 11 million tons in 1988, out of which 4.88 million tons were assumed to have been disposed as waste after processing and use. Out of the waste plastics, 580 thousand tons were recycled in the form of regenerated pellets or fabricated goods. The remaining 4.3 million tons were either incinerated (3.17 million tons) or disposed by landfill (1.13 million tons). Part of the incineration heat, corresponding to 720 thousand tons of the waste, was recovered as electricity.

1.3 million tons of the waste plastics were recycled or re-utilized as energy source, which corresponded to 27% of the whole waste volume.

Proposed directions of waste management are as follows:

A)   Material-Saving

Down-sizing of wall thickness, volume reduction in processing charge and prolongation of product life for the purpose of minimizing the volume of plastics wastes.

B)   Recycling and Re-utilization of the Waste

1)   Increase of the energy recovery rate:

- Installation of power generation facilities based on the plastics waste incineration
- Development of an efficient heat recovery technology
- De-regulation and de-control in respect of legal restrictions on the power generation

2)   Promotion of re-utilization:

Today's difficulties lie in the variety and numerous grades of the plastics, the immature collection systems of the wastes, and the limited quality of the regenerated products. Both of the social infra-structure to collect the wastes and the re-utilization technology should well be developed in parallel.

Proposed Collection System

- Marking on plastic parts to be recycled
- Re-designing of product to allow an easy collection
- Information network with respect to the details of plastic
  wastes
- Cooperative linkage between the plastic manufacturers and
  the plastic users to promote collection and re-utilization
  of the plastic wastes.
- Establishment of an efficient collection system covering
  the public consumers, local governments and
  regeneration/re-utilization business undertakers.

Improvement of Re-generated Product Value and Market Development

- Product development based on regenerated plastics
- Improvement of regeneration processes

Development of Regeneration Business

- Expansion of favorable financial arrangements

Already having this subject in mind as a common
understanding in the plastic industries, respective plastic
companies are conducting various studies for the collection
and re-utilization of the plastic wastes.

A big topic in this respect is a possible recycling system
of automotive plastics. Plastics used in cars have
increased year by year, as was 4.7% of the total car weight
in 1980, 5.7% in 1984, and 7.5% in 1989.

Together with the light weight metals such as aluminum and
magnesium, plastics are expected as a promising material for
automotive applications for the fuel-saving light weight
structure.

However, if a recycling possibility is incorporated in the
criteria of the material selection, automotive plastics may
greatly be affected. Plastic companies are now involved in
joint studies with automotive companies in respect of the
recycling possibility.

(4) Feedstocks

Almost all raw materials for Japan's chemical industries come
from overseas sources. The largest among them is petrochemical
feedstocks. In 1990, 97.5% of the feedstock for ethylene
production was naphtha, while the remaining 2.5% was LPG. 72.5%
of the naphtha was imported from overseas sources, out of
which more than 70% came from the Middle East region. An
extremely high naphtha ratio in the feedstock structure and its
high dependence on the overseas sources constitute our concern
for the future.

Traditionally, the imported naphtha price was linked with the crude price, and was considered as crude price plus 1$/bbl. However, recently the price margin has become as much as 4 to 6$/bbl due to the increasing demand, especially, in the Asian regions.

In the light of the tightening naphtha market, highlightened as an important subject of the petrochemical industries was a diversification of the feedstocks into LPG, heavy condensate, gas oil and others. Currently, certain studies are underway including such legislative arrangements as taxation and reserve mandate. Already there is a movement toward the exemption of heavy condensate from the oil tax.

To take ammonia as an example of the basic chemicals, the production was 1.86 million tons in 1990. However, its feedstocks were already diversified as follows:

Coal, Petrocoke	20%	Naphtha	13%
LPG	14%	LNG, NG	21%
Off-Gas	22%		

Due to the recent price hike of naphtha, the economic preference is presently in the order to coal and petrocoke in the first place, LPG and LNG in the second place and naphtha in the third place in terms of the variable production cost.

A high operating rate has been maintained, as is currently 93.5%, after the capacity reduction was made some years ago. It will not be in the long distant future when the importation of ammonia will be started.

In the case of methanol, already 93.7% of the domestic demand is supplied from the overseas sources. In pursuit of such economical feedstocks as natural gas and oil well gas, off-shore production of methanol and ammonia will be actively developed by way of direct investment.

It appears that the power rate is relatively high in Japan when compared with those of the other countries. Previously, an average power rate for a large account was not as much as Yen 10/KWH (7.7 Cent/KWH at the exchange rate of Yen 130/$) before it went up to Yen 17.07/KWH (13.1 Cent/KWH) due to the sharp increase of the crude oil price in 1980. Since then, that rate had been maintained until 1985.

After that, the power rate has been revised four times because of the substantial drop of the crude price together with the sharp appreciation of the Japanese currency. Thus, it went down to Yen 13.30/KWH (10.2 Cent/KWH) in 1990.

As we think this level still high, many Japanese chemical companies operate coal-fired power stations in order to minimize the energy cost. Although it depends on a steam/power ratio, a saving of around Yen 5/KWII (3.8 Cent/KWH) is expectable.

As I explained in the foregoing parts, our Japanese chemical industries have no advantages at all as far as feedstocks and energy are concerned. This characteristic has already been embodied in the business structure of methanol, because the feedstock cost differential is most significant. I think the production of ammonia and fertilizers will more or less follow the same shift as the methanol industries experienced.

The following had a high import ratio in 1990 among the petrochemical products:

Product	Import Ratio
Ethylene Glycol	24.8 %
EDC	18.4 %
Acrylonitrile	14.8 %

While the import ratio of these commodity products further increases, Japanese chemical companies will step forward into the area of specialty products.

In line with the globalization, overseas investments will be developed not only for the marketing purposes but also for the upward integration in pursuit of competitive feedstocks.

(5) Restructuring

I am of opinion that there are three key principles for business management, namely, they are:

Clean-Up Operation, Fix-Up Operation, and Innovation.

Clean-Up Operation means to withdraw from marginal and loss-making business, or to withdraw from those operations which are not in line with the long-term corporate strategy, and to concentrate thus obtained managerial resources on the development of the key operations.

Fix-Up Operation is to provide the selected remaining operations with thorough reinforcement so as to bring them up to so-called 'a top group runner in a marathon race', who is constantly highlightened in the live TV program. For this purpose, we need incessant improvements in both products and processes. In particular, I emphasize the importance of catalyst development, process simulation and computer controlling when process development is conceived.

In parallel with technical studies, we should also develop strong marketing powers, a functionally organized structure, an efficient information network, and a workable logistic system.

Successful innovation also requires a close connection in the both aspects of products innovation and process innovation.

Although this kind of restructuring is effectively promoted in the U.S.A. or in Europe by means of M&A, it is not the case in Japan because of the difference in the social structures. There are several explanations why M&A is hindered in our country. One reason may be the antimonopoly law which prohibits a holding company system. Entangled company grouping arrangements and the life-time employment system are also considered as hindering factors of M&A.

The degree of business concentration is by far lower in the Japan's chemical business, when compared with those of the other sectors. A number of companies of the similar size are severely competing with each other in the same business domain. As a consequence, Japanese chemical industries have so far repeatedly followed a vicious cycle, namely, to follow a route of -- over capacity -- price erosion -- business loss -- weak financial structure.

Having experienced so many negative consequences of the excessive competitions, Japanese chemical companies are now well prepared for a self-control in expanding production facilities by carefully taking account of the supply and demand situations.

Another new attitude is to organize a joint investment by establishing an alliance between several companies. This type of cooperative arrangement may increase importance in the future of the Japanese chemical industries because of the difficulties of M&A.

(6) Improvement of the R&D Efficiency

I would like to point out that the R&D expenditure has a large share in the corporate budget. The following are the 1989 figures of the ratios of R&D expenditures against the total turnover at several sectors in the chemical industry of Japan:

Chemical Industry (in general)	4.84%
- Chemicals, Petrochemical and Textile	4.09%
- Fatty Oil, Paint	3.93%
- Pharmaceutical	7.50%
- Others	4.11%

(Source: Statistics Bureau, Prime Minister's Office)

Although this level itself may be comparable to those of American or European companies, the absolute amount is by far smaller because the size of company is small. Also, in reality, there are many overlapped R&D programs being conducted in the same domains and in the same directions.

Key factors for the successful R&D may lie in such concepts as:

Focused	Committed	Management Support
Professional	Adapt to the Market	Patient

An essential element to the success is to try to diversify into the neighboring areas of one's own strong fields by eliminating so called Me-too-ism.

Also important in the R&D activities is to promote alliances in the form of joint studies with technical colleges, governmental institutions and other private companies.

3.   TARGETS FOR THE FUTURE DEVELOPMENT

Recently, a number of Japanese chemical companies worked out and publicly released their respective visions toward the 21st century. The following are the common features of these visions (1):

1)   Global activities are strongly intended as advocated as a 'World Enterprise', or an 'International Company'.

2)   To double the sales amount within 10 years.

3)   To make a clear determination of strategies as a whole group including related companies, to improve the efficient use of the business resources and to expand the business size of the whole group.

4)   To expand the business scale, while promoting a restructuring in terms of business line and product-mix --- to increase the market share in the commodity business, and to promote positive developments in the processing business together with diversification into new products.

5)   To promote internationalization by developing, as a core approach, direct investment in the overseas countries.

6)   To attain a pre-tax profit level of more than 10% of the turnover.

7)   To raise an owned capital ratio to a level of 50% to 60%.

8)   To promote a positive increase of R&D budget to minimum 5%, preferably to a level of 10% of the turnover.

All of the above points constitute a future image of a Japanese chemical company, which is commonly conceived by the management members. It is, however, very unlikely that all companies will be successful in materializing above rosy pictures. Most probably, we may face with a harsh era of selection, and the differences in company competence may become further great among the chemical industries.

(1) Yamamoto, K. ; Kagaku Keizai (Aug.) 1991

The Next Thirty Years - Will it Fulfill our Expectations?

Robert M. Nowak

Central Research & Development, 1776 Building,
The Dow Chemical Company, Midland, Michigan, USA 48674

There is an old English saying, "There's nothing new under the sun." We, as scientists, know that this is not true. In fact, companies who embrace that philosophy usually find themselves economically disadvantaged and often bankrupt due to the technological innovations of their competitors. A company called Photoprotective Technologies, Inc. uses as a slogan on one of their brochures, "There is something new under the sun." And that new emanates from science and technology. The future for discovery, for the development of enabling technologies, for paradigm shifts, has never been brighter. Science and technology will never let us down. It will always be exciting, fast moving, and highly competitive.

A lot has happened since 1899, when Charles Duell resigned as director of the United States Patent Office because he is claimed to have said, "Everything that can be invented has been invented." By 1899, science had barely awakened.

The technological world has had its greatest achievements since World War II.

World War II was a terrible event. It laid waste to the industries of Europe and the Far East. With Europe and Japan in ruins, and American industry geared to mass production as the world had never known it, America had an almost insurmountable lead in the world of commerce. But surmount it the Europeans and Japanese did. For they had a significant advantage not available to American industry. And that advantage was to work with, as we Americans call it, "a clean slate."

Who does not recall the story of the construction of a new, modern Japanese steel industry unsurpassed in the world after World War II. Who does not know the story of the growth of the Korean steel industry that suddenly gave Japan serious competition? Who does not know the story of the Japanese automotive industry, the envy of the United States and the countries of Europe. Suddenly, a new concept was introduced -- that of quality and the importance of satisfying the customer -- while the U.S. auto industry was still caught up in the mass production syndrome. As the competitive world changed, it became apparent that the eighties and the nineties would be characterized by intense competition, global competition. Thomas Merton, a noted Catholic theologian, once wrote, "No Man Is An Island." No country can any longer remain an island unto itself.

And, of course, as we progressed through the eighties, other Far Eastern countries began or are beginning to emerge as economic powers -- Korea, Taiwan, the countries of Southeast Asia. What does the future hold for Eastern Europe and Russia and China? History has shown that a democratic free market approach will win every time in the world of science and the world of commerce. Creative, innovative, hard-driving people must be allowed to follow their initiatives and dreams. They must be allowed to build, to create, and to accumulate as one way of tangibly demonstrating their accomplishments. As the countries of Eastern Europe and Russia and China understand and embrace that principle, these great nations, too, will join the competitive world.

The future has never been brighter for science and the commercialization of new science. Never have more scientists been at work continuing to build on already established scientific principles and bases and establishing completely new scientific principles and bases from which whole new industries will be born.

Y. Imanishi (Ed.)
Progress in Pacific Polymer Science 2
© Springer-Verlag Berlin Heidelberg 1992

The future will be characterized by:

1. A continually growing world scientific community.

2. More intense global competition.

3. A greater difficulty in establishing and maintaining unique technological/commercial positions by any single company.

4. The emergence of new democracies developing capabilities to compete with the well-established major economic powers.

5. More multinational alliances to take advantage of the unique positions of each of the partners.

6. More cooperation multinationally in very expensive, highly scientific areas which could include such areas as the superconducting supercollider or mapping the human genome.

7. The emergence of whole new areas of science and the development of enabling materials to allow new degrees of freedom in new commercial applications.

New areas of science and technology are exploding all around us at this very moment and will continue to accelerate at such a phenomenal rate that it will be difficult for companies to choose which paths they should take to build competitive advantages for their futures. One thing is certain -- no company can become or remain excellent in too many areas. Each must choose where it will try to become technologically excellent and commercially sound.

During the last sixty years, there have been a number of major scientific revolutions, revolutions that have changed the face of science and allowed for the development of major new commercial opportunities. Four of the most important scientific revolutions are understanding the nature of atoms and molecules, the computer, biotechnology, and advanced engineering materials.

The Bohr/Pauling era allowed us to understand the nature of atoms and molecules and laid the basis for many important discoveries. The groundwork for chemistry and physics was being laid.

The computer ushered in the electronics world as we know it today and envision it tomorrow. The electronics business has become in a few short decades the largest business in the world today. The computer has become enabling technology for many other scientific areas.

The exciting new science of biotechnology was born from the research carried out in American and English universities in the 1970s. The rapid evolution of new techniques in DNA replication and mapping, hybridoma synthesis, monoclonal antibody specificity, chimerization, and, recently, polymerase chain reaction (PCR) promises to change significantly the pharmaceutical industry and eventually the agricultural industry.

A fourth major scientific revolution taking place today is that of providing enabling materials that are allowing some of the most exciting large growth industries to fulfill further their visions of the future. This scientific revolution is in the world of advanced engineering materials. This area promises to deliver to the electronics industry, to the aerospace industry, to the machine and machine tool industry, even to the automotive and construction industries those materials necessary to free the designers from the limits now restricting them from moving forward to ever more exciting innovations. As always, depending on the industry and the application, it will be the balance of price/performance that dictates where and when such materials will be used.

The world of advanced engineering materials is already very large, for it encompasses such materials as advanced composites in their many forms and possibilities. It involves fine ceramics as well as improved metals, but, perhaps most of all, it involves new polymer chemistry. The Department of Commerce issued a report in the spring of 1990 called "Emerging Technologies, A Survey of Technical and Economic Opportunities." Twelve emerging technologies were identified and analyzed for potential commercial impact in the year 2000. These technologies in four major categories feature a combined U.S. market potential of about $350 billion in annual product sales and a world market potential approaching $1 trillion.

The major categories and market potential are shown in the following two slides.

Major Emerging Technology Categories

Emerging Materials

- Advanced Materials
- Superconductors

Emerging Electronics & Information Systems

- Advanced Semiconductor Devices
- Digital Imaging Technology
- High Density Data Storage
- High Performance Computing
- Optoelectronics

Emerging Manufacturing Systems

- Artificial Intelligence
- Flexible Computer-Integrated Manufacturing
- Sensor Technology

Emerging Life Sciences Applications

- Biotechnology
- Medical Devices and Diagnostics

Emerging Technologies and Market Potential
(USA by the Year 2000)

	$ Billion
Materials	155
Electronics & Information Systems	148
Manufacturing Systems	30
Life Sciences	23
Total	356

The area of advanced materials is almost half of that vision and polymers will be a significant portion of that half.

There has never been a more exciting time to be working in the area of advanced polymeric materials. Choose your material area:

Advanced Polymeric Materials

New Engineering Thermoplastics
New High Performance Thermosets
Thermotropic Liquid Crystals
Lyotropic Liquid Crystals
Molecular Composites
Advanced Composites
Non-Linear Optical Polymers
Electronically Conducting Polymers

The list of exciting new engineering materials is constantly expanding. Unfortunately, the discovery of a new polymer does not necessarily mean a commercially viable new business can be formed.

The area of advanced engineering thermoplastics is a good example. Companies have been inventing new engineering thermoplastics for the last twenty years.

Beginning with Union Carbide's polysulfone around 1972, no less than ten to twelve advanced engineering thermoplastics have been commercialized by as many as fifteen companies throughout the world. While total world capacity may have reached 140 million pounds, world consumption for 1989 has been estimated to be only sixty million pounds.

Our analysis says that no company is making an acceptable profit in this business today. The problems are too many competitors, too many materials, products too expensive to produce, and markets too small. And yet new products keep emerging. Polyphenylene sulfide's volume is about one-quarter of that total. Will polyphenylene sulfide emerge from the pack to become a profitable business? The next few years should answer the question. It's been a slow start for polyphenylene sulfide. It's taken over twenty years from the initial commercialization of polyphenylene sulfide to reach a global sales volume of around 20 million pounds. There are already eight competitors globally and a great deal of excess capacity.

Will thermotropic liquid crystal polymers repeat the problems of the engineering thermo-plastics market? Thermotropic liquid crystal polymers may prove to be the next super property thermoplastic. However, here too, though discovered in 1972, volume has not yet reached twenty million pounds with more than half the total being Amoco's Xydar, a poly-mer passed from Carborundum to Dart to Amoco. More than fifteen companies are samp-ling materials in the marketplace, and no company has a strong patent position. The prob-lem of too many competitors for each new technology area is one that we will have to learn to handle more appropriately. I am convinced that these problems are solvable but not without very hard work, ingenuity, and imagination.

There never seems to be a dampening of enthusiasm for new materials, for there is no limit to the imagination of the organic polymer chemists to continue to invent thrilling new materials. The problem will never be in the invention, it will continue to be in the profitable commercialization of these new materials. Market opportunities will be the key. Existing markets must grow; new opportunities must be found. We must not only be creative technically, we must also be creative commercially.

As bright as the future looks for developing new technologies, a number of significant problems do exist. The problem of too many fierce competitors was already mentioned. The question is, in this era of intense competition, can any company carve out its special profitable niche for long? Can a superior patent position be strong enough to allow sufficient payback for the research before competitors step in? This worry has not deterred many companies from increasing R&D budgets and chasing many of the same objectives. A partial answer will be global alliances between companies on various continents sharing research and the business. Another answer is that many of the new materials brought to the market will not be commercially viable and must be, will be, abandoned. Will PEEK survive? Will Ultem survive? Will PEK survive? Only time will tell. Those remaining may make the "good business" list.

While the future, indeed, is bright, chemistry has created problems for the environment and new chemistry will create more problems for the environment. More attention than ever before needs to be given to air and water quality, to plastic recycle, to toxicity of new materials. Certainly, the world has now awakened to the ecological and health safety problems scientific progress can create. Responsible action must walk hand in hand with new discoveries and commercialization.

There is a long list of problems that continually need to be addressed. Problems created by PCB's, lead additives, pesticide residues, SO_2 emissions, plastic recycle, fluorocarbon blowing agents, chlorinated hydrocarbons, solvent emissions, and the list goes on and on and will continue to grow. Is ozone depletion a problem that chemistry can solve? How serious is the greenhouse effect? Was the Montreal Protocol necessary? Do we now jump too soon on scanty data to solve perceived problems or must this become our mode of operation whether true or perceived? The answer is environmental concerns will be raised to the legislative level in countries around the world long before the last pieces of data are gathered to determine the actual size of the environmental insult. That being the case, we must examine each new product we bring to the marketplace to make sure we understand and cope with any negative environmental impact. Our goal must always be zero negative environmental impact.

We scientists have a tremendous obligation to assure that when our discoveries are commercialized they are produced, distributed and used in a safe manner. The world is awash with environmental zeal. Environmental groups everywhere are forcing new legislation, some thoughtful, some alarming, that is shaping our industry. We must work in the political arena in our respective countries to strive for responsible legislation that preserves the environment without unreasonably stifling our industry. The environmental wave will cause new product development to weigh heavily in favor of durables versus disposables, of utility versus convenience. While this trend will certainly limit the growth of some of our commodity plastics, our ingenuity will develop new applications to replace those loses.

There are other problems that could be enumerated. In the United States, the Chemical Manufacturers Association (CMA) has recently completed an attitude survey of the general public. This survey showed that the chemical industry ranked near the bottom in favorability -- only ahead of the tobacco industry. These results showed that the public generally thinks that the chemical industry lacks credibility, cannot be trusted, is foot dragging in its efforts to improve itself, especially on environmental issues, and is neither honest nor ethical in its dealings with others. In short, the chemical industry in the U.S.A. has a tarnished image which we must work very hard to erase. What is the public attitude about the chemical industry in your country? Probably very similar to the attitude in the U.S. This is a problem we must address. This problem must be challenged and corrected. If left unchallenged, it can have severe negative consequences on our industry.

We could also talk about a growing shortage of well-trained scientists for the future which appears to be emerging not just in the U.S. but other countries as well, but I'll leave that for another day.

There are no limits to the creativity of man. There are no existing problems that cannot be solved, no ecological problems arising from new technologies that can't be solved.

We are held back only by those whose imagination does not allow them to advance beyond where we are today, by those who are satisfied with the way things are, by those who abhor even the slightest risk that will take us into ever more exciting and new horizons that the unknown holds for us. These people cannot stop progress, they can only impede it.

It is man's vision and the ability to make that vision come true that is so exciting for us as scientists.

The promise of the future makes the discoveries of the past seem insignificant.

Rigid Chain Vinyl Polymers from Multi-Substituted Ethylenes

Takayuki Otsu*, Akikazu Matsumoto, and Masahiro Yoshioka

Department of Applied Chemistry, Faculty of Engineering, Osaka City University, Sumiyoshi-ku, Osaka 558, Japan

Abstract: Novel vinyl polymers were synthesized from 1,1- or 1,2-disubstituted and 1,1,2-trisubstituted ethylenes by radical polymerization. The polymers obtained consist of a rigid chain structure on account of the bulky side groups compared with flexible poly(monosubstituted ethylene)s. The substituted polymethylenes obtained from 1,2-disubstituted ethylenes such as fumaric and maleic derivatives were revealed to have new properties different from ordinary vinyl polymers. Radical polymerization behaviors of these multi-substituted ethylenes and some properties of the resulting polymers were investigated.

1. INTRODUCTION

A number of monosubstituted ethylenes homopolymerize in the presence of an initiator to give high molecular weight polymers consisting of a substituted polyethylene structure, which has a methylene group in the recurring unit [1]. Since the existence of such a methylene group makes these polymers flexible and processable, a number of plastic materials have been produced industrially. Some 1,1-disubstituted ethylenes such as alkyl methacrylates can homopolymerize to give high molecular weight polymers consisting of the substituted polyethylene structure, but many of other 1,1-disubstituted ethylenes do not give high polymers because of steric effect of the substituents in propagation, i.e., a low ceiling temperature.

On the other hand, cis or trans and cyclic 1,2-disubstituted ethylenes have been well-known till 1980 to be not homopolymerized with a radical initiator owing to the much more increased steric effect of the substituents except for a few exceptions, e.g., fluoro-substituted ethylenes and cyclic derivatives, vinylene carbonate (VCa), maleic anhydride (MAn), N-substituted maleimides (RMI), and acenaphthylene. However, if 1,2-disubstituted and tri- or tetra-substituted ethylenes can homopolymerize, the polymers with a substituted polymethylene structure, which is different from the above polyethylenes, would be obtained (eqs. 1 and 2), i.e., the polymers would be less flexible (more rigid), and have higher glass transition temperatures than those for the polyethylenes, although their processabilities would decrease.

Y. Imanishi (Ed.)
Progress in Pacific Polymer Science 2
© Springer-Verlag Berlin Heidelberg 1992

$$n \quad \underset{\substack{| \\ X}}{CH} = \underset{\substack{| \\ Y}}{CH} \quad \xrightarrow{\quad R\cdot \quad} \quad \left(\underset{\substack{| \\ X}}{CH} - \underset{\substack{| \\ Y}}{CH} \right)_n \tag{1}$$

(cis or trans)

$$n \quad \underset{\substack{\rule[0.3ex]{2em}{0.4pt} \\ Z}}{CH = CH} \quad \xrightarrow{\quad R\cdot \quad} \quad \left(\underset{\substack{\rule[0.3ex]{2em}{0.4pt} \\ Z}}{CH - CH} \right)_n \tag{2}$$

It has been accepted for a long time that dialkyl fumarates (DRF) and dialkyl maleates (DRM), which are trans and cis 1,2-disubstituted ethylenes, respectively, do not give any high polymers in the presence of radical initiators. In 1981, however, Otsu and coworkers [2,3] found that DRF could homopolymerize in bulk with 2,2'-azobisisobutyronitrile (AIBN) as the most typical radical initiator of which concentration is much higher (about ten times) than that for ordinary polymerization, as shown in eq. 3.

$$n \quad \underset{\substack{| \\ COOR}}{\overset{\substack{COOR \\ |}}{CH = CH}} \quad \xrightarrow{\quad R\cdot \quad} \quad \left(\underset{\substack{| \\ COOR}}{CH} \right)_{2n} \tag{3}$$

DRF

Recent our investigations as to polymerization of 1,2-disubstituted ethylenes including DRF have revealed the feature of this polymerization in detail, i.e., polymerization rate, absolute rate constants, and reactivities of the monomer and the polymer radical, as well as the polymer structures and some properties such as tacticity of the polymer, rigidity of the chain, thermal properties, and some applications. Radical polymerization of 1,1-disubstituted and 1,1,2-trisubstituted ethylenes was also investigated. The detailed results are described and discussed in this article.

On the other hand, some internal olefins such as cis- and trans-2-butenes, 1,2-disubstituted ethylenes, were found in 1965 by Otsu and coworkers [4-7] to polymerize in the presence of Ziegler-Natta catalysts to give high molecular weight polymers consisting of 1-butene recurring unit through a monomer-isomerization polymerization mechanism. Thereafter, it was found that a number of internal olefins could undergo monomer-isomerization polymerization and copolymerization in the presence of Ziegler-Natta catalysts. Some cyclic olefins such as norbornene and cyclobutene have been well-known to polymerize with tungsten or molybdenum chloride catalyst systems to give unsaturated linear polymers via a methathesis polymerization mechanism [8]. Because the polymers derived from these polymerizations do not consist of a rigid chain structure, the detailed results will be not undertaken in this article.

Moreover, tetra- and trisubstituted ethylenes bearing any substituents other than fluorine atom are well-known to be not homopolymerized, although those with some strong electron-accepting substituents were found to copolymerize with electron-donating styrene derivatives. As an example, cyanofumarates copolymerized spontaneously with p-methoxystyrene to give an alternating copolymer via the formation of a tetramethylene diradical [9,10]. Although the results are very interesting, these will be also not described in this article.

2. POLYMERIZATION CHARACTERISTICS OF DRF

Polymerization Reactivity of DRF

The structure of ester substituents of DRF as a sterically hindered 1,2-disubstituted ethylene influences seriously the polymerization reactivities. The reactivities of DRF deduced from the yield and the molecular weight of the resulting polymers increased when the bulkiness of the ester alkyl groups increased, as shown in Figure 1. The observed results are understood as the characteristics of radical polymerization which proceeds via chain reactions, i.e., because the polymer chain with bulky substituents becomes more rigid, bimolecular termination between such rigid propagating radicals having a bulky group occurs less frequently, relative to

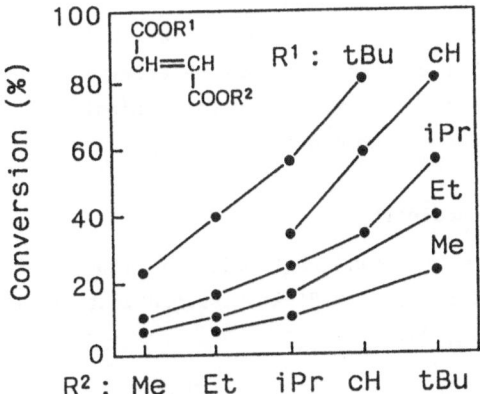

Figure 1.   Relationship between conversion and ester alkyl groups of DRF for radical polymerization in bulk initiated with AIBN (0.02 mol/L) at 60°C for 10 h.

Table 1. Absolute Rate Constants of Propagation and
Termination for Polymerization of DRF

Monomer	Ester Alkyl Group	k_p (L/mol·s)	k_t (L/mol·s)	Method[a]	Ref.
DtBF	C(CH$_3$)$_3$/C(CH$_3$)$_3$	0.46	11.2	A	13
iPtBF	CH(CH$_3$)$_2$/C(CH$_3$)$_3$	0.39	17.5	A	12
MtBF	CH$_3$/C(CH$_3$)$_3$	0.23	–	A	14
DiPF	CH(CH$_3$)$_2$/CH(CH$_3$)$_2$	0.31	0.84	B	16
DEF	C$_2$H$_5$/C$_2$H$_5$	0.015	164	C	15
DEF	C$_2$H$_5$/C$_2$H$_5$	0.029	8.0	B	17
DMF	CH$_3$/CH$_3$	0.058	–	A	14
MMA	–	367	1.87×10^7	C	18
St	–	176	7.24×10^7	C	19

a) A: ESR, B: scavenge of the radical, C: rotating sector.

the propagation reaction, thus giving high molecular weight and faster polymerization rates [3].

In polymerization of DRF, the concentration of the propagating radical is quite high, hence the ESR spectra due to the propagating poly(DRF) radicals are easily detected under usual polymerization conditions [11]. It may be utilized for determination of absolute rate constants of propagation (k_p) and termination (k_t) [12-14]. Table 1 summarizes k_p and k_t of some DRFs determined by the ESR spectroscopy as well as other methods, i.e., a rotating-sector method [15] and a scavenge of the propagating radicals [16,17]. It is clear that the both values of k_p and k_t for DRF are much smaller than those for methyl methacrylate (MMA) and styrene (St) [18,19], because of the low reactivities of sterically hindered DRF monomers and of propagating poly(DRF) radicals as a rigid substituted polymethylene. These results indicate that the polymerization reactivity is dominated by the ratio of rate constants for propagation to termination ($k_p/k_t^{0.5}$) rather than the absolute values of the constants.

Reactivity of Primary Radicals

More recently, we have found that polymerization reactivities of DRFs increase by the use of dimethyl 2,2'-azobisisobutyrate (MAIB) as an initiator instead of AIBN [12,13].

AIBN   MAIB

Table 2. Radical Polymerization of DRFs and Ordinary Vinyl
Monomers with MAIB and AIBN in Benzene at 60°C[a]

Monomer	MAIB			AIBN			R_p(MAIB)
	$R_p \times 10^5$ (mol/L·s)	$M_n \times 10^{-4}$	M_w/M_n	$R_p \times 10^5$ (mol/L·s)	$M_n \times 10^{-4}$	M_w/M_n	R_p(AIBN)
DtBF	9.00	5.2	2.5	1.15	2.1	2.9	7.83
iPtBF	8.01	4.7	2.2	1.07	3.9	2.0	7.49
DiPF	3.78	2.0	2.6	0.53	1.3	2.5	7.13
DEF	0.68	–	–	0.15	–	–	4.53
MMA	7.09	2.3	2.3	6.82	2.3	2.3	1.04
AN	3.98	–	–	3.67	–	–	1.08
VAc	3.06	1.5	2.0	1.79	1.2	2.2	1.71
St	1.32	0.7	2.0	1.27	0.8	1.8	1.04

a) [Monomer] = 1.5 mol/L, [Initiator] = 0.02 mol/L.

Table 2 shows the results of radical polymerization of various monomers with MAIB and AIBN in benzene at 60°C [20]. In the case of DRF, polymerization rate (R_p) with MAIB was much higher than that with AIBN in spite of the almost same rates of primary radical formation from both initiators. The ratio of R_p for both the initiators, i.e., R_p(MAIB)/R_p(AIBN), was found to be 4.5-7.8 for DRFs, and this value increases along with increase in bulkiness of the ester substituent of DRFs, whereas R_p(MAIB)/R_p(AIBN) in polymerization of ordinary vinyl monomers were much smaller (1.0-1.7) than those of DRFs. For example, in polymerization of conjugative monomers such as MMA, acrylonitrile (AN), and St, the polymerization reactivity changed hardly with both initiators, and non-conjugative vinyl acetate (VAc) showed a relatively large R_p(MAIB)/R_p(AIBN). The M_n of poly(DRF)s was also found to vary according to the initiator, but the ratios in M_n were smaller than those in R_p, suggesting that the primary radical of AIBN inactivates outside the cage and the real initiator efficiency decreases.

These kinetic studies conclude that the increase in polymerization reactivities is due to the difference in the reactivity of both primary radicals generated from MAIB and AIBN toward DRF, i.e., in the polymerization with AIBN primary radical termination and inactivation of the primary radical outside the cage occur because the primary radical of AIBN has a lower initiation reactivity than that of MAIB in DRF polymerization.

In the cases that AIBN was used as an initiator, all DRFs showed lower polymerization reactivities than the vinyl monomers employed in this study, but the R_p and M_n for polymerization of DtBF and iPtBF initiated with MAIB were comparable to or larger than those of other vinyl monomers.

Stereoregularity of Poly(DRF) and Propagation Mechanism

The stereoregularity of polymers obtained from 1,2-disubstituted ethylenes has been investigated by many workers since the discovery of stereo-
regular polymers by Natta and coworkers [21,22]. In radical polymerization
of ordinary vinyl monomers, because stereochemical structure of a penulti-
mate unit at the propagating chain end is not determined until the addition
of an attacking monomer to a propagating radical, the direction of monomer
addition and the opening mode of the carbon-to-carbon double bond cannot be
discriminated each other. On the other hand, in the case of polymerization
of 1,2-disubstituted ethylenes like DRFs, the direction of monomer addition
dominates the configuration of two carbons, suggesting that the determination of stereoregularity of substituted polymethylenes enable us to discuss
the direction of monomer addition and the opening mode independently.

We have already studied the stereoregularity of poly(DRF)s by ^{13}C NMR
spectroscopy in order to clarify the propagation mechanism of DRFs in radical polymerization [23-25]. The tacticity of poly(DRF), i.e., poly(alkoxy-
carbonylmethylene) should be considered on the basis of a methylene repeating unit, and it is dominated by both opening mode of a carbon-to-carbon
double bond and direction of monomer addition to the propagating radical.
When an opening of the double bond is restricted to either mode, tacticity
of poly(DRF)s can be determined with solely the direction of addition,
i.e., P_m and P_r, which represent the probability of *meso* and *racemo* additions, respectively (Figure 2). Thus, we determined P_m values for some

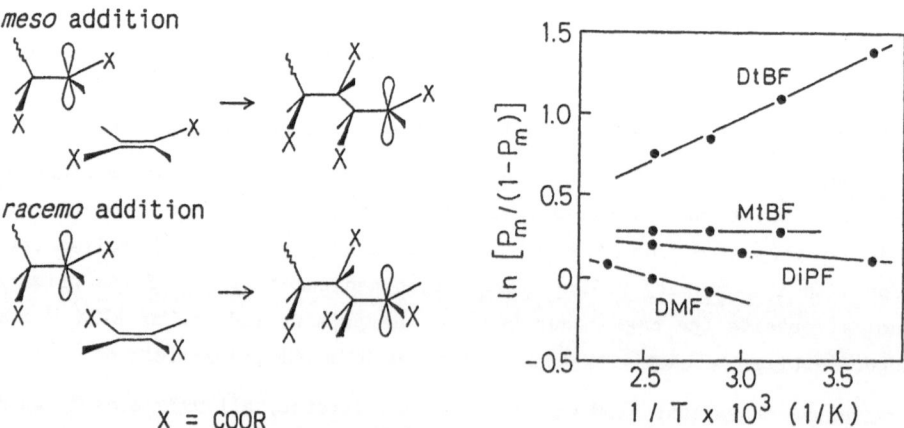

meso addition

racemo addition

X = COOR

Figure 2. *meso* and *racemo* additions
of DRF to a poly(DRF) radical.

Figure 3. Arrhenius plots of
$P_m/(1-P_m)$ for polymerization of DRF.

DRFs by means of restriction of the trans opening, and revealed that the polymerization temperature dependence of P_m is drastically changed with the structure of ester alkyl groups (Figure 3) [25]; the difference in activation enthalpies for *meso* and *racemo* additions is positive for DMF, but negative for DtBF. Moreover, the absolute rate constants for *meso* and *racemo* additions were evaluated by the use of the P_m and k_p values [14].

Properties of Poly(DRF)

Some characteristic properties of poly(DiPF) as a typical poly(DRF) are summarized [3]: It is colorless powder, soluble in many organic solvents, and a transparent film is obtained by casting of the solution. Specific gravity: 1.12 g/cm^3; Refractive index: 1.4698. Miscible with poly(MMA). Hydrolysis is difficult. $T_{init} = 223°C$, $T_{max} = 300°C$, T_g and T_m are not detected below the decomposition temperature. Viscosity relationship: $[\eta]$ = 7.53 x 10$^{-6}M^{0.98}$ dL/g in benzene at 30°C [26]. Persistence length: 7.6 nm [26]. Oxygen permeability: $P = 9.0$ x 10^{-10} cm^3(STP)·cm/cm^2·sec·cmHg, selectivity of O$_2$ to N$_2$: 4.1. For these properties, poly(DRF) can be used for optical or contact lens, and LB [27] or O$_2$ enrichment membranes [28].

3. POLYMETHYLENES FROM MALEIC DERIVATIVES

Polymerization of Amide Derivatives of Fumaric and Maleic Acids

DRM does not homopolymerize under the same conditions as that for polymerization of DRF, but it can homopolymerize in the presence of a radical initiator and morpholine as an isomerization catalyst via a monomer-isomerization radical polymerization mechanism [3,29]. However, since morpholine also serves as a retarder, the polymerization reactivities do not exceed those for DRF in the absence of morpholine.

On the other hand, it has recently been found that some amide derivatives of maleic acid can homopolymerize without isomerization to the fumaric derivatives [30-32]. The radical polymerization of some amide derivatives below, i.e., N,N,N',N'-tetraalkylfumaramides (TRFAm), N,N,N',N'-tetraalkylmaleamides (TRMAm), alkyl N,N'-diethylfumaramates (FAE), and alkyl N,N'-diethylmaleamates (MAE), was investigated.

TRFAm   TRMAm   FAE   MAE

Table 3. Radical Polymerization of Amide Derivatives of
Fumaric and Maleic Acids in Bulk at 120°C for 10 h[a])

Monomer	Geometry	Substituent[b])		Yield	M_n
		X^1	X^2	(%)	$\times 10^{-3}$
TRFAm	trans	CONEt$_2$	CONEt$_2$	57.0[c])	2.1[c])
TRMAm	cis	CONEt$_2$	CONEt$_2$	44.0	4.2
FAE	trans	COOiPr	CONEt$_2$	67.2	5.0
MAE	cis	COOiPr	CONEt$_2$	49.4	5.7
DRF	trans	COOiPr	COOiPr	85.6	11.6

a) [DTBPO] = 0.05 mol/L. b) CHX^1=CHX^2. c) In toluene, [TRFAm] = 1 mol/L.

The yield and M_n of the resulting polymers for bulk polymerization of these fumaric and maleic derivatives at 120°C with di-*tert*-butyl peroxide (DTBPO) are listed in Table 3 [30]. The polymerization reactivities of these derivatives were in the following order; DRF > TRFAm > FAE for the fumaric derivatives, and MAE > TEMAm > DRM for the maleic derivatives. The respective trans isomers always show higher reactivity than the cis isomers, agreeing well with the previous results on reactivities of various cis and trans isomers in the literature [33-36].

Polymerization of *N*-Substituted Maleimides

As described in introduction, some cyclic 1,2-disubstituted ethylenes such as VCa [37-39], MAn [40], and RMI [41-43] are known to be homopolymerized in a radical mechanism. For example, radical polymerization of VCa leads to a high molecular weight polymer, which is then hydrolyzed to poly(hydroxymethylene) [39]. Whereas homopolymerization of MAn, a strong electron-accepting monomer, gives hardly a high molecular weight polymer, but an alternating copolymerization with a number of electron-donating monomers undergoes easily [44]. Similar copolymerization tendency is also observed for RMI, but RMI homopolymerizes quite readily in the presence of radical initiators to give high molecular weight and thermally stable polymers (eq. 4), as described below.

Table 4.   Radical Polymerization of RMI in Benzene at 60°C[a]

RMI	R	Yield (%)	$[\eta]$[b] (dL/g)	M_n × 10^{-4}	M_w/M_n
MI	H	34.3	0.29[c]	–	–
EMI	C_2H_5	33.4	0.24[c]	2.1	2.3
IPMI	$CH(CH_3)_2$	74.0	0.58[c]	8.8	6.0
nBMI	$(CH_2)_3CH_3$	78.4	0.76	15.9	2.6
IBMI	$CH_2CH(CH_3)_2$	66.6	0.50	5.9	3.0
sBMI	$CH(CH_3)CH_2CH_3$	70.9	0.53	9.5	2.5
tBMI	$C(CH_3)_3$	82.8	0.68	14.9	2.5
tAMI	$C(CH_3)_2CH_2CH_3$	71.2	0.61	13.0	2.1
CHMI	cyclo-C_6H_{11}	39.1	0.27[c]	4.3	4.6
nOMI	$(CH_2)_7CH_3$	40.1(1h)	0.79	17.3	2.0
tOMI	$C(CH_3)_2CH_2C(CH_3)_3$	26.5	0.27	5.0	1.8
DodMI	$(CH_2)_{11}CH_3$	46.5(1h)	0.69	16.2	2.0
ODMI	$(CH_2)_{17}CH_3$	44.6(1h)	0.60	17.0	2.0
BzMI	CH_2Ph	20.4	0.13	0.9	2.7
MBzMI	$CH(CH_3)Ph$	47.5	0.30	7.4	2.0
DPhMMI	$CHPh_2$	20.4[d]	0.12	2.2	1.9
TrMI	CPh_3	4.9[e]	–	0.3	1.3
PhMI	Ph	78.9	0.10[c]	–	–
2-MPhMI	2-(CH_3)-Ph	27.6	0.17	3.3	1.9
2,6-DMPhMI	2,6-$(CH_3)_2$-Ph	20.4	0.10	1.9	1.6

a) [RMI] = 1 mol/L, [AIBN] = 0.005 mol/L, for 5 h.   b) In benzene at 30°C.
c) In DMF at 30°C.   d) [DPhMMI] = 0.5 mol/L.   e) [TrMI] = 0.2 mol/L.

The results of radical polymerization of various RMI are summarized in
Table 4 [45]. The polymerization reactivities, i.e., yield and molecular
weight of the polymers depended on the structure of the *N*-substituents.
From the kinetic study of polymerization of *N-tert*-butylmaleimide (tBMI)
[46], which is one of the monomers with the highest polymerization reactivity among RMIs in spite of the bulky *N*-substituent, it was revealed that
the R_p was expressed as $R_p = k$ [AIBN]$^{0.5}$[tBMI]$^{1.4}$, and the overall activation energy was 99.6 kJ/mol. The high polymerization reactivity of tBMI
was assumed to result from the decrease in the rate of bimolecular termination between rigid polymer radicals bearing a bulky substituent, on the
basis of the results of ESR studies and the data from the viscometric and
light scattering measurements.

The poly(RMI)s were confirmed to consist of substituted polymethylene
structure by IR and NMR spectroscopies. The poly(RMI)s are soluble in many
organic solvents, and a thin and tough film is obtained from the chloroform
solution. It was confirmed that poly(RMI) with various length *n*-alkyl
groups showed a high T_g from 97°C to 185°C respective to the length of the
alkyl chain [47], but the polymers with bulky *N*-substituents such as *tert*-
alkyl groups did not melt and did not show any transition temperatures

below their decomposition temperatures. In thermogravimetric analysis
(TGA) of poly(RMI)s, no weight-loss was observed at temperature below
300°C, and the maximum decomposition temperature was 400-440°C [45,48],
except for N-$tert$-alkyl substituted derivatives of which decomposition
started at 240-280°C via a two-step reaction [49], i.e., formation of poly-
maleimide (poly(MI)) by quantitative olefin elimination at the first step
and decomposition of the resulting poly(MI).

From the results for measurements of permeability of O_2 and N_2 through
the poly(RMI) membranes, which were prepared by casting of the chloroform
solution, it was revealed that the high permeability coefficient was great-
er than 10^{-9} $cm^3(STP) \cdot cm/cm^2 \cdot sec \cdot cmHg$, and depended on the length of the
n-alkyl groups of the side chain [47]. The excellent permeability of gases
through these polymaleimides may be ascribed to high diffusion of pene-
trants in the membranes [50].

Radical polymerization of N-(alkyl-substituted phenyl)maleimides
(RPhMI) was also investigated to synthesize thermally stable vinyl polymers
soluble in organic solvents [51], since thermally stable poly(N-phenylmale-
imide) (poly(PhMI)) is soluble in a few polar organic solvents. The radical polymerization and copolymerization reactivities of RPhMI were revealed
to depend on the position, number, and bulkiness of the alkyl substituents;
The introduction of alkyl groups into the ortho position decreased the
yield and molecular weight of the polymers by steric hindrance. The meta
and para substitution increased the homopolymerization reactivities because
of the introduction of electron-donating alkyl groups. The polymers produced were soluble in common organic solvents including benzene,
chloroform, and THF, dependent of the position, number, and structure of
the alkyl groups introduced. These alkyl-substituted polymers showed
excellent thermal stability as well as poly(PhMI).

4. POLYMERS FROM ITACONIC AND CITRACONIC ACID DERIVATIVES

Itaconic and Related Derivatives

Formation of such less-flexible polymers is expected from polymerization of not only 1,2-disubstituted ethylenes but also 1,1-disubstituted
ethylenes when the substituents are bulky. It has been generally recognized that 1,1-disubstituted ethylenes with bulky substituents polymerize
hardly because of their low ceiling temperatures due to the steric effect
of the substituents, but itaconic acid (IA), α-(hydroxycarbonylmethyl)-

acrylic acid and its derivatives such as diesters homopolymerize radically [52]. Therefore we have re-examined the polymerization reactivities of IA derivatives from the present viewpoints, because the itaconic polymers are expected to be applied in many industrial fields. The IA derivatives including citraconic acid (CA) and mesaconic acid (MA) derivatives as structural cis and trans isomers consist of monoester, diester, amic acid, amic ester, and diamide [53]. The derivatives of IA and CA also include anhydride and imide as a cis cyclic monomer.

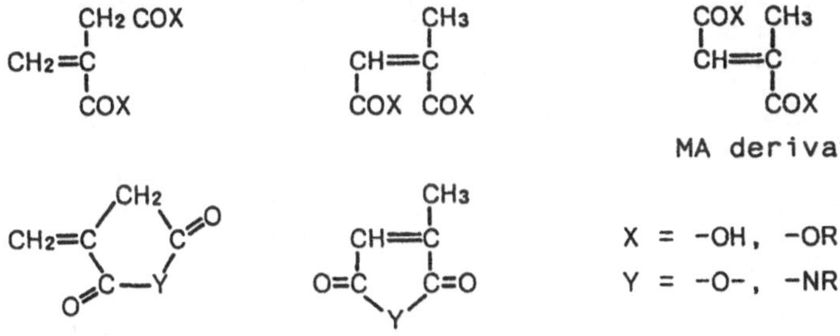

$$CH_2=C\begin{array}{c}CH_2\,COX\\|\\COX\end{array} \qquad CH=C\begin{array}{c}CH_3\\|\\COX\ COX\end{array} \qquad CH=C\begin{array}{c}COX\ CH_3\\|\\COX\end{array}$$

MA derivative

IA Derivative   CA Derivative

X = -OH, -OR, -NR$_2$

Y = -O-, -NR-

IA is a 1,1-disubstituted ethylene, and CA and MA are 1,1,2-trisubstituted ethylenes. The steric effects of the substituents influence importantly not only polymerization reactivity but also properties of the polymers. IA derivatives homopolymerize radically, but CA and MA derivatives do not at all except for the imide derivatives of CA as described later.

Radical Polymerization of DRI

DRIs undergo radical polymerization easily to give high molecular weight polymers (eq. 5) [53,54].

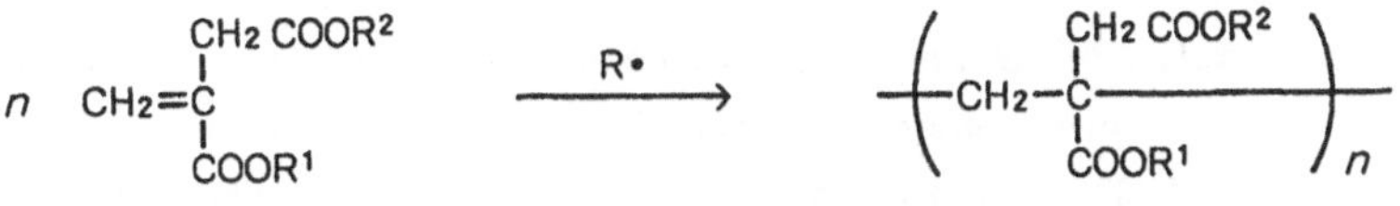

$$n\ \ CH_2=C\begin{array}{c}CH_2\,COOR^2\\|\\COOR^1\end{array} \xrightarrow{\ R\bullet\ } \left(\!\!\begin{array}{c}CH_2\,COOR^2\\|\\-CH_2-C-\\|\\COOR^1\end{array}\!\!\right)_n \qquad (5)$$

DRI

Table 5 shows the results of radical polymerization of DRI with various ester alkyl groups initiated with AIBN in bulk at 60°C. The increase in bulkiness of the ester alkyl groups was found to decrease polymerization reactivity, i.e., the polymerization rate and M_n of the resulting polymer decreased in the following order: DnPI > DiPI > DsBI > DtBI. Similar

Table 5. Radical Polymerization of DRI in Bulk with AIBN
at 60°C for 10 h[a]

DRI	Ester Alkyl Group[b]		Yield	M_n x	M_w/M_n
	R^1	R^2	(%)	10^{-4}	
DMI	CH_3	CH_3	33.5	3.6	2.0
DEI	C_2H_5	C_2H_5	31.0	2.8	2.3
DnPI	$n-C_3H_7$	$n-C_3H_7$	40.7	2.8	2.8
DiPI	$i-C_3H_7$	$i-C_3H_7$	38.1	2.7	2.0
DnBI	$n-C_4H_9$	$n-C_4H_9$	40.9	4.0	2.2
DiBI	$i-C_4H_9$	$i-C_4H_9$	44.3	7.2	1.9
DsBI	$s-C_4H_9$	$s-C_4H_9$	33.3	2.6	1.9
DtBI	$t-C_4H_9$	$t-C_4H_9$	13.7	0.7	1.8
DnAI	$n-C_5H_{11}$	$n-C_5H_{11}$	43.9	3.6	2.6
DiAI	$i-C_5H_{11}$	$i-C_5H_{11}$	54.8	7.4	1.9
DneoAI	$neo-C_5H_{11}$	$neo-C_5H_{11}$	33.2(5h)	9.2	1.8
DcAI	$c-C_5H_9$	$c-C_5H_9$	63.0	6.1	2.1
DcHI	$c-C_6H_{11}$	$c-C_6H_{11}$	44.2(5h)	10.7	2.4
DcHMI	$CH_2-c-C_6H_{11}$	$CH_2-c-C_6H_{11}$	73.3	11.0	1.9
DnOI	$n-C_8H_{17}$	$n-C_8H_{17}$	27.5	5.3	2.3
DEHI	2-EH[c]	2-EH	62.1	4.4	3.1
DMAdI	DMAd[d]	DMAd	15.5	0.7	2.2
MEI	C_2H_5	CH_3	24.5	2.5	1.8
MnPI	$n-C_3H_7$	CH_3	27.1	2.5	1.8
MiPI	$i-C_3H_7$	CH_3	21.5	2.2	1.8
MnBI	$n-C_4H_9$	CH_3	25.6	3.0	1.7
MiBI	$i-C_4H_9$	CH_3	27.1	4.1	1.9
MsBI	$s-C_4H_9$	CH_3	18.9	1.8	1.4
MtBI-I	$t-C_4H_9$	CH_3	10.5	1.6	1.5
MtBI-II	CH_3	$t-C_4H_9$	38.6	3.3	2.4

a) [AIBN] = 0.02 mol/L.   b) $CH_2=C(COOR^1)CH_2COOR^2$.   c) 2-Ethylhexyl.
d) 3,5-Dimethyl-1-adamantyl.

effect was also observed for polymerization of methyl alkyl itaconates
(MRI) [55].   These results are different from those for polymerization of
DRF, suggesting that the reactivities seem to be controlled by the steric
effect of the 1,1-disubstituents on the propagation rather than the
termination.

In polymerization of β-monoalkyl itaconates and IA [56], the polymerization reactivities were higher than those of DRI and MRI, and independent
of the structure of R^2.   It indicates that the effect of the alkyl groups
on the polymerization reactivity appears more importantly in R^1 than R^2.
In fact, it was found that the polymerization reactivity of methyl α-
(tert-butoxycarbonylmethyl)acrylate (MtBI-II) was higher than that of tert-
butyl α-(methoxycarbonylmethyl)acrylate (MtBI-I).   Recently, from the
kinetic studies by means of ESR spectroscopy [57-60], the substituent
effects of the ester alkyl groups on k_p and k_t were investigated [60] and

the rate constants were determined to be lower than those for ordinary vinyl monomers.

The poly(DRI)s obtained are powdery or rubbery materials, depending on their structure of the side chain, and soluble in many organic solvents. Transparent films can be obtained from the solutions, and the refractive indexes are similar to that of poly(MMA). Poly(DRI)s except for poly(DtBI) underwent thermal degradation via a one-step reaction; T_{init} = 220-260°C, T_{max} = 300-350°C. Poly(DtBI) decomposed exceptionally via a quantitative elimination of isobutene and water; T_{init} = 205°C and T_{max} = 220°C.

Polymerization of N-Substituted Itaconimides

Recently, it has been revealed that N-alkylitaconimides (RII) and N-(alkyl-substituted phenyl)itaconimides (RPhII) polymerize in the presence of a radical initiator to give a high molecular weight polymer (eq. 6) [53].

The results of polymerization of RII in benzene at 60°C are shown in Table 6. All monomers gave polymers with 10^4-10^5 of M_n in high yields. From bulk polymerization, polymers with a higher molecular weight more than 10^6 were obtained. It was found that the polymerization rate of RPhII was larger than that of RII, although the alkyl substitution at ortho positions

Table 6. Radical Polymerization of RII in Benzene at 60°C[a]

Monomer	R	Time (h)	Yield (%)	M_n x 10^{-4}	M_w/M_n
EII	C_2H_5	5	76.1	7.3	2.6
iPII	$i\text{-}C_3H_7$	5	78.6	2.9	3.6
nBII	$n\text{-}C_4H_9$	5	72.8	18.4	2.0
tBII	$t\text{-}C_4H_9$	5	76.2	8.7	2.1
cHII	$c\text{-}C_6H_{11}$	5	87.7	13.3	2.8
nOII	$n\text{-}C_8H_{17}$	5	63.0	7.1	1.9
ODII	$n\text{-}C_{18}H_{37}$	5	92.1	20.4	3.4

a) [Monomer] = 1 mol/L, [AIBN] = 0.005 mol/L.

decreased polymerization reactivities. Kinetic study on polymerization of RII by means of ESR spectroscopy was carried out as well as DRI [61], and the chain flexibility of poly(nBII) was also elucidated by intrinsic viscosity and light scattering in dilute solution to discuss with relation to the decrease in k_t [62].

Poly(RII) and poly(RPhII) are soluble in many organic solvents and transparent film can be obtained. These polyimides show high decomposition temperatures; i.e., T_{init} = 250-300°C and 300-350°C for poly(RII) and poly(RPhII), respectively, which are higher than those of poly(DRI).

Other IA Derivatives

It has been also reported sicne 1958 [52] that itaconic anhydride (IAn), which is an another cyclic derivative of IA, polymerizes as well as other IA derivatives, but its reactivity is less than those of RII and RPhII. Poly(IAn) may be derived into some poly(IA) derivatives by reaction with amines and alcohols [63,64]. The polymerization reactivities of several itaconamides (IAm) and itaconamates (IAE) were also investigated [48]. It was found that N,N'-dialkyl substituted IAm homopolymerize in the presence of a radical initiator, but N,N,N',N'-tetraalkyl-substituted one did not give a polymer, because the latter consists of an α,N,N-trisubstituted acrylamide structure resemble to N,N-disubstituted methacrylamides. Similar results were obtained for polymerization of N-mono- and N,N-disubstituted IAE.

Polymerization Reactivity of CA and MA Derivatives

Dialkyl citraconates and dialkyl mesaconates, α-methylmaleic and α-methylfumaric esters, respectively, do not homopolymerize because of steric hindrance due to a trisubstituted ethylene structure, but undergo copolymerization with vinyl monomers. The alternating copolymerization of these diesters and citraconic anhydride with some electron-donating monomers such as St, VAc, and isobutyl vinyl ether has been reported [65-68].

Recently, radical polymerization reactivity of N-alkylcitraconimides (RCI) and N-(alkyl-substituted phenyl)citraconimides (RPhCI) was investigated. It was revealed that RCI and RPhCI polymerized to give polymers with molecular weight of 10^3-10^4 (eq. 7), although they are one of sterically hindered 1,1,2-trisubstituted ethylenes [69]. However, the reactivity of RCI or RPhCI is less than RII, RPhII, RMI, and RPhMI.

$$\begin{array}{c} CH_3 \\ CH=C \\ n \quad O=C \quad C=O \\ N \\ R \\ RCI \end{array} \xrightarrow{R\cdot} \left(\begin{array}{c} CH_3 \\ CH-C \\ O=C \quad C=O \\ N \\ R \end{array}\right)_n \quad (7)$$

The decomposition temperatures were determined by TGA as follows: T_{init} = 309 and 265°C, T_{max} = 365 and 355°C for N-isopropyl and N-n-butyl derivatives of RCI, respectively. These values are lower than those for poly(RMI)s.

References

1. T. Otsu, *"Radical Polymerization (I) Elementary Reaction Mechanism"*, Kagaku Dojin, Kyoto, 1971
2. T. Otsu, O. Ito, N. Toyoda, and S. Mori, *Makromol. Chem., Rapid Commun.*, **2**, 725 (1981)
3. T. Otsu, T. Yasuhara, and A. Matsumoto, *J. Macromol. Sci.-Chem.*, **A25**, 537 (1988), and references cited therein
4. A. Shimizu, T. Otsu, and M. Imoto, *J. Polym. Sci.*, **B3**, 449 (1965)
5. T. Otsu, A. Shimizu, and M. Imoto, *J. Polym. Sci.*, **A1**, 4, 1579 (1966)
6. J. P. Kennedy and T. Otsu, *Adv. Polym. Sci.*, **7**, 369 (1970)
7. K. Endo and T. Otsu, *"Monomer-Isomerization Polymerization"*, in Handbook of Mass and Heat Transfer, Ed., N. P. Cheremisinoff, Vol.3., pp.553-581, Gulf Publishing, 1988
8. K. J. Ivin, *"Cycloalkenes and Bicycloalkenes"*, in Ring-Opening Polymerization, Eds., K. J. Ivin and T. Saegusa, Chapter 3, Elsevier, 1984
9. H. K. Hall, Jr., A. B. Padias, A. Pandya, and H. Tanaka, *Macromolecules*, **20**, 247 (1987)
10. C. Lee and H. K. Hall, Jr., *Macromolecules*, **22**, 21 (1989)
11. T. Otsu, *Makromol. Chem., Macromol. Symp.*, **10/11**, 235 (1987)
12. M. Yoshioka and T. Otsu, *Macromolecules*, in press
13. T. Otsu and M. Yoshioka, *Macromolecules*, submitted
14. M. Yoshioka, A. Matsumoto, and T. Otsu, *Polym. J.*, **23**, 1249 (1991)
15. T. Otsu, B. Yamada, and T. Ishikawa, *Macromolecules*, **24**, 415 (1991)
16. B. Yamada, E. Yoshikawa, H. Miura, K. Shiraishi, and T. Otsu, *Polymer*, **32**, 1892 (1991)
17. B. Yamada, E. Yoshikawa, H. Miura, and T. Otsu, *Polym. Bull.*, **26**, 543 (1991)
18. M. S. Matheson, E. E. Auer, E. B. Bevilacqua, and E. J. Hart, *J. Am. Chem. Soc.*, **73**, 1700 (1951)
19. M. S. Matheson, E. E. Auer, E. B. Bevilacqua, and E. J. Hart, *J. Am. Chem. Soc.*, **71**, 497 (1949)
20. M. Yoshioka and T. Otsu, *Macromolecules*, submitted
21. G. Natta, P. Pino, P. Corradini, E. Danusso, E. Mantica, G. Mazzanti, and G. Moraglio, *J. Am. Chem. Soc.*, **77**, 1708 (1955)
22. G. Natta, *Makromol. Chem.*, **16**, 213 (1955); *J. Polym. Sci.*, **16**, 143 *J. Polym. Sci.*, **35**, 94 (1960)
23. X. Wang, T. Komoto, I. Ando, and T. Otsu, *Makromol. Chem.*, **189**, 1845 (1988)
24. M. Yoshioka, A. Matsumoto, T. Otsu, and I. Ando, *Polymer*, **32**, 2741 (1991)
25. M. Yoshioka, A. Matsumoto, and T. Otsu, *Polym. J.*, **23**, 1191 (1991)

26. A. Matsumoto, T. Tarui, and T. Otsu, *Macromolecules*, 23, 5102 (1990)
27. K. Shigehara, M. Hara, H. Nakahama, S. Miyata, Y. Murata, and A. Yamada, *J. Am. Chem. Soc.*, 109, 1237 (1987)
28. S. B. Choi, A. Takahara, N. Amaya, Y. Murata, and T. Kajiyama, *Polym. J.*, 21, 433 (1989)
29. N. Toyoda, M. Yoshida, and T. Otsu, *Polym. J.*, 15, 255 (1983)
30. A. Matsumoto and T. Otsu, *Polym. Commun.*, in press
31. A. Matsumoto, K. Fukushima, and T. Otsu, *J. Polym. Sci., Part A: Polym. Chem.*, 29, 1697 (1991)
32. A. Matsumoto, R. Kotaki, and T. Otsu, *J. Polym. Sci., Part A: Polym. Chem.*, 29, 1707 (1991)
33. F. M. Lewis and F. R. Mayo, *J. Am. Chem. Soc.*, 70, 1533 (1948)
34. A. R. Bader, R. P. Buckley, F. Leavitt, and M. Szwarc, *J. Am. Chem. Soc.*, 79, 5621 (1957)
35. B. Giese, *Angew. Chem. Int. Ed. Engl.*, 22, 753 (1983)
36. O. Ito and M. Matsuda, *J. Polym. Sci., Polym. Chem. Ed.*, 28, 1947 (1990)
37. M. S. Newman and R. W. Addor, *J. Am. Chem. Soc.*, 75, 1263 (1953)
38. K. Hayashi and G. Smets, *J. Polym. Sci.*, 27, 626 (1958)
39. N. D. Field and J. R. Schaefgen, *J. Polym. Sci.*, 58, 533 (1962)
40. N. G. Gaylord, *J. Macromol. Sci.-Rev.*, C13, 235 (1975)
41. P. O. Tawney, R. H. Snyder, R. P. Conger, K. A. Leibbrand, C. H. Stiteler, and A. R. Williams, *J. Org. Chem.*, 26, 15 (1961)
42. R. C. P. Cubbon, *Polymer*, 6, 419 (1965)
43. M. Yamada, and I. Takase, *Kobunshi Kagaku*, 22, 626 (1965); M. Yamada, I. Takase, and T. Mishima, *Kobunshi Kagaku*, 26, 393 (1969)
44. J. M. G. Cowie, *"Alternating Copolymers"*, Plenum, New York, 1985
45. T. Otsu, A. Matsumoto, T. Kubota, and S. Mori, *Polym. Bull.*, 23, 43 (1990)
46. A. Matsumoto, T. Kubota, and T. Otsu, *Polym. Bull.*, 24, 459 (1990)
47. A. Matsumoto, Y. Oki, and T. Otsu, *Polym. J.*, 23, 201 (1991)
48. T. Otsu, A. Matsumoto, and T. Kubota, *Polym. International*, 25, 179 (1991)
49. T. Otsu, A. Matsumoto, and A. Tatsumi, *Polym. Bull.*, 24, 467 (1990)
50. A. Matsumoto, Y. Oki, and T. Otsu, *Polym. J.*, 23, 1371 (1991)
51. A. Matsumoto, T. Kubota, and T. Otsu, *Macromolecules*, 23, 4508 (1990)
52. S. Nagai, T. Uno, and K. Yoshida, *Kobunshi Kagaku*, 15, 550 (1958); S. Nagai and K. Yoshida, *Kobunshi Kagaku*, 17, 77, 79, 748 (1960); S. Nagai, *Bull. Chem. Soc. Jpn.*, 36, 1459 (1963); 37, 369 (1964)
53. T. Otsu, H. Watanabe, J. -Z. Yang, M. Yoshioka, and A. Matsumoto, *Makromol. Chem., Macromol. Symp.*, in press
54. T. Otsu, M. Okuo, H, Watanabe, and J. -Z. Yang, *Chem. Express*, 5, 953 (1990)
55. J. -Z. Yang and T. Otsu, *Makromol. Chem., Rapid Commun.*, 12, 205 (1991)
56. J. -Z. Yang and T. Otsu, *Polym. Bull.*, 25, 145 (1991)
57. T. Sato, S. Inui, H. Tanaka, T. Ota, M. Kamachi, and K. Tanaka, *J. Polym. Sci., Polym. Chem. Ed.*, 25, 637 (1987)
58. T. Sato, N. Morita, H. Tanaka, and T. Ota, *J. Polym. Sci., Part A: Polym. Chem.*, 27, 2497 (1989)
59. T. Sato, K. Morino, H. Tanaka, and T. Ota, *Makromol. Chem.*, 188, 2951 (1987)
60. T. Otsu, K. Yamagishi, and M. Yoshioka, to be published
61. T. Otsu and H. Yamazaki, unpublished data
62. A. Matsumoto, S. Umehara, and T. Otsu, to be published
63. J. -Z. Yang, O. Nakatsuka, and T. Otsu, *Chem. Express*, 5, 805 (1990)
64. T. Otsu and J. -Z. Yang, *Polym. International*, 25, 245 (1991)
65. J. -Z. Yang and Otsu, *Makromol. Chem., Rapid Commun.*, 11, 549 (1990)
66. J. -Z. Yang and Otsu, *Macromolecules*, in press
67. J. -Z. Yang and Otsu, *Polym. International*, 26, 636 (1991)
68. J. -Z. Yang and Otsu, *Polym. Bull.*, 26, 509 (1991)
69. T. Otsu and J. -Z. Yang, *Polym. Bull.*, 24, 475 (1990)

Silicon-Mediated Synthesis of Poly(Fluorovinyl Ethers) and Poly(Perfluoroalkenes)

B. E. Smart, W. B. Farnham and M. J. Nappa

Du Pont Central Research and Development, Experimental Station
Wilmington, DE 19880-0328 USA

ABSTRACT

Under the influence of anionic catalysts, silyl ethers react with perfluorinated olefins to produce exceptionally high yields of partially fluorinated vinyl ethers. The number of vinyl fluorines substituted can be controlled for a wide variety of fluorinated olefins and silyl ethers. Application of this reaction to difunctional starting materials affords quantitative yields of condensation polymers of moderate molecular weight.

Perfluoroalkylsilanes (R_fSiMe_3) have been found to be useful reagents for construction of new C-C bonds. Activated by anionic catalysts, R_fSiMe_3 react with fluorinated olefins to give perfluroalkylated olefins and Me_3SiF. Application of this new reaction to polymer syntheses leads to a variety of otherwise inaccessible structures. Only low molecular weight oligomers have been made so far, but the process is notable as a unique example of an anion-initiated perfluoroalkene oligomerization that does not produce rearranged, branched structures.

INTRODUCTION

The preparation of high molecular weight fluorinated polyethers of well-defined structure is a long-standing goal in fluoropolymer synthesis. Conventional ring-opening, anionic polymerization of fluorinated epoxides suffers from complicating chain-transfer processes[1]. Other approaches to fluorinated polyethers also suffer from adequate control of molecular architecture[2]. A fundamentally different approach seemed warranted, and a new method for C-O bond construction was investigated. Anionic activation of tetracoordinate organosilicon species has been utilized for a variety of C-C and C-heteroatom bond formation strategies[3], but organosilicon reagents have seldom been employed in syntheses of highly fluorinated materials[4].

We have discovered an anion-catalyzed reaction between silyl ethers and perfluoroalkenes that gives partially fluorinated vinyl ethers in high yields. This process has been extended to difunctional derivatives to produce condensation polymers in quantitative yields (Scheme 1). This paper summarizes the polymer forming aspects of our work on the new C-O bond forming reaction[5].

Y. Imanishi (Ed.)
Progress in Pacific Polymer Science 2
© Springer-Verlag Berlin Heidelberg 1992

Scheme 1

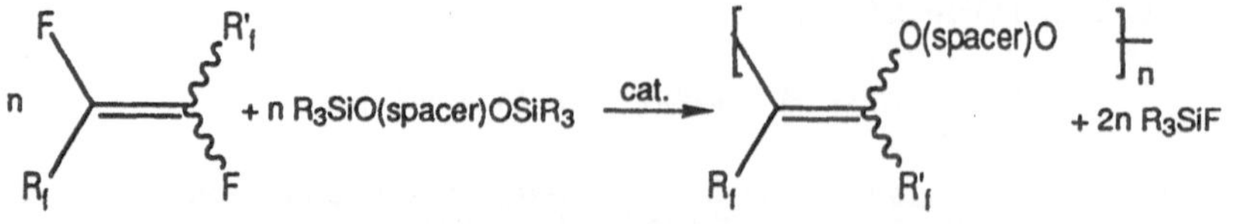

RESULTS

The difunctional silyl ether starting materials are obtained conveniently in quantitative yield by controlled treatment of the corresponding diols with hexamethyldisilazane in the presence of catalytic trimethylsilyl chloride (Scheme 2). The reactivity of the silyl ethers (Scheme 1) depends upon the silicon ligands, so a judicious choice of catalyst and fluoroolefin coreactant must be made to obtain optimum results in the substitution reaction. As shown in Scheme 3, fluorosilicate catalysts activate silyl ethers containing simple hydrocarbon alkoxy ligands, and good control over the degree of substitution on the fluorinated double bond is realized. Partially fluorinated alkoxy silanes require milder conditions to achieve comparable selectivity. It should be noted that some of the catalysts which activate silyl ethers cause side reactions with terminal fluoroolefins such as hexafluoropropene. Nevertheless, the degree of fluorine substitution on hexafluoropropene can be controlled using sufficiently reactive silyl ethers and carefully controlled conditions.

Scheme 2

PREPARATION OF SILYL ETHERS

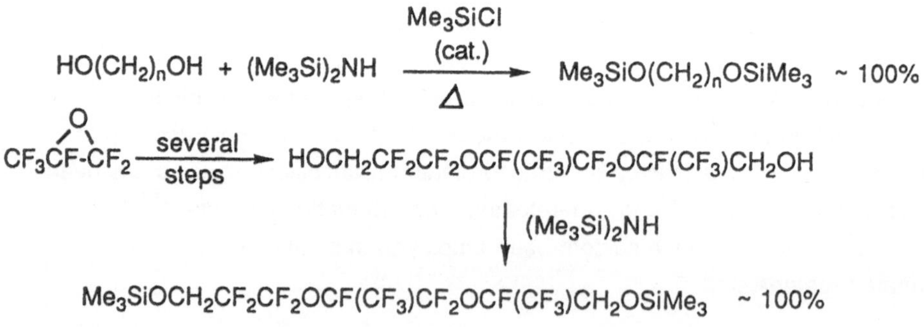

Scheme 3

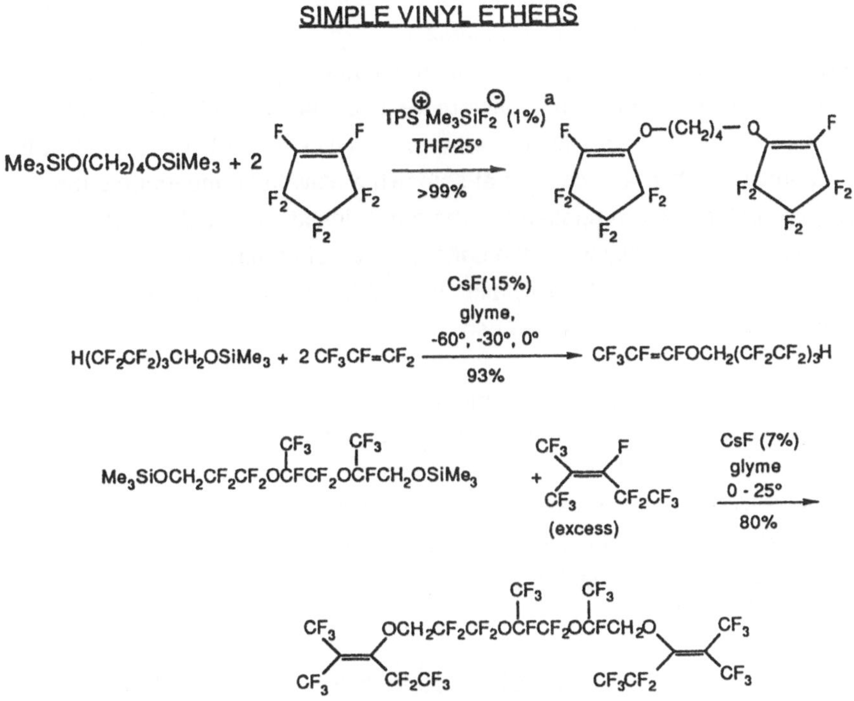

SIMPLE VINYL ETHERS

Me₃SiO(CH₂)₄OSiMe₃ + 2 [cyclopentene with F substituents] $\xrightarrow[\text{>99%}]{\substack{\text{TPS Me}_3\text{SiF}_2 \text{ (1%)} \\ \text{THF/25°}}}$ a [product: two fluorinated cyclopentene rings linked by O—(CH₂)₄—O]

H(CF₂CF₂)₃CH₂OSiMe₃ + 2 CF₃CF=CF₂ $\xrightarrow[\text{93%}]{\substack{\text{CsF(15%)} \\ \text{glyme,} \\ -60°, -30°, 0°}}$ CF₃CF=CFOCH₂(CF₂CF₂)₃H

Me₃SiOCH₂CF₂CF₂OĊFCF₂OĊFCH₂OSiMe₃ (with CF₃ substituents) + [(CF₃)₂C=CF(CF₂CF₃)] (excess) $\xrightarrow[\text{80%}]{\substack{\text{CsF (7%)} \\ \text{glyme} \\ 0 - 25°}}$

[product structure with two perfluoro vinyl ether groups]

a) TPS = tris(piperidino)sulfonium

A variety of anionic catalysts, including fluorosilicates, bifluoride, carboxylates and other oxyanions are effective for the substitution reaction. Similar catalysts have been found useful for activating other silicon species[6]. Bulk polymerizations using difunctional silyl ethers and vinyl ethers can be carried out without solvent, provided that very reactive catalysts are used.

Since the conversion of fluorinated olefins to vinyl ethers is known[7], it should be emphasized that the principal benefit from the silyl ether chemistry is increased selectivity and yield. Base-catalyzed reaction of alcohols and fluoroolefins almost always gives a mixture of substitution and addition products, but clean substitution occurs with silyl ethers. Homogeneity, which persists during bulk polymerization, is an additional attractive feature. Substitution of the first fluorine in a 1, 2-disubstituted perfluorinated olefin proceeds more rapidly than subsequent fluorine substitution in the product vinyl ether. The 2/1 adducts shown in Scheme 4 are therefore conveniently prepared in excellent yield. On a laboratory scale, these difunctional vinyl ethers permit good control of stoichiometry required for polymer formation.

Using model difunctional silyl ethers and vinyl ethers, number average molecular weights up to 64,000 have been achieved, but attainment of exceptionally high molecular weight material remains problematic. This, at least in part, is due to a minor crosslinking reaction involving polymeric vinyl ether chains and difunctional silyl ethers which takes place in the presence of the usual anionic catalysts under forcing conditions. Thus, treatment of vinyl ether-ended oligomer ($\overline{M}n$ ca. 12,000) with excess difunctional silyl ether and catalyst at somewhat elevated temperatures (65-100°C) produced material which appeared to be cross-linked. This side reaction seems to be important especially when the concentration of terminal vinyl ether groups becomes quite small. Protodesilylation is a likely termination process, but we do not have convincing evidence that this occurs.

Preparation of the corresponding perfluoropolyethers requires a subsequent fluorination step to saturate double bonds and replace hydrogens in the "spacer" groups with fluorine. We have found that the partially fluorinated vinyl ethers produced in the polycondensation reaction undergo reaction with elemental fluorine to give the desired perfluorinated polyethers. Ultraviolet irradiation is required to replace the least reactive hydrogens[8].

Scheme 4

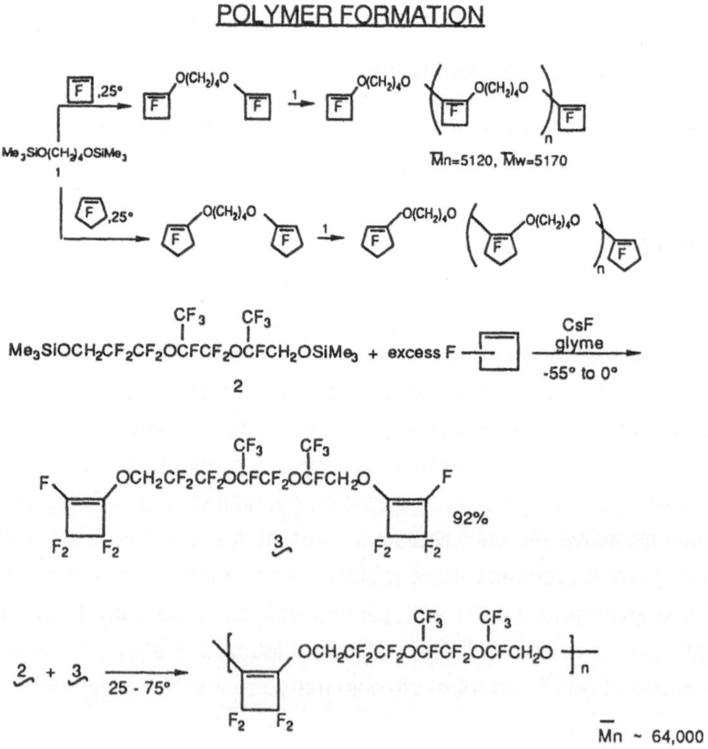

POLYMER FORMATION

Previous mechanistic studies of another silicon-mediated reaction, "group transfer polymerization" of methacrylates[6a,9] indicated the involvement of pentacoordinate silicon species. Our working hypothesis for the subject reaction of silyethers is that formation of pentacoordinate silicon species (Scheme 5) should increase reactivity and promote nucleophilic character in the alkoxy ligands.

Scheme 5

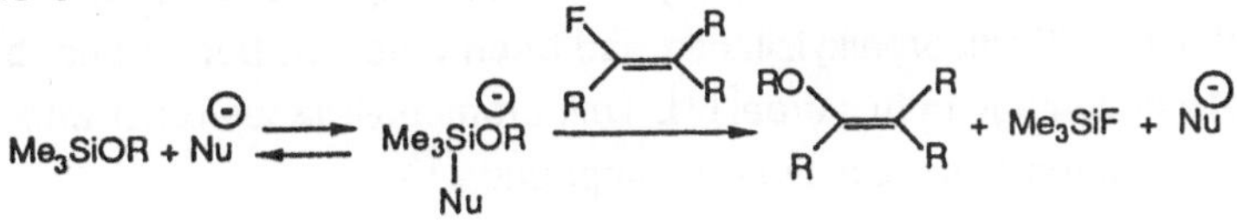

Other groups have described increased reactivity in pentacoordinate silicates for different processes[3a,10]. Demonstrating that silicon remains bound to the oxygen during the product-forming step, however, is not straight-forward. We have prepared and isolated a pentacoordinate silicate which contains a partially fluorinated alkoxy ligand (Scheme 6). This complex (M+= Na, TAS) undergoes the desired reaction with perfluorocyclobutene to give a vinyl ether product, but this process may involve

Scheme 6

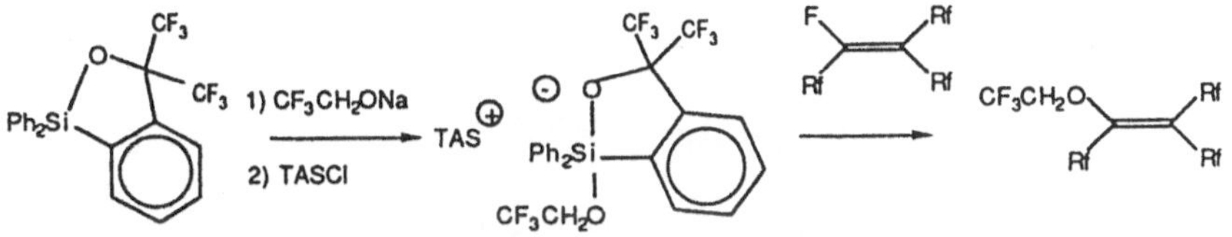

TAS = tris(dimethylamio)sulfonium

Table 1

Regioselectivity with $F\!\!-\!\!\langle\text{cyclobutene}\rangle$

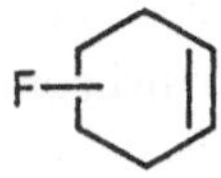

reagent	products $F\!\!-\!\!\langle\rangle\!\!-\!\!OR$	+	$F\!\!-\!\!\langle\rangle\!\!-\!\!OR$
$H(CF_2)_6CH_2OSiMe_3$ + CsF	69.2%		30.8%
$CF_3CH_2OSiMe_3$ + CsF	70.0%		30.0%
$H(CF_2)_6CH_2OH$ + NaH	95.0%		5.0%
$H(CF_2)_6CH_2OH$ + Cs_2CO_3	70.5%		29.5%

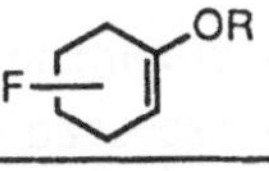

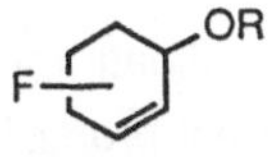

predissociation (especially with M+=Na). Indirect evidence has been sought by examining the regiochemistry of substitution in perfluorocyclohexene. As shown in Table 1, a counterion effect is operative, but this does not directly address the question whether silicon is present at the product-determining step.

In view of the beneficial influence of silicon for the C-O bond forming reactions, we have investigated silicon reagents for utility in C-C forming processes for fluorocarbon syntheses. Perfluoroalkylsilanes had been prepared but had not been utilized to synthetic advantage in this area[11]. Trifluoromethylsilanes react with aldehydes to give trifluoromethyl-substituted compounds[12].

We have reinvestigated Gilman's preparation of perfluoroalkylsilanes[11] and find somewhat better yields (80-85%) and more forgiving conditions than those reported previously (Scheme 7).

Scheme 7

Synthesis of R_fSiMe_3

$$R_fBr + TMSCl \xrightarrow[-30° \text{ to } 25°]{\text{Mg, THF}} R_fSiMe_3$$

$$R_f = CF_3(CF_2)_n\text{-} \quad n = 5,7$$

$$\text{-}(CF_2)_m \quad m = 6, 8,10$$

Activated by anionic catalysts (e.g., CsF, TAS fluorosilicates, carboxylates, fluoroalkoxides, fluorocarbanions, etc.), R_fSiMe_3 react readily with fluorinated acceptor molecules to give perfluoroalkylated products and trimethylfluorosilane. We have focused on acid fluorides and olefins as acceptors, although certain reactive aromatics and other carbonyl compounds function satisfactorily.

As a candidate for a polymer-forming process, reactions with fluoroolefins seemed the most promising. As discussed earlier, fluorinated olefins react readily with many nucleophilic species. These acceptors react with R_fSiMe_3 to give either mono- or bis(perfluoroalkylated) products (Scheme 8). Product specificity is enhanced for cyclic olefins whose higher reactivity and thermodynamically uphill isomerization pathways account for this tendency.

Scheme 8

OLEFIN SYNTHESIS

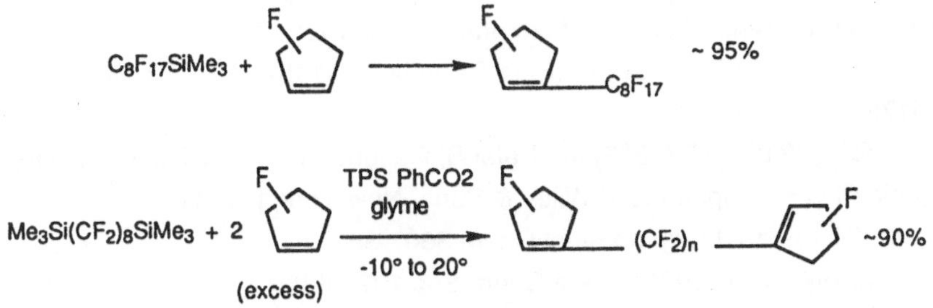

$C_8F_{17}SiMe_3$ + (cyclopentene with F) ⟶ (product with C_8F_{17}) ~ 95%

$Me_3Si(CF_2)_8SiMe_3$ + 2 (cyclopentene with F) (excess) →[TPS PhCO2, glyme, -10° to 20°] (product) $(CF_2)_n$ (product with F) ~90%

Because the yields of perfluoroalkylated olefins in some cases are quite good and principal side reactions involve coupling of the perfluoro-alkylene fragments, we have attempted to use this chemistry to make perfluorocarbon polymers. Using a model system (Scheme 9), preliminary reactions have provided low molecular weight oligomers ($\overline{Mn}$ ca. 3200). Improved selectivity is obviously required to produce high molecular weight polymer from these reactive species, and we are attempting to achieve this by judicious choice of metalloid ligands.

Scheme 9

Application to Polymer Synthesis

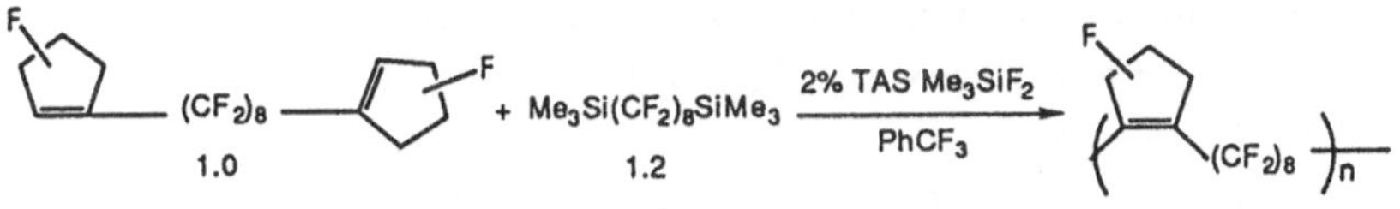

(cyclopentene with F) $(CF_2)_8$ (cyclopentene with F) 1.0   + $Me_3Si(CF_2)_8SiMe_3$ 1.2 →[2% TAS Me_3SiF_2, $PhCF_3$] (polymer with F, $(CF_2)_8$)$_n$

$\overline{M}_n$ ~ 3200, $\overline{M}w$ ~ 4300

Silicon-based reagents are well suited to the construction of C-O and C-C bonds systems. The ready availability of acceptors which contain fluorine as a potential leaving group makes such reagent/substrate combinations especially attractive. These and cognate systems should find wide applicability in fluorocarbon syntheses.

REFERENCES

1   Hill JT, and Erdman, JP (1977) In: Saegusa T, Goethals E (eds) Ring-Opening
    Polymerization, ACS Symposium Series 59, Wash DC, p269
2   a) Sianesi D, Tontanelli R (1967) Macromol Chem 102: 115
    b) Faucitano A, Buttafava A, Caporiccio G, Viola CT (1984) J Am Chem Soc
    106: 4172
3   a) Corriu RJP, Young JC (1989) In: Patai S, Rappoport Z (eds) The Chemistry
    of Organicsilicon Compounds, J Wiley & Sons, New York, p 1241
    b) Larson GL (1989) J Organometal Chem 360: 39.
4   a) Fujita M, Hiyama T (1985), J Am Chem Soc 107: 4085
    b) Fujita M, Obayashi M, Hiyama T (1988) Tetrahedron 44: 4135
    c) Yamazaki T, Ishikawa N (1984) Chem Letters 521
    d) Boutevin B, Pietrasanta Y (1985) Progress in Organic Coatings 13: 297
5   Formation of Macrocycles and the use of these materials as anion hosts has
    been described: Farnham WB, Roe DC, Dixon DA, Calabrese JC, Harlow RL
    (1990) J Am Chem Soc 112: 7707
6   a) Webster OW, Hertler WR, Sogah DY, Farnham WB, RajanBabu TV (1983)
    J Am Chem Soc, 105: 5706
    b) Dicker IB, Cohen GM, Farnham WB, Hertler WR, Laganis ED, Sogah DY
    (1990), Macromolecules 23: 4034
7   Chambers RD (1973) Fluorine in Organic Chemistry, J Wiley & Sons, New York
8   Nappa MJ, Sievert AC, Tong WR, (1991) J Fluorine Chem 53: 397
9   Sogah DY, Farnham WB (1985) In: Organosilicon and Bioorganosilicon
    Chemistry, J Wiley & Sons, New York, Chapter 20
10 Sakurai H (1989) Synlett 1: 1
11 a) Jukes AE, Gilman HJ (1969) Organometal Chem 18: 33
    b) Smith MR Jr, Gilman HJ (1972) Organometal Chem 46: 251
12 a) Prakash GKS, Krishnamurti R, Olah GA (1989) J Am Chem Soc 111: 393
    b) Stahly GP, Bell DR (1989) J Org Chem 54: 2873

Synthesis and Characterization of Perfectly Alternating Segmented Fully Cyclized Polyimide Siloxane Copolymers

M. E. Rogers, D. Rodrigues, A. Brennan*, G. L. Wilkes and J. E. McGrath**
Departments of Chemistry and Chemical Engineering;
NSF Science and Technology: High Performance Polymeric Adhesives and Composites
Virginia Polytechnic Institute and State University
Blacksburg, Virginia 24061-0212

* University of Florida
Dept. of Mat. Sci. and Engr.
317 MAW
Gainesville, Florida 32611

** To whom correspondence should be addressed.

Abstract

Perfectly alternating segmented, fully cyclized, poly(siloxane-imide) copolymers were synthesized and characterized as new materials. The reaction utilized the transimidization of aminopropyl terminated polydimethyl siloxane oligomers with aminopyridine capped polyimides based on oxydiphthalic anhydride and bisaniline P. The reaction was conducted in refluxing chlorobenzene at about 130°C. Transimidization processes appear to be quite rapid under these conditions, and high molecular weight copolymers were achieved in times as short as 10 minutes. The resulting copolymers could be cast or melt pressed into transparent films whose mechanical properties reflected the compositions and block molecular weights. In general, the reactions proceeded smoothly under homogeneous conditions in chlorobenzene and produced materials with glass temperatures somewhat depressed from the 240°C value of the controlled polyimide. The method allows excellent control of the microphase morphology, which is a function of block size and interaction parameters. The synthesis and characterization of these materials will be discussed and compared to the more widely studied randomly segmented copolymers which had been prepared in our labs and elsewhere.

1. INTRODUCTION

Polyimides are of particular interest for many engineering applications due to their excellent thermal and mechanical properties in the aerospace and microelectronics industries. Many important applications have been developed (1,2) The polyimide siloxane segmented copolymers have been of particular interest since incorporation of amine functionalized polysiloxanes into polyimides have resulted in improved solubility and processability, decreased water absorption, atomic oxygen resistance (3), lower dielectric constants (4) and enhanced adhesion (5). The copolymers are generally synthesized as shown in Scheme 1 from a combination of a dianhydride and

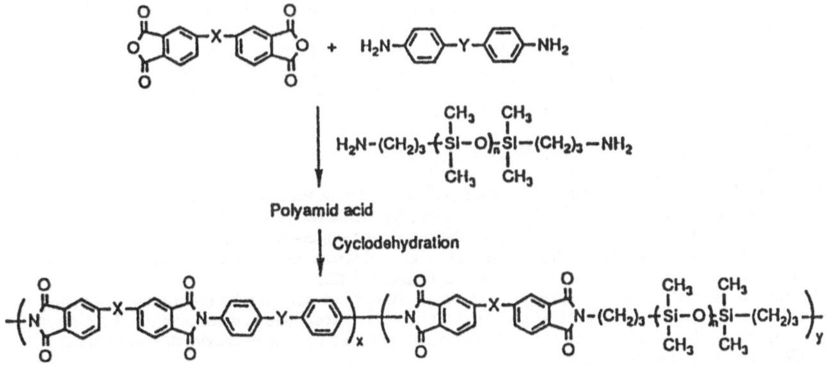

Scheme 1

diamine monomer and an aminopropyl terminated polysiloxane to form the segmented polyamic acid. The polyamic acid is then cyclodehydrated by either thermal imidization or solution imidization to give randomly segmented polyimide siloxane copolymers (3). This is indeed the classical approach of reacting one oligomer with two monomers which react to form the "hard" segment (4).

Y. Imanishi (Ed.)
Progress in Pacific Polymer Science 2
© Springer-Verlag Berlin Heidelberg 1992

Corresponding perfectly alternating segmented copolyimides would have desirable properties similar to randomly segmented copolyimides. However, the perfectly alternating segmented copolyimides should have higher structural regularity giving better defined microphase separation. Thus, depending on their composition, well defined perfectly alternating segmented copolyimides may afford higher upper glass transition temperatures, improved tensile strength and modulus as compared to analogous randomly segmented copolyimides.(4)

Takekoshi, et al., (7) demonstrated that an amine-imide exchange or transimidization method could be employed for the synthesis of polyetherimide homopolymers by reacting bisphthalimide monomers capped with 2-aminopyridine and appropriate diamine monomers in the presence of a transition metal catalyst. In view of our extensive experiences, polyimide oligomers endcapped with 2-aminopyridine were postulated to be also reactive toward amine terminated oligomers through a transimidization reaction. Amino alkyl (and possibly aryl) terminated oligomers such as polydimethylsiloxane and polyarylene ethers of controlled molecular weight combined with polyimides of a predetermined molecular weight and endcapped with 2-aminopyridine should afford perfectly alternating segmented polyimide copolymers. Aspects of the synthesis and characterization of perfectly alternating segmented polyimide-siloxane copolymers have been investigated and will be discussed in this paper.

2. EXPERIMENTAL

2.1. Oligomer Synthesis. The polyimides reported herein were largely based on oxydiphthalic anhydride (ODPA), and the aromatic diamine bisaniline P (Bis P). The ODPA was provided by Occidental Chemical Company and the Bis P was provided by Air Products and Chemicals, Inc. Both were of high purity and used as received. 2-aminopyridine (2AP) was purchased from Aldrich and recrystallized from a mixture of 75% (by volume) chloroform and 25% petroleum ether. The polymerization solvents were N-methylpyrrolidone (NMP), o-dichlorobenzene (DCB) and chlorobenzene (ΦCl). They were distilled over phosphorous pentoxide and stored in sealed flasks under nitrogen until use. The poly(amic acid) preparation was performed in a four-necked flask equipped with a mechanical stirrer, nitrogen inlet and a condenser with drying tube. Thus, the dianhydride was dissolved in NMP with slight heating, then cooled to ambient temperature. A calculated amount of the reactive endcapping agent 2-aminopyridine was added to the reaction mixture and allowed to react with the ODPA while stirring for 15 to 20 minutes. The diamine was then added as a powder and rinsed with NMP to bring the final reaction solids content to ~17-20 percent. The reaction was allowed to proceed for 20 hours to allow for the generation of the required molecular weight and for equilibration to a most probable molecular weight distribution.

The imidization was conducted as previously described.(8) A reverse Dean Stark trap with a condenser filled with DCB was fitted to the flask. An additional amount of DCB was added to the reaction mixture to bring the solvent ratio to 80%NMP/20%DCB. The reaction mixture was heated to 165°C by immersion in a hot silicone oil bath. The reaction was allowed to heat for a total of 24 hours to ensure complete imidization. The solution was cooled to ambient temperatures, and precipitated in methanol in a high speed blender. Upon cooling, the ODPA-Bis P solutions became turbid. The polyimide oligomers were collected and dried in a vacuum oven for 18 hours at 200°C and for 1 hour at 300°C.

The synthesis of amine terminated polysiloxane (PSX) (3) has been reported elsewhere. Aminopropyl terminated polydimethylsiloxanes with <Mn>'s of 1070 and 2670 g/mole, as determined by potentiometric titration, were used. The synthesis of high molecular weight polyimides endcapped with phthalic anhydride (PA) has also been reported(3).

2.1.2 Copolymer Synthesis

The copolymer synthesis was carried out in a three neck round bottom flask equipped with a mechanical stirrer, nitrogen inlet, thermometer and a condenser with drying tube. The stoichiometry was offset by using an excess of the polysiloxane oligomer. In every case, the calculated <Mn> was 40,000 g/mole. The polyimide oligomer was dissolved in chlorobenzene and heated to ~125°C. The amine terminated oligomer was added slowly and rinsed with chlorobenzene to bring the final reaction solids content to 15 percent. The reaction was stirred at ~125°C for 2 hours. The solution was cooled to ambient temperatures and cast into films. The films were heated slowly in a vacuum oven and dried for at least 3 hours at 250°C.

2.2 Characterization

2.2.1 Intrinsic viscosity measurements: Intrinsic viscosity measurements were performed in chloroform at 25°C using a Cannon-Ubbelohde viscometer.

2.2.2 Proton NMR Analysis: All proton spectra were measured on a Varian Unity 400 MHz NMR. Samples were dissolved in deuterated chloroform. TMS was used as an internal reference standard for the polyimide oligomers. Chloroform was used as a reference for the polyimide-siloxane copolymers.

2.2.3 Potentiometric titration: The <Mn>'s of the aminopropyl terminated polydimethylsiloxanes were determined by potentiometric titration with HCl on a MCI Automatic Titrator GT-05.

2.2.4 Thermal Analysis: Glass transition temperatures, Tg's, were obtained by differential scanning calorimetry on a Seiko DSC 210. Scans were run at 10°C per minute and the reported values were obtained from a second heating after quick cooling. The thermooxidative stability was determined by thermogravimetric analysis in air on a Perkin-Elmer TGA-7 at 10°/minute heating rate.

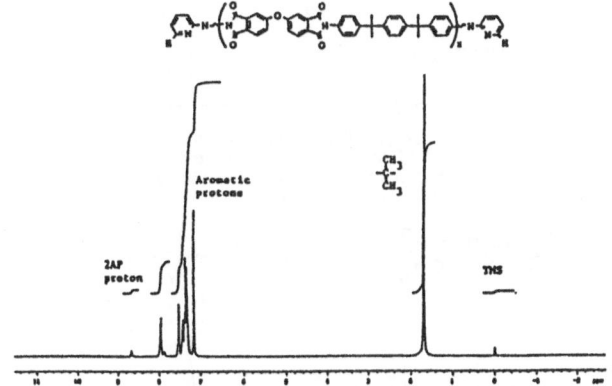

Scheme 2. Polyimide Oligomer Synthesis

3. RESULTS AND DISCUSSION

The synthesis of the polyimides (PI) is illustrated in Scheme 2. The polymerizations were conducted in a single "one pot" reactor, which minimizes solution transfer steps. Number average molecular weights, <Mn>, of the oligomers were evaluated by end group analysis using ^{1}H-NMR from the integral ratio of the proton for 2AP, at 8.7ppm, to the Bis P methyl protons at 1.7 ppm, see Figure 1.

^{1}H NMR of ODPA-Bis P-2AP Polyimide with a theoretical <Mn> = 4000 g/mole
Figure 1

Table 1 shows the comparison of the theoretical <Mn> to the <Mn> determined by ^{1}H NMR.

Table 1

ODPA-BIS P-2AP Polyimide Oligomers

<Mn> THEORETICAL	<Mn> BY ^{1}H NMR	[η] dl/g
4000	4900	0.21
6000	7300	0.23
8000	9700	0.33

The low intensity of the 2AP proton at 8.7 ppm limits the accuracy of the integration, but nevertheless, the <Mn> determined by ^{1}H NMR is in reasonable agreement with the theoretical <Mn>.

The copolymer synthesis was carried out as shown in Scheme 3.

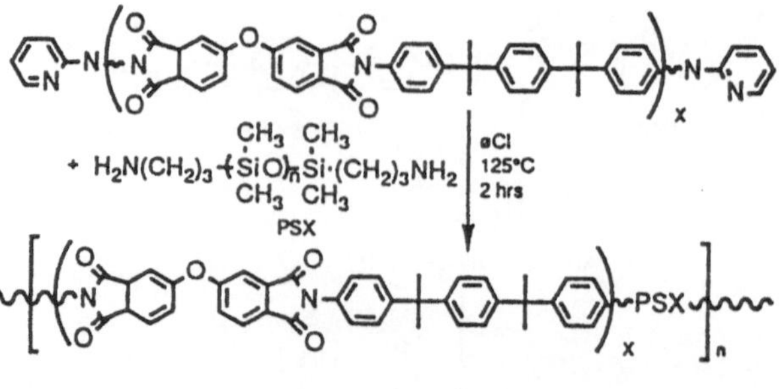

Scheme 3. Perfectly Alternating Segmented Copolymer Synthesis by the Transimidization Route

Table 2 shows that high molecular weight is achieved within 10 minutes and increases slightly after 2 hours.

Table 2

Copolymerization Results of ODPA-Bis P-2AP*
and H₂N-PSX-NH₂** in Refluxing Chlorobenzene

CATALYST Zn(OAc)₂·2H₂O	TIME minutes	[η] 25°C, CHCL₃
100PPM	10	0.79
	20	0.80
	120	0.96
no catalyst	10	0.89
	20	0.81
	120	0.87

The use of a catalyst is not necessary in the copolymerization of polyimides and aminopropyl terminated polydimethylsiloxanes. Intrinsic viscosities after 2 hours reaction time are comparable for copolymerization either with or without a catalyst. Solution cast films from the reaction flask were tough, flexible and transparent. Figure 2 shows a representative ¹H NMR of a polyimide-siloxane copolymer.

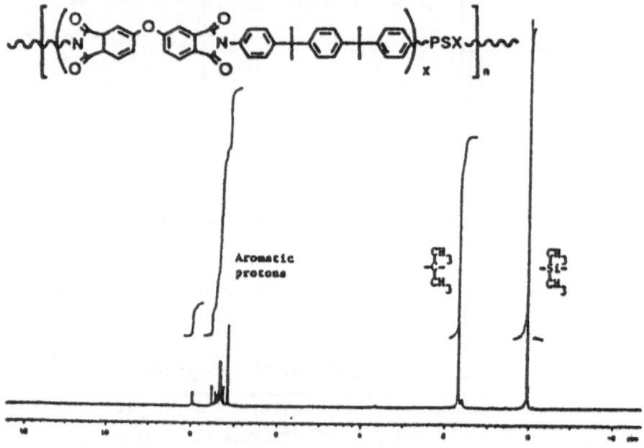

¹H NMR of a ODPA-Bis P Polyimide-siloxane containing 31 wt% PSX
Figure 2

Two series of polyimide-siloxane copolymers were investigated. In the PI-PSX1070 series, an aminopropyl terminated polydimethylsiloxane oligomer having a molecular weight of 1070 g/mole was reacted with three polyimide oligomers of ODPA-Bis P-2AP with molecular weights of 4000, 6000, and 8000 g/mole. In the PI-PSX2670, an aminopropyl terminated polydimethylsiloxane oligomer with a molecular weight of 2670 g/mole was reacted with the same polyimide oligomers used in the PI-PSX1070 series. Intrinsic viscosity data of the PI-PSX1070 and PI-PSX2670 series is given in Table 3.

Table 3

Intrinsic Viscosities

[η] of PI* Oligomer (dl/g)	[η] of PI-PSX1070 Copolymer (dl/g)	[η] of PI-PSX2670 Copolymer (dl/g)
1. 0.21	0.71	0.64
2. 0.23	0.87	0.71
3. 0.33	1.05	0.73

*The theoretical <Mn> if the PI oligomers in 1, 2 and 3 are 4000, 6000 and 8000 g/mole respectively.

After a reaction time of 30 minutes, high viscosities of the copolymers relative to the polyimide oligomers were achieved. Compared with the ODPA-Bis P-PA homopolymer having a theoretical <Mn> of 40,000 g/mole and an intrinsic viscosity of 0.54 dl/g, the polyimide-siloxane copolymers all showed higher intrinsic viscosities. The incompatibility of the polyimide and siloxane segments in the copolymer gives a more extended chain conformation as compared to the polyimide homopolymer of a similar molecular weight, thus, explaining the higher intrinsic viscosities observed for the copolymers. Increasing the molecular weight of the PI oligomers results in a corresponding decrease in the wt% siloxane incorporated and an increase in the intrinsic viscosity of the copolymers in each series. The intrinsic viscosities and wt% PSX in the PI-PSX1070 were higher than the intrinsic viscosities of a copolymer using the same polyimide oligomer in the PI-PSX2670 series. The intrinsic viscosity of the copolymer would be expected to increase since adding siloxane decreases the amount of flexible chains in the copolymer.

DSC data of the two polyimide-siloxane series and a high molecular ODPA-Bis P-PA homopolymer is presented in Table 4.

Table 4

Thermal Analysis

Theoretical <Mn> of PI Oligomer (g/mole)	Tg of PI oligomer (°C)	wt% PSX Theoretical	Upper Tg of PI-PSX(°C)	5% TGA loss (°C)
PI-PSX2670 Series				
4000	219	31	212	434
6000	232	22	229	446
8000	240	17	235	468
PI-PSX1070 Series				
4000	219	16	201	472
6000	232	11	227	
8000	240	8	233	
ODPA-Bis P-PA				
40,000	267	--	--	502

All of the copolymers exhibited two Tg's indicating that the copolymers were phase separated. The Tg's of the polyimide oligomers range from 219°C to 240°C and the high molecular weight polyimide shows a Tg of 267°C. The upper Tg's, which result from the polyimide phase, of the polyimide-siloxane copolymers range from 201°C to 233°C for the PI-PSX2670 series. In both series, the Tg of the copolymers is less than the Tg of the corresponding polyimide oligomers indicating that there is some mixing between the polyimide and PSX phases. The depression of the Tg is even more pronounced when the lower molecular weight polyimide oligomer (4000 g/mole) is used. The lower molecular weight PSX oligomer used in the PI-PSX1070 series results in a greater depression of the upper Tg's than the higher molecular weight PSX used in the PI-PSX2670 series. The depression of the upper Tg is not dependent on the wt% PSX incorporated but depends on the size of the polyimide and PSX oligomers with the lower molecular weight oligomers giving lower upper Tg's and thus increased phase mixing.

Also included in Table 4, is the 5% weight loss measured by TGA. The thermooxidative stability of the copolymers is less than the polyimide homopolymer due to the presence of the PSX oligomers. The 5% weight loss is dependent on the wt% PSX incorporated since the thermooxidative stability of the copolymers decreases as greater weights of the PSX are incorporated into the copolymer. Two thermograms from the PI-PSX2670 series are shown in Figure 3. The copolymer with the higher weight percent PSX begins to degrade first but it also gives a greater char yield.

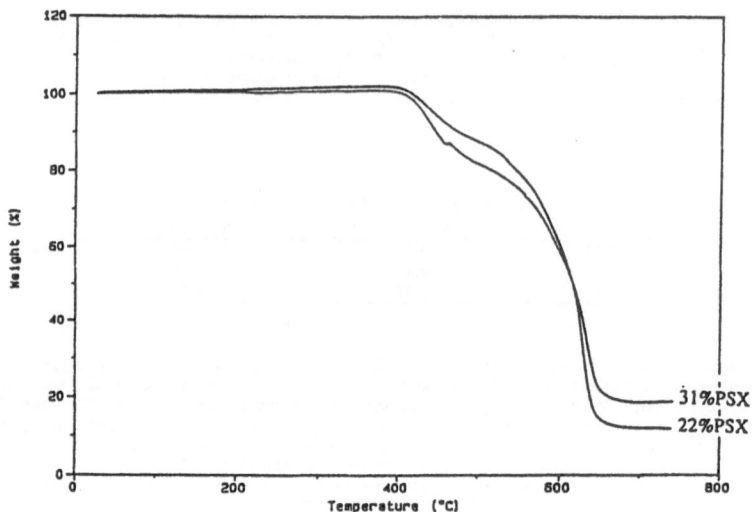

Thermogram of Polyimide-siloxane Copolymers
Figure 3

4. CONCLUSIONS

Controlled molecular weight polyimides oligomers endcapped with 2-aminopyridine were successfully synthesized by a convenient "one pot" method. Utilizing transimidization chemistry polyimides endcapped with 2-aminopyridine were reacted with aliphatic aminopropyl terminated polydimethylsiloxane oligomers to afford perfectly alternating segmented copolymers of high molecular weight. These copolymers could be transformed into tough, transparent films by either solution or melt cast fabrication. The copolymers exhibit two Tg's indicating microphase separation. However, a significant depression of the upper Tg suggests partial phase mixing occurred.

5. ACKNOWLEDGEMENTS

We appreciate the support of this research by the NSF Science and Technology Center for High Performance Polymeric Adhesives and Composites. An IBM fellowship to Martin E. Rogers is also gratefully acknowledged.

6. REFERENCES

1. K. L. Mittal, Editor, "Polyimides", Volumes 1 and 2, Plenum Press, 1984; C. Feger, M. M. Khojasteh and J. E. McGrath, Polyimides: Chemistry Materials, and Characterization, Elsevier (1989).
2. M. Bowden and S. R. Turner, Editors, "Polymers for High Technology, Electronics and Photonics", ACS Symp. Series 46 (1987).
3. C. A. Arnold, J. D. Summers, Y. P. Chen and J. E. McGrath, Polymer (London), 30(6), 986 (1989); J. D. Summers, Ph.D. Thesis, VPI and SU ,1987; C. A. Arnold, Ph.D. Thesis VPI and SU, 1989.
4. A. Noshay and J. E. McGrath, "Block Copolymers Overview and Critical Survey", Academic Press (1977).
5. C. A. Arnold, Y. P. Chen, D. H. Chen, M. E. Rogers and J. E. McGrath, Mat. Res. Soc. Symp. Proc., Vol. 154, 149(1989).
6. T. Yoon, C. A. Arnold and J. E. McGrath, J. Adhesion, (accepted 1991).
7. T. Takekoshi, J. L. Webb, P. P. Anderson and C. E. Olsen, IUPAC 32nd International Symposium on Macromolecules, 464, 1988; T. Takekoshi, in Polyimides, D. Wilson, P. M. Hergenrother and H. Stenzenberger Editors, Chapman and Hall (1990).
8. R. O. Waldbauer, M. E. Rogers, C. A. Arnold, G. A. York, Y. Kim and J. E. McGrath, Polymer Preprints, 31(2), 430 (1990).

Tailored Design of New Polyphosphazenes with Special Properties

Harry R. Allcock

Department of Chemistry, The Pennsylvania State University,
University Park, Pennsylvania 16802, U.S.A.

Abstract: Polymers that contain heteroelements such as phosphorus and nitrogen in the skeletal structure offer a means for extending the properties normally associated with synthetic macromolecules into areas hitherto typical of inorganic materials such as ceramics, metals, or electroactive or electro-optical solids.   The polymeric molecular structure provides materials strength and flexibility, while the heteroelements provide flame resistance, radiation resistance, and access to new reactions for the linkage of electroactive or biologically interesting units to the polymer structure.   In this paper synthetic pathways that provide access to such polymers are described, together with a review of structure-property relationships that provide a high level of molecular and materials design.

BACKGROUND AND PURPOSE

Polymers As A Branch Of Materials Science

Polymer science is a branch of the broader field of materials science which also includes ceramics, metals, and electroactive or electro-optical solids.   Of these four areas, traditional polymer science is derived mainly from organic chemistry via petrochemicals and biological polymers, whereas the other three areas are rooted in inorganic chemistry.   These different chemical origins become manifest in different properties and limitations.   For example, the long-chain structure of linear or branched organic polymers gives rise to strength, toughness, and ease of fabrication, while the inorganic composition of ceramics is responsible for their high temperature stability.

The purpose of our research program is to design and synthesize new polymer systems that contain both organic and inorganic components in order to generate materials with hybrid properties.   These can be visualized as new substances that occupy the area connecting the four classical materials fields and which possess new combinations of properties.

Hybrid Organic-Inorganic Polymers

A wide range of alternatives exists for the incorporation of inorganic elements into macromolecular structures.   For example, inorganic or organometallic units can be incorporated into the side groups of a classical vinyl organic polymer.   However, the greatest differences in properties from organic polymers are likely to result from the introduction of inorganic elements into the *backbone* of a macromolecule [1-3], and this is the approach described in this paper.

The main precedent for the work to be described here was the development of the poly(organosiloxanes) (1), starting in the 1940's and continuing to the present day. The high thermo-oxidative stability that resulted from the incorporation of alternating oxygen and silicon atoms into the backbone was an indication of the broader possibilities that could be envisaged with the use of other elemental combinations.   Our work has been focussed on the design and synthesis of polymers with alternating phosphorus and nitrogen atoms in the skeleton (2) [3, 4-8], and more recently with an extension of these ideas to poly(carbophosphazenes) (3) and poly(thiophosphazenes) (4) [9-11].

As will be evident from the following discussion, the variety of polymer structures based on a phosphorus-nitrogen skeleton far exceeds those that are accessible in the

Y. Imanishi (Ed.)
Progress in Pacific Polymer Science 2
© Springer-Verlag Berlin Heidelberg 1992

polysiloxane series, although the commercial development of polyphosphazenes is still at an early stage compared to the siloxane systems.

$$\left[-O-\underset{\underset{R}{|}}{\overset{\overset{R}{|}}{Si}}-\right]_n \qquad \left[-N=\underset{\underset{R}{|}}{\overset{\overset{R}{|}}{P}}-\right]_n$$

1 $\qquad\qquad\qquad\qquad$ 2

$$\left[-N=\underset{\underset{R}{|}}{\overset{\overset{R}{|}}{C}}-N=\underset{\underset{R}{|}}{\overset{\overset{R}{|}}{P}}-N=\underset{\underset{R}{|}}{\overset{\overset{R}{|}}{P}}-\right]_n \qquad \left[-N=\underset{\underset{R}{|}}{\overset{\overset{R}{|}}{S}}-N=\underset{\underset{R}{|}}{\overset{\overset{R}{|}}{P}}-N=\underset{\underset{R}{|}}{\overset{\overset{R}{|}}{P}}-\right]_n$$

3 $\qquad\qquad\qquad\qquad$ 4

Polyphosphazenes - Historical Perspective

The chemical origins of this field can be traced back over 150 years to the reports by Liebig, Wohler, and Rose [12, 13] that phosphorus pentachloride and ammonia react to yield a white, crystalline, "organic"-type compound since shown to be hexachlorocyclotriphosphazene (5). Later, in 1897, Stokes [14] described the conversion of this compound to an insoluble, hydrolytically sensitive elastomer subsequently known as "inorganic rubber". The intractibility of inorganic rubber is now known to be due to its crosslinked structure, and its hydrolytic sensitivity to the presence of phosphorus-chlorine bonds.

This material was viewed as a mere scientific curiosity until the mid 1960's when, following a detailed study of the thermal behavior of 5, Allcock, Kugel, and Valan found a way to control the ring-opening polymerization of 5 to yield an *uncrosslinked*, essentially linear high polymer, and developed reaction conditions that allowed the chloro- side groups to be replaced by organic side units to form hydrolytically stable high polymers [4-6]. Since that time a very wide range of more than 300 different organic derivative polymers has been synthesized. The number of fundamental papers and patents has grown steadily over the years, reaching roughly 200 per year in 1990, and a total of approximately 2000 reports by mid-1991. Some of this growth in interest can be traced to the commercial development of several polyphosphazene elastomers [15-18], but the unique fundamental chemistry and materials science of these polymers underlies a large part of the growth in interest.

POLYMER SYNTHESIS METHODOLOGY

The Ring-Opening-Polymerization / Macromolecular Substitution Route

The main method of synthesis for polyphosphazenes [4-6] involves the thermal ring opening polymerization of monomer 5 to uncrosslinked high polymer 6, followed by solution state nucleophilic replacement of the chlorine atoms in 6 by organic or organometallic side groups. The overall process is illustrated in Scheme I.
The ring-opening polymerization may proceed via an uncatalyzed reaction, in which case the polymerization temperature is in the range of 250°C, or it may be catalyzed by a Lewis acid initiator such as triphenylphosphate-boron trichloride [19], in which case the polymerization temperature may be as low as 210°C. Uncrosslinked poly(di*fluoro*phosphazene), $(NPF_2)_n$, can be prepared by a similar process (at 350°C) and is also used as a reactive intermediate [20]. The macromolecular substitution reactions are carried out with the reactants in a solvent such as tetrahydrofuran, benzene, or toluene, with the insolubility of the sodium chloride or amine hydrochloride providing a strong force for driving the reactions to 100% completion. The only serious limits to this process are connected with a

Scheme I

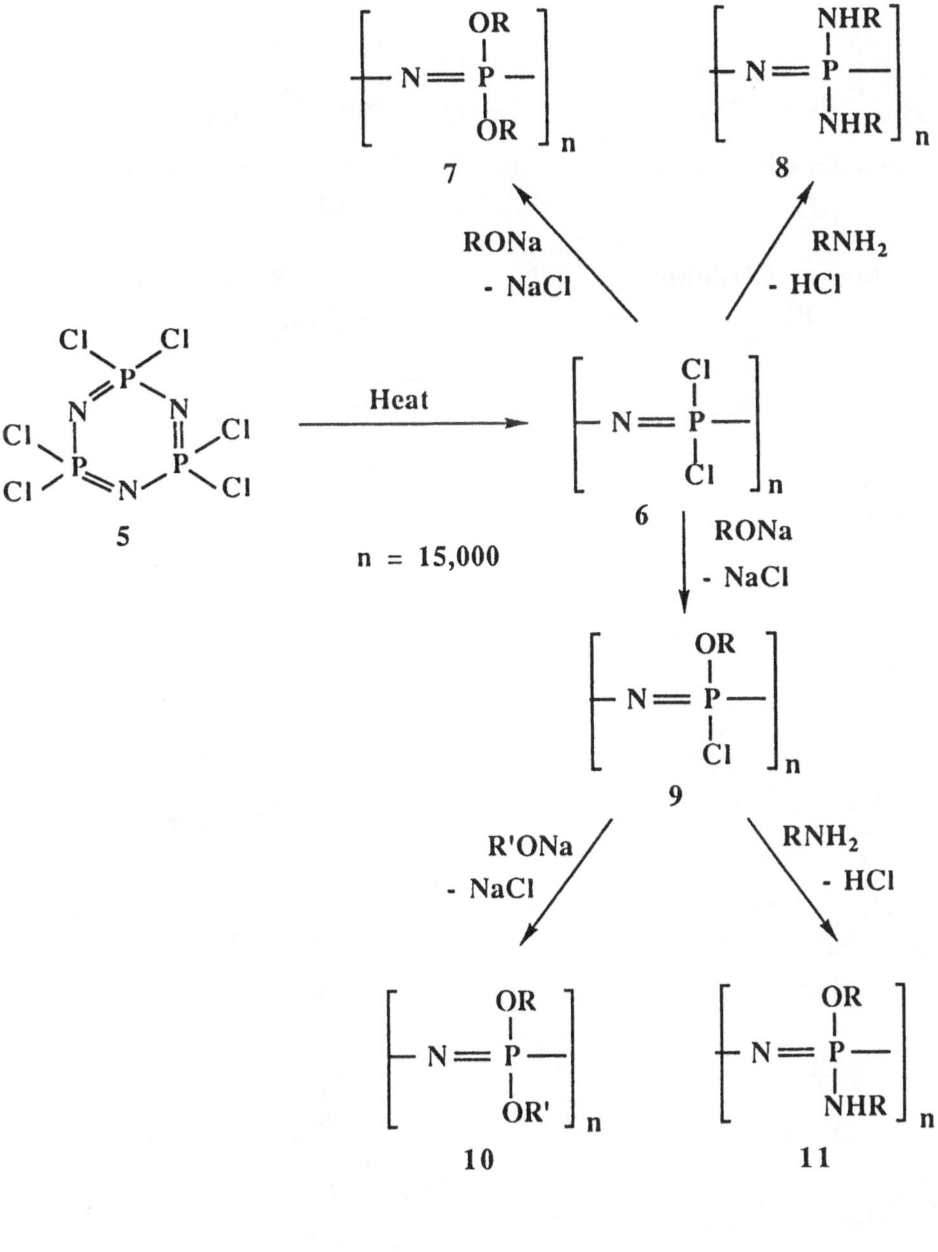
OR
N=P
OR
n
7
NHR
N=P
NHR
n
8
RONa
- NaCl
RNH2
- HCl
Cl
Cl
P
N
N
Cl
Cl
P
N
P
Cl
Cl
5
Heat
n = 15,000
Cl
N=P
Cl
n
6
RONa
- NaCl
OR
N=P
Cl
n
9
R'ONa
- NaCl
RNH2
- HCl
OR
N=P
OR'
n
10
OR
N=P
NHR
n
11

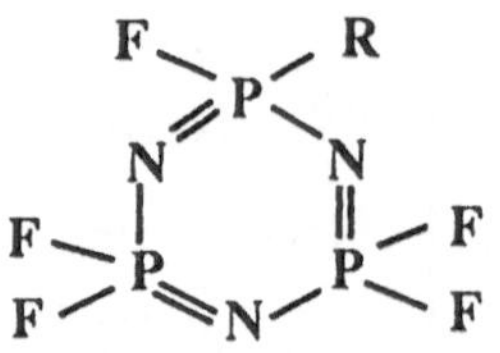

12

R = Me, Et, t-Bu, or Ph

13

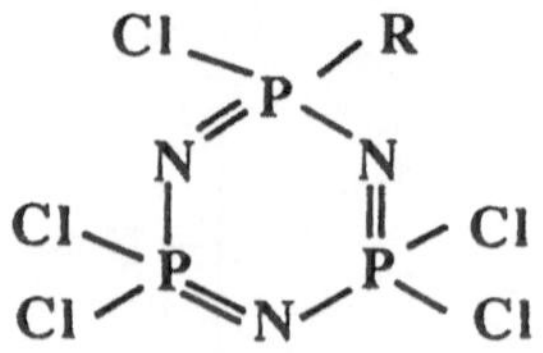

R = Me, Et, n-Pr, n-Bu, or Ph

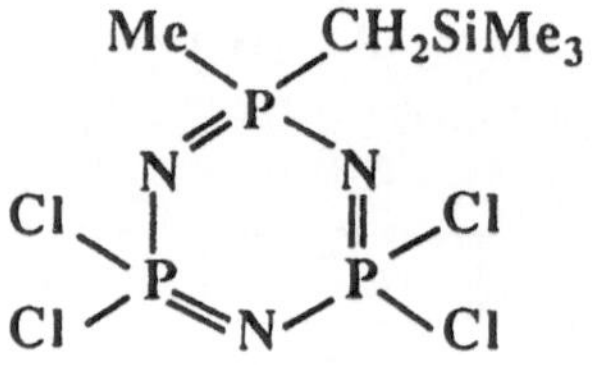

14

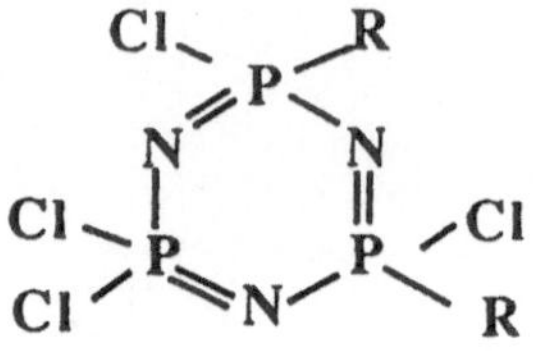

15

R = Me or Et

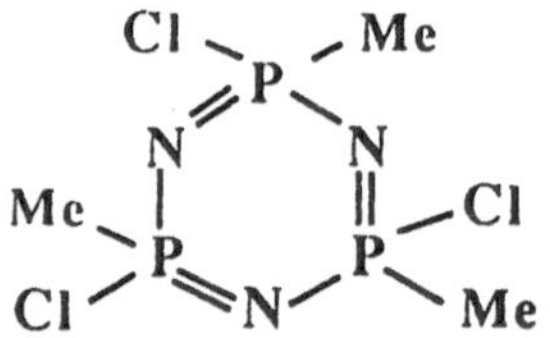

16

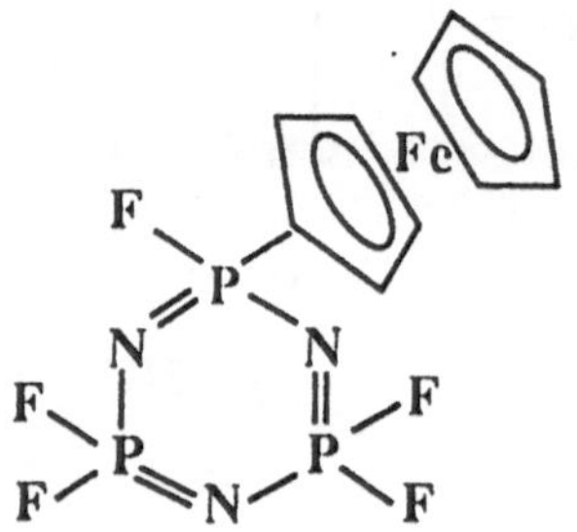

17

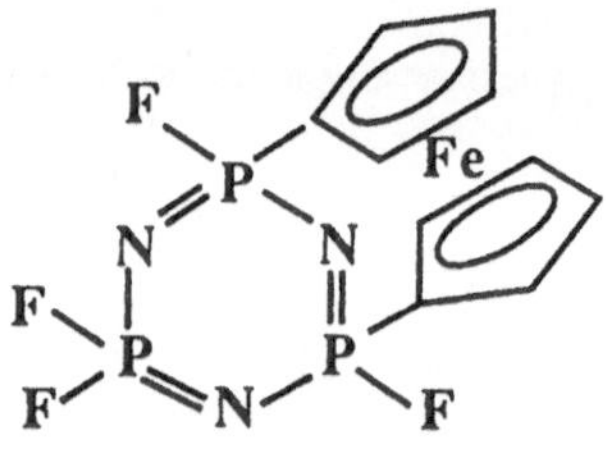

18

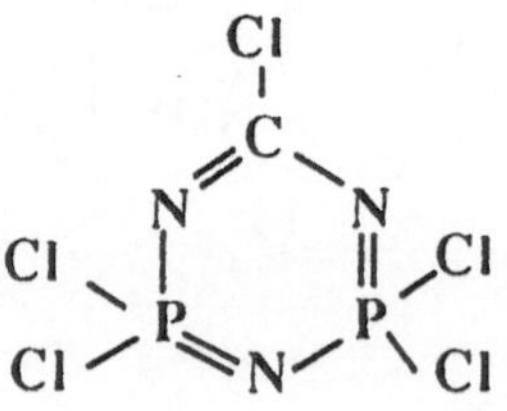

19

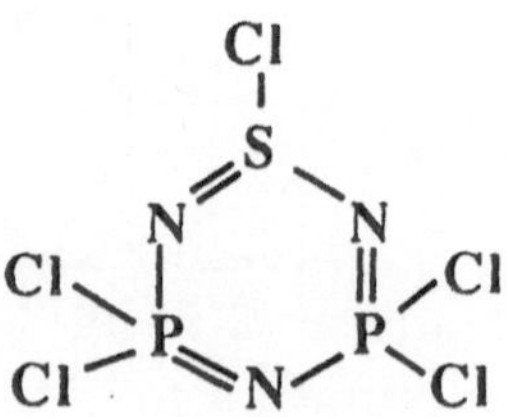

20

retardation of the chlorine replacement reaction when bulky nucleophiles are employed, in which case more forcing reaction conditions, such as temperatures above 100°C and the use of pressure reactors [21] usually ensure complete halogen replacement.

An additional strong advantage of the macromolecular substitution process is the option that two or more different types of side groups can be introduced into each polymer molecule by sequential or simultaneous co-substitution reactions. This is illustrated in Scheme I. Occasionally, as when bulky diethylamino units are introduced first, this technique permits the synthesis of polymers in which each phosphorus bears only one bulky unit and one less hindered substituent (non-geminal structure) [22] rather than structures that contain both geminal and non-geminal side group arrangements. Sequential or simultaneous co-substitution reactions are used for the preparation and manufacture of commercial high performance phosphazene elastomers of the type shown in structure 10 [15-17].

Ring-Opening-Polymerization Of Organo-Substituted Cyclic Phosphazenes

An alternative method for polyphosphazene synthesis involves carrying out the halogen replacement reactions at the cyclic trimer stage, followed by polymerization of the organo-substituted trimers. Two alternatives have been developed, (a) polymerization of trimers that bear both organic and halogen side groups, and (b) polymerization of those that have organic side groups only. Use of the latter (less common) alternative clearly avoids the need for a subsequent polymer substitution step.

Systems with halogen as a cosubstituent. The phosphazene cyclic trimers shown as 12-18 polymerize when heated to give high polymers that can be subjected to polymer substitution reactions to replace the remaining halogen atoms. In general, bulky organic side groups attached to the trimer molecule, or more than three organic groups per phosphazene ring, lower the tendency for polymerization and favor ring-ring equilibration (ring expansion) reactions instead [23-25]. Nevertheless, this is a powerful route to mixed-substituent polymers that cannot be prepared by the conventional route. In particular it provides access to polymers with alkyl- or aryl groups bonded directly to phosphorus, species that are difficult to prepare by the direct reactions of organometallic reagents with poly(dichlorophosphazene) [6]. Many of the known polymers made by this route are elastomers with low glass transition temperatures [< -60°C].

Systems without halogen side groups. Normally, phosphazene cyclic trimers that bear organic side groups only do not polymerize. Instead they undergo ring expansion reactions when heated. However, if the phosphazene ring is spanned by a transannular bridge, such as the ferrocenyl unit shown in structure 21, the ring strain imparted to the molecule provides a driving force for phosphazene ring-opening polymerization to give polymer 22 directly [26, 27]. For the system shown, a trace of trimer 5 is needed as an intiiator, but related systems are known where this is not a requirement.

OR = OCH_2CF_3 Trace of $(NPCl_2)_3$ as initiator

21   22

Synthesis of Poly(carbophosphazenes) and Poly(thiophosphazenes)

Two recent developments have illustrated the opportunities that exist for expanding the methodology discussed above to the synthesis of other, related inorganic-organic polymers. These developments involve the first preparation of phosphazene-like polymers that contain heteroelements such as carbon or sulfur in the backbone as well as phosphorus and nitrogen. The synthetic pathways involve the newly discovered ring-opening polymerization of cyclocarbophosphazenes and cyclothiophosphazenes **19** and **20** [9-11]. Macromolecular chlorine replacement reactions then yield derivative polymers. A sulfur-VI thiophosphazene system, based on similar chemistry has also been reported recently [28]. The development of these heterophosphazene polymer systems is still in its infancy, but their influence on our understanding of structure-property relationships is already important (see later).

The Condensation Route To Polyphosphazenes

An alternative approach to the synthesis of polyphosphazenes has also been reported [29, 30]. In this process, the *condensation* reactions of N-silylphosphor-animines yield medium molecular weight poly(alkyl- and aryl-phosphazenes). The advantage of this route is that it yields polymers in which all the side groups are alkyl- or aryl- units bonded to phosphorus through carbon-phosphorus bonds - precisely those structures that are the most difficult to prepare by the classical route. Catalysis of these condensation polymerizations by Bu_4NF has also been described [31]. It has been demonstrated [32] that alkyl side groups can be lithiated, and these sites can form the basis of a lithium-replacement macromolecular substitution chemistry.

GENERAL STRUCTURE-PROPERTY RELATIONSHIPS

The principal advantage of polyphosphazene synthesis chemistry is the ease with which different side groups can be incorporated into the macromolecular structure. The inorganic backbone controls some of the properties (see later), but variations in side group structure allow the development of an almost unprecedented range of polymer properties superimposed on those generated by the skeleton. For example, fluoroalkoxy side groups yield hydrophobic, solvent resistant properties coupled with low glass transition temperatures and either elastomeric or film or fiber-forming qualities, depending on crystallinity. Many of these polymers have surface properties that are reminiscent of poly(tetrafluoroethylene), but they are more easily fabricated. Aryloxy side groups raise the glass transition temperatures and change the solubility behavior. Solubility in water can be accomplished by the use of side groups such as alkyl ether, carboxy-aryloxy, methylamino, glucosyl, or glycery side groups. Side groups such as alkylamino, borazine, or carborane units yield polymers that, when heated to 1000°C, undergo condensation crosslinking to ceramics. Other side groups generate biological activity. A major purpose for current work in polyphosphazene chemistry is to develop structure-property relationships that will lead to a predictive scheme for planning future synthetic work.

APPLICATION TO BIOLOGY AND MEDICINE

Different Aspects

The use of synthetic polymers in biology and medicine is a rapidly developing subject. Solid materials comprised of polymers have many features in common with living tissues, including flexibility, elasticity, strength and, in some cases, the ability to fuction as hydrogels and membranes. Other polymers are of biomedical interest because of their ease of degradation in a biological environment or their biological activity. Biologically active, water-soluble polymers have potential uses as non-migrating pharmaceutical molecules.

Polyphosphazenes are of particular interest in this respect because the synthetic versatility allows different biologically-related properties to be designed into the molecule by changes in side group structure. Moreover, the phosphorus-nitrogen backbone offers the possibility that, in the presence of specific side groups, the polymer can be broken down by a biological environment to phosphate, ammonia, and the side group.

Modification Of Surfaces For Biomedical Compatibility Or Bioactivity

An ideal biomaterial would have certain specific bulk properties, such as elasticity, combined with biocompatible surface characteristics. Only rarely, if at all, is this combination of properties found in an unmodified polymer. More frequently a material is chosen for its bulk properties, and the surface is then modified by reaction chemistry. Polyphosphazenes are especially appropriate for surface modification because the side groups at the surface can be chosen for their ease of chemical reactions. A number of examples of surface reactions have been developed in our program, including the replacement of hydrophobic fluoroalkoxy surface groups by hydrophilic hydroxy, alcohol, or alkyl amino units [33, 34], the nitration or sulfonation of surface aryloxy units [35], the oxidation of surface alkylaryloxy units to aryloxycarboxylate groups [36], quaternization of surface alkyl halide units and coupling to heparin [37], and the immobilization of dopamine and enzymes to polyphosphazene surfaces [38, 39]. All these surface reactions provide access to unique materials for biomedical applications.

Membranes

Amphiphilic polymers are of special interest for biomedical membrane uses. The introduction of hydrophobic and hydrophilic side groups into the same polyphosphazene allows the development of amphiphilic bulk and surface properties. Moreover, the macromolecular substitution method of polymer synthesis allows the ratios of the two types of side groups to be controlled by changes in the reactant ratios. Thus, a complete sequence of different membrane materials can be prepared with properties that range from highly hydrophobic to strongly hydrophilic. For example, polymers that bear trifluoroethoxy or phenoxy and methylamino or methoxyethoxy-ethoxy side groups have been prepared and their semipermeability behavior studied [40]. In addition, the strength of the membranes can be increased by radiation crosslinking methods [40, 41]. In one case, dialysis membranes with higher permeabilities than standard cellulose dialysis tubing have been prepared [40].

Hydrogels

A hydrogel is a formed when a crosslinked, previously water-soluble polymer absorbs water. Synthetic hydrogels have many characteristics in common with soft living tissues. Two different types of polyphosphazenes have shown promise for use as biomedical hydrogels. These are poly[di(methoxyethoxyethoxy)phosphazene] ("MEEP") shown as structure 23, and poly[di(carboxylatophenoxy)phosphazene] (24). MEEP is an excellent precursor for the formation of neutral hydrogels because it readily undergoes gamma- or ultraviolet radiation-induced crosslinking through the side groups [41, 42]. This converts a water-soluble polymer to a hydrogel, the properties of which can be altered by the radiation dose. Small doses yield a loose, flexible hydrogel with a high water content, while high doses give a heavily crosslinked, rubbery hydrogel.

The second system, based on polymer 24 is water-soluble as its sodium salt, but crosslinks ionically in the presence of di- or tri-valent cations such as calcium or aluminum ions [43], to form hydrogels. This polymer is under development as a component for the microencapsulation of hybridoma mammalian cells to allow their use in prototype artificial liver or pancreas devices, or in biotechnology [44-46].

Bioerodible Polymers

A few organic side groups sensitize the polymers to hydrolytic breakdown. These side groups are amino acid esters, imidazolyl units, glyceryl, and glucosyl groups [47-51]. These polymers bioerode to phosphate, ammonia, and the protonated product from the side group, and are thus good candidates for temporary tissue replacement uses or controlled drug delivery matrices. Several of these systems have been examined for this last application [48].

Biologically Active Polyphosphazenes

It will be clear from the foregoing discussion that organic molecules with hydroxyl or amino functionalities can also be linked as side groups to a polyphosphazenes chain. When these structures are combined with the hydrolytically labile side groups discussed above, or the water-solubilizing side units mentioned earlier, the system provides access to a wide range of macromolecular drug molecules with controllable rates of hydrolytic degradation. Steroids, antibacterial agents, antitumor agents, and other bioactive units have been utilized in this way [52-55], and it is clear from the known results that a broad possibilities for pharmaceutical design are possible by the use of this system.

23

24

25

26

27

28

POLYMERS WITH ELECTROACTIVE AND ELECTRO-OPTICAL PROPERTIES

Recent studies have extended the structure-property relationships in polyphosphazenes to include polymers that show interesting electrical or optical behavior in the solid state. These fall into three categories - (1) polymers that are good solid solvents for salts and which function as solid ionic conductors, (2) species that bear electronically active side groups, and (3) polymers that bear rigid organic units that generate liquid crystalline or non-linear-optical behavior.

Solid Ionic Conductors

The polymer MEEP, mentioned earlier, has been found to be an excellent solid solvent for salts such as lithium triflate or silver triflate [42, 56-59]. The solid solvent properties of this polymer are connected with the high loading of etheric oxygen atoms in the side group structure which provides coordination sites for the cations and facilitates ion pair separation. This same mechanism allows ion movement under the influence of an electric current, a process that is assisted by the high molecular mobility of the side groups and the polymer backbone. Room temperature ionic conductivities are two to three orders of magnitude higher than those of comparable solid electrolytes based on poly(ethylene oxide). MEEP solid electrolytes can be crosslinked by radiation techniques, a process that stabilizes the shape and dimensions of the material without lowering the conductivity [42].

Polyphosphazenes With Electronically Active Side Groups

We have devoted a major effort in recent years to the synthesis of polyphosphazenes that bear transition metals in the side group structure [60]. One purpose of these studies has been to examine the possibility that metal-containing side groups might superimpose electroactive, catalytic, or magnetic propeties onto the solid state or surface properties normally associated with polyphosphazenes. Only one example will be given here.

Earlier, it was mentioned that polymers can be prepared by the ring opening polymerization of cyclic phosphazenes that have metallocenyl side groups (for example, polymer **22**) By varying the type of metallocene (ferrocenyl or ruthenocenyl), the mode of linkage of the metallocene to the skeleton (single link or transannular) and by altering the cosubstituent groups it is possible to prepare a wide range of different polymers. The metal-containing units are electroactive, and the oxidation/reduction potentials depend on all the factors just mentioned [61, 62]. Thus, the electochemical behavior of a polymer can be tailored in ways that are of interest in electrode mediator catalysis. Specific polymers of this type also exhibit weak semiconductivity when doped with iodine, a phenomenon that is ascribed to the jumping of electrons from one metallocenyl group to another.

Polymers With Liquid Crystalline Or Non-Linear-Optical Side Groups

The macromolecular substitution route for polyphosphazene synthesis allows a wide variety or simple or complex organic side groups to be linked to the backbone. This includes side groups that, at the small- molecule level, might be expected to generate liquid crystallinity or non-linear optical properties.

Species **25** and **26** are side chain liquid crystalline polymers [63-65], while polymers **27** and **28** show second order non-linear optical characteristics [66-68]. Note that in all these cases the active side unit must be separated from the main polymer chain by a flexible spacer group to allow orientation of the active unit independently of the disposition of the backbone. Species **25** generates nematic-type liquid crystallinity at temperatures between 118 and 126°C. Polymer **28**, after being poled by exposure to a 10-20 kV electric field, showed d_{33} value2 of 34-40 pm/V This is one of the highest second order NLO values reported for a polymer, although the value falls quickly to zero after removal of the electric field due to thermal randomization of the side groups. High refractive index polyphosphazene glasses have also been synthesized by the use of side groups with high loadings of pi-electrons [69].

THEORY AND MOLECULAR STRUCTURE

As first postulated by Craig and Paddock [70], and more specifically by Dewar [71], a plausible valence bond description of phosphazene cyclic trimers and tetramers assumes that the backbone contains d_{pi}-p_{pi} bonding that explains the disposition of the one electron each from phosphorus and nitrogen not accounted for by formation of the sigma-bond framework and the lone pair orbital at each nitrogen. When applied to the high polymers this model explains why the backbone bonds have a high torsional flexibility, because each nitrogen $2p_z$ orbital can overlap several of the five different phosphorus 3d orbitals as the bond undergoes torsion. Thus the "delocalized double bonds" in these polymers are not comparable to classical organic double bonds, either with respect to long-range delocalization or torsional barrier. The unusual backbone bonding also explains the transparency to radiation from the near infrared to the short wavelength ultraviolet (220 millimicrons). It also provides a rationale for the resistance of the backbone to free radical degradation, especially under photolytic or gamma-irradiation conditions.

In these terms, the molecular flexibility of the backbone should be affected by the introduction into the skeleton of elements other than phosphorus, and this is found to be the case. For example, the replacement of one phospphorus atom in three along the chain by carbon raised the glass transition temperature by about 20°C for each different side group studied [10], and this is ascribed to the introduction of organic-type

p_{pi}-p_{pi} carbon-nitrogen bonds which raise the torsional barrier. The preliminary results indicate that the replacement of phosphorus by sulfur does not have nearly so dramatic an effect.

Only a few studies have been made of the chain conformations of well-characterized polyphosphazenes. With the exception of $[NPF_2]_n$, which assumes a spiral helical conformation, most polyphosphazenes adopt a cis-trans planar conformation in the microcrystalline state [6, 72-77]. Although it is tempting to explain this behavior in terms of the pi-bonding in the backbone, in fact it appears to be mainly a response to side group repulsions which are minimized by the large side group - side group intramolecular distances that are generated by this conformation [78, 79]. In any event, it can be shown that the influence of bulky side groups, which raise the glass transition temperatures, is via intramolecular side group -side group interactions. This provides an exceptionally useful means for the correlation of solid state properties with side group molecular structure in a way that is difficult to achieve in classical polymer chemistry.

REFERENCES

1 Allcock HR (1967) Heteroatom Ring Systems and Polymers. Academic Press, New York
2 Allcock HR, Lampe FW (1990) Contemporary Polymer Chemistry (2nd ed). Prentice Hall, Englewood Cliffs, New Jersey
3 Mark JE, Allcock HR, West RC (1991) Inorganic Polymers. Prentice Hall, Englewood Cliffs, New Jersey
4 Allcock HR, Kugel RL (1965) J Am Chem Soc 87:4217
5 Allcock HR, Kugel RL, Valan KJ (1966) Inorg Chem 5:1709
6 Allcock HR, Kugel RL (1966) Inorg Chem 5:1716
7 Allcock HR (1972) Phosphorus-Nitrogen Compounds. Cyclic, Linear, and High Polymeric Systems. Academic Press, New York
8 Allcock HR (1985) Chem. & Eng. News 63:22
9 Manners I, Renner G, Nuyken O, Allcock HR (1989) J Am Chem Soc 111:5478
10 Allcock HR, Coley SM, Manners I, Renner G, Nuyken O (1991) Macromolecules 24:2024
11 Dodge JA, Manners I, Allcock HR, Renner O, Nuyken O (1990) J Am Chem Soc 112:1268
12 Liebig J (1834) Ann Chem 11:139
13 Rose H (1834) Ann Chem 11:131
14 Stokes HN (1897) Am Chem J 19:782
15 Rose SH (1968) J Polymer Sci B6:837
16 Tate DP (1974) J Polym Sci Polym Symp 48:33
17 Singler RE, Schneider NS, Hagnauer GL (1975) Polym Sci Eng 15:321
18 Penton HR In: Zeldin M, Wynne KJ, Allcock HR (eds) (1988) Inorganic and Organometallic Polymers. ACS Symp Ser 360:277
19 Sennett MS, Hagnauer GL, Singler RE (1983) Polym Mat Sci Eng 49:397; Fieldhouse JW, Graves DF (1980) US Patent 4,226,840
20 Evans, TL, Allcock HR (1981) J Macro Sci - Chem A16(1):409
21 Allcock HR, Mang MN, Dembek AA, Wynne KJ (1989) Macromolecules 22:4179
22 Allcock HR, Cook WJ, Mack DP (1972) Inorg Chem 11:2584
23 Allcock HR, Brennan DJ, Dunn BS (1989) Macromolecules 22:1534
24 Allcock HR, McDonnell GS, Desorcie JL (1990) Macromolecules 23:3873
25 Allcock HR, McDonnell GS, Desorcie JL (1990) Inorg Chem 29:3839
26 Manners I, Riding GH, Dodge JA, Allcock HR (1989) J Am Chem Soc 111:3067
27 Allcock HR, Dodge, JA, Manners, I, Riding GH (in press) J Am Chem Soc
28 Manners I, Liang M (1991) J Am Chem Soc 113:4044
29 Wisian-Neilson P, Neilson RH (1980) J Am Chem Soc 102:2848
30 Neilson RH, Wisian-Neilson P (1988) Chem Rev 88:541
31 Montague RA, Matyjaszewski K (1990) J Am Chem Soc 112:204
32 Wisian-Neilson P, Ford RR, Neilson RH, Ray AK (1986) Macromolecules 19:2089

33 Allcock HR, Rutt JS, Fitzpatrick RJ (1991) Chemistry of Materials 3:442
34 Allcock HR, Fitzpatrick RJ (1991) Chemistry of Materials 3:450
35 Allcock HR, Kwon S (1986) Macromolecules 19:1502
36 Allcock HR, Fitzpatrick RJ, Salvati L (in press) Chemistry of Materials
37 Neenan, TX, Allcock HR (1982) Biomaterials 3,2:78
38 Allcock HR, Hymer WC, Austin PE (1983) Macromolecules 16:1401
39 Allcock HR, Kwon S (1986) Macromolecules 19:1502
40 Allcock HR, Gebura M, Kwon S, Neenan TX (1988) Biomaterials 19:500
41 Allcock HR, Kwon S, Riding GH, Fitzpatrick RJ, Bennett JL (1988) Biomaterials 19:509
42 Bennett JL, Dembek AA, Allcock HR, Heyen BJ, Shriver DF (1989) Chemistry of
    Materials 1:14
43 Allcock HR, Kwon S (1989) Macromolecules 22:75
44 Cohen S, Bano MC, Visscher KB, Chow M, Allcock HR, Langer R (1990) J Am Chem
    Soc 112:7832
45 Bano MC, Cohen S, Visscher KB, Allcock HR, Langer R (1991) Biotechnology 9:468
46 Cohen S, Cima LG, Allcock HR, Vacanti JP, Langer R (in press) Clinical Materials
47 Allcock HR, Fuller TJ, Mack DP, Matsumura K, Smeltz KM (1977) Macromolecules
    10:824
48 Laurencin C, Koh HJ, Neenan TX, Allcock HR, Langer R (1987) J Biomed Mater 21:1231
49 Allcock HR, Scopelianos AG (1983) Macromolecules 16:715
50 Allcock HR, Kwon S (1988) Macromolecules 21:1980
51 Allcock HR, Pucher SR (1991) Macromolecules 24:23
52 Allcock HR, Fuller TJ (1980) Macromolecules 13:1338
53 Allcock HR, Austin PE, Neenan TX (1982) Macromolecules 15:689
54 Allcock HR, Allen RW, O'Brien JP (1977) J Am Chem Soc 99:3984
55 Allcock HR, Austin PE (1981) Macromolecules 14:1616
56 Blonsky PM, Shriver DF, Austin PE, Allcock HR (1984) J Am Chem Soc 106:6854
57 Allcock HR, Austin PE, Neenan TX, Sisko JT, Blonsky PM (1986) Macromolecules
    19:1508
58 Blonsky PM, Shriver DF, Austin PE, Allcock HR (1985) Polym Mater Sci Eng 53:118
59 Blonsky PM, Shriver DF, Austin PE, Allcock HR (1986) Solid State Ionics 18 & 19:258
60 Allcock HR, Desorcie JL, Riding GH (1987) Polyhedron 6:119
61 Saraceno RA, Riding GH, Allcock HR, Ewing AG (1988) J Am Chem Soc 110:980
62 Saraceno RA, Riding GH, Allcock HR, Ewing AG (1988) J Am Chem Soc 110:7254
63 Kim C, Allcock HR (1987) Macromolecules 20:1726; Singler RE, Willingham RA, Lenz
    RW, Furakawa A, Finkelman H (1987) Macromolecules 20:1727
64 Allcock HR, Kim C (1989) Macromolecules 22:2596
65 Allcock HR, Kim C (1990) Macromolecules 23:3881
66 Dembek AA, Kim C, Allcock HR, Devine RLS, Steier WH, Spangler CW (1990)
    Chemistry of Materials 2:97
67 Dembek AA, Allcock HR, Kim C, Steier WH, Devine RLS, Shi Y, Spangler (1991) In:
    Marder S, Sohn JE, Stucky GS (eds) Materials for Nonlinear Optics. ACS Symp Ser 455,
    Washington, DC
68 Allcock HR, Dembek AA, Kim C, Devine RLS, Shi Y, Steier WH, Spangler CW (1991)
    Macromolecules 24:1000
69 Allcock HR, Mang MN, Dembek AA, Wynne KJ (1989) Macromolecules 22:4179
70 Craig DP, Paddock NL (1958) Nature 181:1052
71 Dewar MJS, Lucken EAC, Whitehead MA (1960) J Chem Soc 2423
72 Allcock HR, Konopski GF, Kugel RL, Stroh EG (1970) J Chem Soc Chem Commun
    16:985
73 Allcock HR, Arcus RA, Stroh EG (1980) Macromolecules 919
74 Allcock HR, Tollefson NM, Arcus RA, Whittle RR (1985) J Am Chem Soc 107:5166
75 Allcock HR, Ngo DC, Parvez M, Whittle RR, Birdsall WJ (1991) J Am Chem Soc
    113:2628

76 Bishop S, Hall IV (1974) Brit Polymer J 6:193
77 Chatani Y, Yatsuyanagi K (1987) Macromolecules 20:1042
78 Allcock HR, Allen RW, Meister JJ (1976) Macromolecules 9:950
79 Allen RW, Allcock HR (1976) Macromolecules 9:956

Polymer Degradation Studies by FTIR

Richard G. Davidson,

Materials Research Laboratory,
PO Box 50, Ascot Vale, Melbourne, Victoria 3032, Australia

INTRODUCTION

Polymer degradation can take many forms, chemical, photochemical, thermal and mechanical, and can affect a wide range of polymers in various ways. Although it is usual to assume that 'polymer degradation studies' implies the study of degradation processes that occur in pure polymers when subjected to external stimuli, the topic also includes the examination of degraded polymeric materials recovered from the field and the use of degradation as a characterisation tool. The applied nature of the research work performed at Materials Research Laboratory directs our interests in polymer degradation to the latter two categories, rather than investigation of degradation processes in pure polymers .

Polymer degradation in the economic sense may be defined as any undesirable change in the chemical or physical properties as a result of external applied stimuli such as heat, light etc. This definition is relevant when assessing in-service degradation and failure. For characterisation work, the response of the material to severe degradation, eg. pyrolysis, is moderated by its composition. Thus not only may differences in original formulation be detected, but also changes wrought by degradation.

An essential requirement for both characterisation work and examination of field specimens is a comprehensive data base. This is provided by the published literature on polymer degradation in both academic (pure materials, well defined stimuli) and industrial spheres (formulated materials, poorly defined stimuli - eg. natural weathering). This database must cover all evaluation techniques, not just IR spectroscopy.

The use of FTIR spectrometers is now almost universal. The accuracy, sensitivity and reproducability of the new instrumentation have facilitated the development of a range of new data processing methods such as spectral subtraction, deconvolution, curve resolving etc., and new sampling techniques, but it is essential to remember that the underlying principles of infrared spectroscopy have not changed. The precautions necessary to avoid spectral artefacts are thus an important aspect of the work.

This paper will consider the precautions required in sample preparation and in spectral data processing and interpretation, the examination and evaluation of degraded field specimens and the use of polymer degradation in materials characterisation.

INFRARED SPECTROSCOPY

Most forms of polymer degradation result in destruction of part of the polymer chain and/or introduction of new chemical structures. These may be directly detectable by IR spectroscopy, or indirectly by derivatisation or by their effects on some other property that can be measured by IR spectroscopy.

Y. Imanishi (Ed.)
Progress in Pacific Polymer Science 2
© Springer-Verlag Berlin Heidelberg 1992

The principles and performance of FTIR spectrometers are now sufficiently well known to need no detailed discussion. The FTIR system now allows greater <u>information</u> retrieval from the IR data than was possible with non-digital systems, and a much greater range of sample formats. It is therefore important to ensure that the data are all truly representative of the sample, and when comparisons are used (eg. subtraction) the samples are as far as possible in the same physical state and the same sampling method is used.

Difference spectroscopy is often used to detect species produced by degradation when the absorption bands of the degradation product are masked by those of the base polymer. This technique was one of the earliest to be exploited by FTIR users (1) and has produced valuable results. Although simple in concept, experience has shown that there are a number of conditions that must be met, and that the results must be carefully interpreted.

The difference spectra will contain two kinds of information, real peaks, negative or positive, caused by loss of polymer or creation of degradation products, and artefacts caused by changes in band shape and position. The artefacts may be real, that is, arising from the degradation process, or they may be extraneous, that is, caused by differences in physical condition between the two samples. For example band shapes would not be the same for powders and cast film; a degraded, rough surface will not give the same band shapes as a smooth, fresh surface. In the same way, it is unrealistic to expect good spectral correspondence between spectra of cast films and the same material dispersed in a KBr disk. Indeed, the latter technique may introduce further errors arising from mechanochemical degradation as will be shown later. Absorbed water can give rise to different levels of hydrogen bonded species, and solvents can induce different levels of crystallinity (2). In our experience, absorbed water has caused the most problems. This is desorbed in the dry, warm atmosphere in the spectrometer, and gives rise to significant changes in the spectra. For example, for a study of the effect of abrasion on a polyester surface, it was necessary to establish the reproducibility of the spectra and hence the difference spectra. A thin film of an aliphatic polyester (ca. 10μm) was placed in the spectrometer and spectra recorded at intervals. Difference spectra between successive recordings are shown in fig. 1.

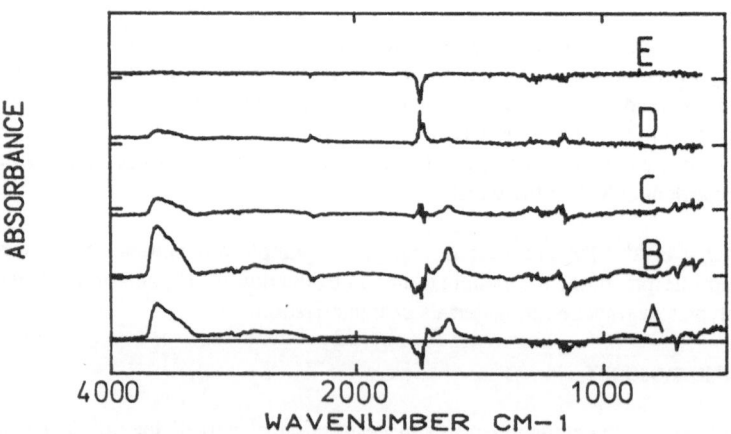

Fig 1. Effect of water desorbed from polyester film.   Difference spectra.   A: 0-2min.
B: 2-5min. C: 5-9min. D: 9-10min. E: 10-11min.

The most prominent features in the first few spectra are absorption bands of water near 3400cm⁻¹ and 1640cm⁻¹, free and bonded carbonyl absorptions near 1740 and 1710cm⁻¹ respectively and features near 1240cm⁻¹ that could be caused by band shape changes, since the shifts are in the wrong direction for free/bonded C-O absorptions. At least 10 minutes is required to reach equilibruim. The influence of sample composition was examined briefly, both by recording spectra from different areas of the sample, and by reducing the aperture. The results showed that composition variations were significant.

Oxidation of nylon parachute fabric results in loss of strength and shortens service life. IR spectroscopy was one of several techniques used to try to detect low levels of oxidation to provide a non-destructive method of evaluation by correlation with changes in tensile strength. The material was examined by internal reflectance spectroscopy, and absorbed water was again found to cause problems. Significant changes occur over a 90 minute period in the carbonyl stretching and N-H bending regions (amide I + II). The removal of water creates a free carbonyl group, absorbing near 1680cm⁻¹, which is apparently unable to associate with N-H protons. Negative bands are associated with loss of hydrogen bonded carbonyl, with corresponding changes in the amide II bands. There does not appear to be much change in the normal amide I band, in which the H-bonding is between C=O and NH groups. These spectral changes are of the same order, or greater than those expected from oxidation products. The loss of water would be much faster when open sampling such as diffuse reflectance or photoacoustic systems are used.

CHEMICAL AND PHOTOCHEMICAL DEGRADATION

Many polymeric components of imported equipment have been formulated for less severe conditions than occur in Australia, especially the northern half of the continent. Consequently, when failures occur, it is necessary to identify the polymer, the cause of the failure and to recommend remedial measures. A recent example involved zip-fastener teeth in army truck canopies in northern Australia where the material temperature may approach 90°C. The teeth exhibited surface crazing, but the failure was by cracking through the body of the teeth. The first problem we faced was choice of an appropriate sampling method. The teeth were small, irregularly shaped and insoluble in all common solvents. The ideal technique in such cases is infrared microscopy. Thin slices of material were taken from the back (protected from UV), centre (virgin material) and at several depths from the surface. The material was identified from the virgin material sample as acetal polymer (polyoxymethylene). The presence of low concentrations of acetate (carbonyl, near 1750cm⁻¹) indicated homopolymer; terminal hydroxyl groups are acetylated to increase thermal stability. In the copolymer with ethylene oxide, the comonomer units block unzipping caused by thermal degradation initiated at the terminal hydroxyl groups.

Spectra of the surface layers were found to contain carbonyl and hydroxyl absorptions consistent with photodegradation (3). This evidence of degradation extended only a short way into the polymer below the level of visible deterioration. Spectra of samples from the centre and back of the teeth contained no evidence of photodegradation, nor were any of the common stabilizers detected. The surface crazing could thus be attributed to photodegradation, but this seemed unlikely to be the cause of the teeth breakage.

According to Grassie and Roche (4), acetal homopolymer may thermally degrade at relatively low temperatures if the protective acetate end group is removed. The degradation product is formaldehyde, which diffuses rapidly out of the polymer and leaves no evidence of degradation. Extensive loss of polymer from this cause could lead to internal stresses high enough to produce cracking. The IR spectra showed that absorbed water was distributed throughout the teeth. In the absence of other evidence, it seemed likely that the high temperatures and humidity, possibly combined with catalytic activity from traces of skin acids had produced conditions conducive to hydrolysis of the acetate end groups and that subsequent thermal degradation over a long period had weakened the material to the point of cracking.

In this case the problem was primarily caused by use of inappropriate material. The local supplier of acetal resins advised that acetal is not recommended for prolonged exposure outdoors, that the copolymer grades (with ethylene oxide) are better than the homopolymer, and a good UV stabiliser is essential.

An example of the potentially practical value of chemical degradation is correlation of physical properties with composition as determined by infrared spectroscopy. Several polyurethane candidates for an application that included prolonged immersion in seawater were tested for resistance to hydrolysis. The samples were immersed in seawater at 70°C and their residual tensile strength measured at intervals. IR spectra of films cast from THF were recorded at the same time (5). The polyesterurethanes, as expected, degraded quite rapidly, by hydrolysis of the ester groups in the soft segments. Carboxylic acid groups are produced, and as the degradation proceeds, the average MW of the polymer decreases and the proportion of -COOH increases. The -COOH group absorbs near 1710cm^{-1} and can be detected either by subtraction, or by curve resolving techniques. Although curve resolving should be more reliable in principle, we have found that it is difficult to achieve uniform results from a diffuse band system such as the one in question (fig 2). Consequently simple subtraction was used, setting the centre of the ester band at 1734cm^{-1} at zero.

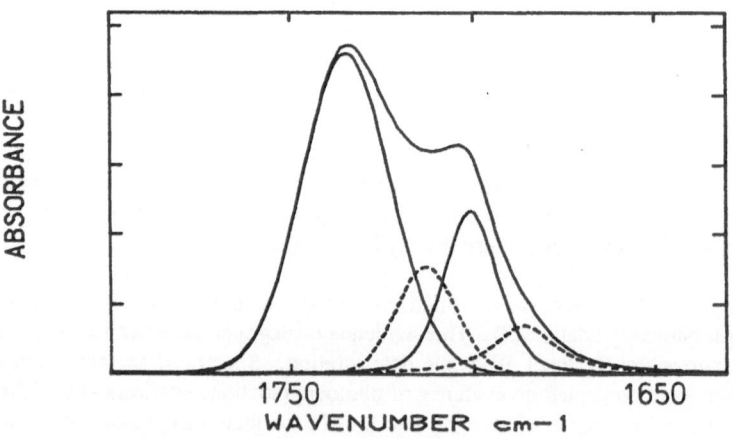

Fig 2. Resolved carbonyl bands - part hydrolysed urethane.

The results show quite good correlation of "acid" absorbance with residual tensile strength (fig 3). This type of relationship is potentially useful for examination of components in the field, since only a small sample is required and the structural integrity of the unit is not compromised. With appropriate care, curve resolving can be used to gain some insights into the hydrolysis process.

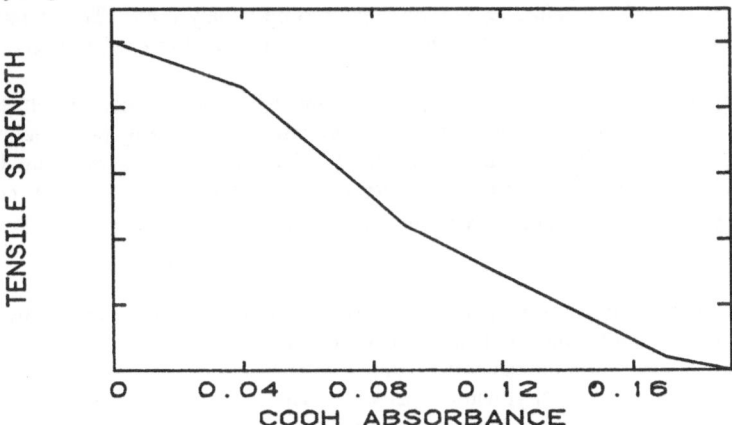

Fig 3. Correlation between carboxylic acid index and tensile strength

THERMAL DEGRADATION

Since FTIR spectroscopy came into widespread use, dynamic analysis of thermal degradation has increased; the spectral data are collected on-the-fly, rather than at discrete intervals.

An early use of the technique was reported by Lin, Bulkin & Pearce in 1979 (6). Those authors investigated thermal degradation of several epoxy resins cured with trimethoxyboroxine. The samples were cast on to a heated, reflective plate in a modified specular reflectance accessory, and spectra were recorded at intervals. The sample temperature was 300°C, and both inert and oxidising atmospheres were used. The spectral data were refined by the use of digital subtraction, but no account appears to have been taken of band shifts and shape changes caused by factors other than direct chemical degradation. The preliminary discussion includes a very extensive table of IR band assignments. The discussion of thermal degradation reaction mechanisms is very comprehensive, and provides a valuable reference for characterisation work on these materials.

This system was later developed to allow analysis of the evolved gases by a slide mechanism that replaced the sample with a clean reflection plate (7). This combined technique was then used to study thermal degradation of poly (ethyleneterephthalate) in conjunction with non-isothermal-pyrolysis-evolved gas - FTIR analysis (FTIR-EGA). (8). More recently, the combined system has been further refined by Hummel et.al. to allow analysis of either the evolved gases alone or the polymer and evolved gases (9). Examples of thermal degradation of poly (methyl methacrylate) and poly (methacrylic acid) were described. This development recognised that the sample mass requirements for the two techniques are different, micrograms for external reflective or transmission work, and milligrams for evolved gas work.

The combination of polymer analysis and evolved gas analysis certainly provides considerable and complementary information about the degradation process, but in many cases pyrolysis-evolved gas - FTIR analysis alone is very useful, and generally simpler. This technique was pioneered by Griffiths (10) but developed to a high level by Lephardt and Fenner (11-13); we have found it to be very useful for characterisation work. In recent years it has been coupled with TGA (TGA-FTIR) but this does not yet appear to be widely used. The manufacturers' data so far suggest that the pyrolysis chamber/gas cell relationship has not yet been optimised.

Lephardt's original work was on the pyrolysis and combustion of tobacco, for which he used 1g samples. We have adapted the techniques to work with 5-10mg samples of and have used it successfully to characterise a variety of polymers. Our standard conditions are 10°C/min ramp rate, 50ml/min purge gas (usually N_2), 10s data collection time (25 scans), continuous (GC mode). We use both identity and evolution profile of the evolved gases to evaluate the sample.

We have examined a number of materials in terms of toxic gas evolution during pyrolysis or combustion. For example, the products from fibreglass insulations included isocyanic acid and hydrogen cyanide, while the support backing and adhesive of a made-up panel produced acetic acid, acetaldehyde and vinyl chloride as well. The effect of a fire in a confined space such as a surface ship or submarine can be imagined. The products come from thermal decomposition of polymer coatings etc, and while the experimental conditions may not duplicate those of a fire, the information is very useful.

During an investigation of the thermal degradation of poly(vinylchloride - vinylacetate) blends, McNeill (14) observed acetyl chloride in the degradation products. We investigated this aspect by FTIR-EGA and found that the amount of acetyl chloride in the effluent decreased with increased residence time in the hot zone. This suggested that acetyl chloride is a primary decomposition product, rather than a product of reaction between HCl and acetic acid from pyrolysis of the copolymer. McNeill's work was performed by thermal volatilisation analysis and did not provide on-the-fly identification of the pyrolysis products.

FTIR-EGA is useful for evaluation of the effects of additives on thermal stability, and conversely can be used to distinguish between similar materials. Chlorine-containing polymers may be readily identified from their hydrogen chloride evolution profiles. Differences in formulations of the same polymer can be detected in the same way. The hydrogen chloride evolution profiles from three samples of PVC compound are shown in fig 4. One of the samples is clearly different from the other two. Confirmatory evidence is usually provided by the profiles for water, carbon dioxide and hydrocarbons. The FTIR-EGA technique has been similarly useful for characterisation of elastomers, epoxy and acrylic adhesive systems, polyurethanes, and acrylic glazing materials. The published literature on thermal degradation has provided a rich base on which to work.

The influence of additives on thermal degradation may be exemplified by the FTIR-EGA study of the effect of zinc oxide on the pyrolysis of polychloroprene (15). This effect (fig 5) has been recognised for some time (16-18) and has generally been attributed to the catalytic action of zinc chloride formed during cross-linking of the polychloroprene (19). We have examined the effect of ZnO, ZnS, $ZnCl_2$ and $ZnSO_4$ on the pyrolysis of polychloroprene and found that only zinc oxide results in the effects seen in fig 5.

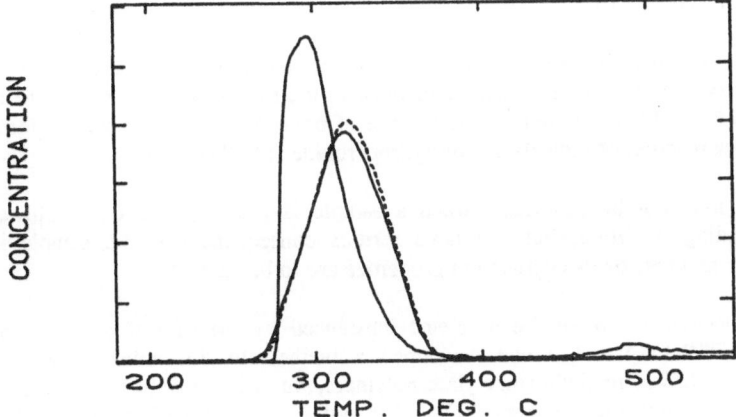

Fig 4. Hydrogen chloride evolution from three samples of PVC compounds.

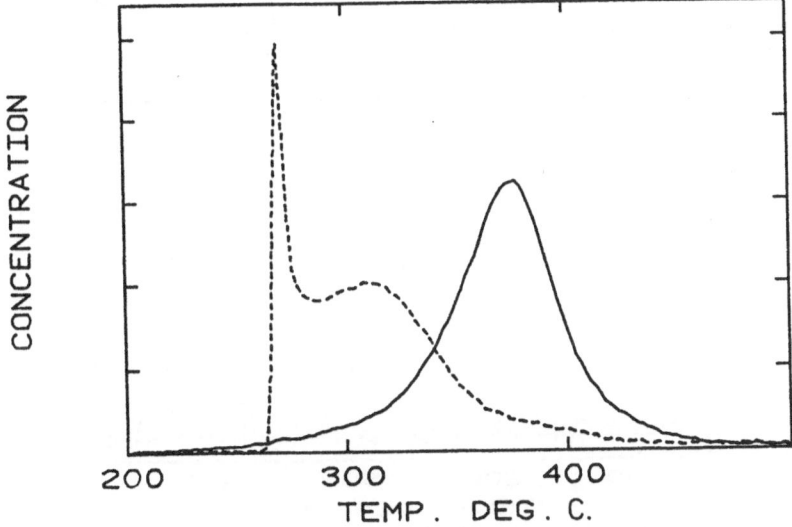

Fig 5. Effect of ZnO on evolution of hydrogen chloride from polychloroprene.
_____ raw polymer; -------- polymer +5phr ZnO.

At the same time, we confirmed that ZnCl$_2$ catalyses the dehydrochlorination of polychloroprene at temperatures below 200°C, but the reaction is very slow, and also accelerates on the evolution of hydrogen chloride by thermal degradation. The effect of ZnO has been observed in other chlorine - containing polymers including polyvinylidene chloride, chlorinated polyethylene and polyepichlorohydrin. The phenomenon thus seems unlikely to be a function of polymer structure.

108

MECHANICAL DEGRADATION

Mechanical degradation of polymers often results in chain scission accompanied by the formation of free radicals. In the presence of oxygen, the final products are complex and may include peroxides and hydroperoxides and a range of oxygen - containing species formed as a result of further reaction of radicals and of hydroperoxides (20,21).

These effects may not be significant when a sample is prepared for IR identification by scraping, grinding or filing, but can have serious consequences if the sample is to be compared with another, or its degradation properties are to be studied.

We recently became aware of the problems introduced by mechanical degradation of our samples for FTIR-EGA. We had developed a method to distinguish between various commercial crosslinked methylmethacrylate polymers, but were unable to achieve sufficient reproducibility for quantitative work.(22). This problem was found to be associated with variable levels of mechanochemical degradation of material sampled by filing or scraping. We have now confirmed this with an investigation of the effect of mechanical degradation on the pyrolysis of polymethylmethacrylate (PMMA) and some of its copolymers (23). Pyrolysis of undamaged PMMA homopolymer yields only a trace of methanol, but mechanically degraded (ie. filed) materials produces significant amounts. (Fig. 6).

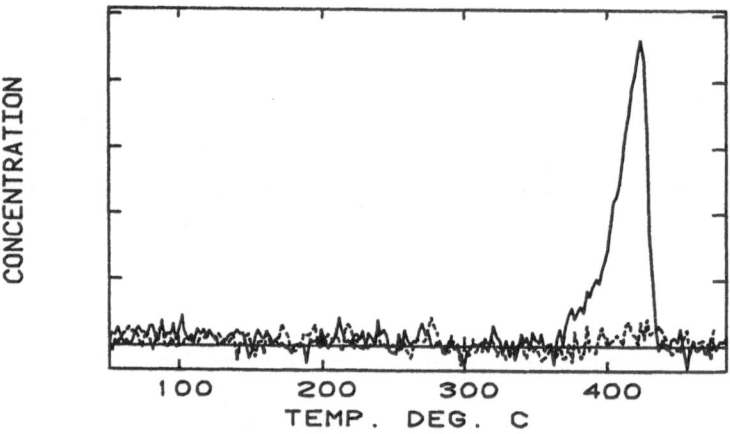

Fig 6. Effect of mechanochemical degradation on the yield of methanol from pyrolysis of PMMA. ------ intact polymer; _____ degraded polymer.

Above 420°C, methane and carbon monoxide are produced from the degraded material but not from the undamaged sample. Carbon dioxide yield is also much higher from the degraded material, and the yield of monomer is significantly reduced. The results are consistent with formation of free radicals by mechanical action, followed by oxidation and chain scission reactions. The extent of the effect is proportional to the amount of mechanical degradation, and that is responsible for our poor reproducibility.

Pyrolysis of mechanically degraded copolymers produces even more striking changes in the pyrolysis behaviour. Not only are the pyrolysis products changed, but also their pattern of evolution. The methanol profile from a methylmethacrylate/ methacrylamide copolymer before and after degradation is shown in fig 7. The two profiles are quite different, and the very large increase in methanol evolution above 400°C is consistent with thermal degradation of residual oxygen-containing species formed by mechanical degradation in air.

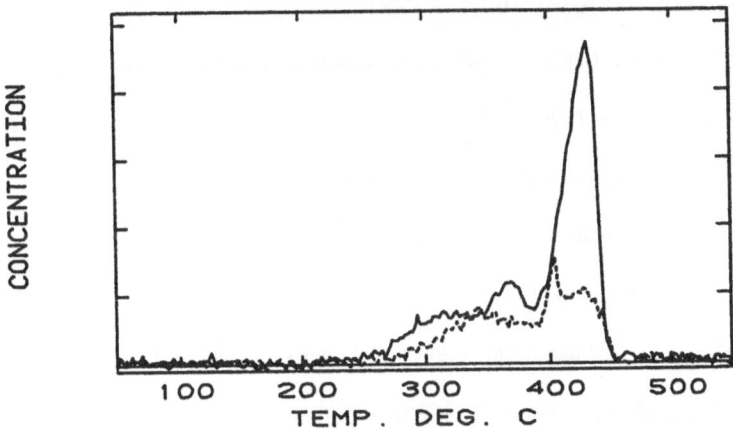

Fig 7. Effect of mechanochemical degradation on methanol yield from pyrolysis of PMMA copolymer. ------ intact polymer; _____ degraded polymer.

Reactions that involve adjacent monomer units have been proposed to account for the appearance of methane, carbon monoxide, methanol and carbon dioxide at high temperatures in the pyrolysis of PMMA polymers.(24,25). Our results suggest a more likely source is mechanochemical degradation products.

Grayson and Wolf (26) reported that propenal was evolved from a cured epoxy resin system at 175°C. The epoxy system was based on a commercial form of tetraglycidylmethylene dianiline (Ciga-Geigy MY720), and diaminodiphenylsulfone. Their samples were prepared by abrading the material to provide shavings less than 10μm thick. We have examined the effect of mechanical degradation on a similar cured epoxy material, and found that propenal is evolved from powder produced by filing, but not from one or two pieces of the same mass. Although propenal has been observed to evolve during cure of epoxy resins (27), it appears that in this case, the source is thermal degradation of products of mechanical degradation.

These two examples demonstrate both the value of FTIR-EGA for studying both mechanical and thermal degradation and the potential for erroneous results arising from the effects of mechanical degradation on subsequent polymer degradation studies.

Acknowledgement

I wish to acknowledge the contribution of Dr D.K.C. Hodgeman of this laboratory, who suggested that we investigate the effects of mechanical degradation on the thermal degradation properties of polymers.

References

1. Keonig JL (1975) Appl Spectrosc 29:293

2. Lin S-B, Koenig JL (1983) J. Polym Sci, Polym Phys Ed 21:1539

3. Grassie N, Roche RS (1968) Makromol Chem 112:34

4. Grassie N, Roche RS (1968) Makromol Chem 112:16

5. Oldfield D, Materials Research Laboratory, unpublished results

6. Lin SC, Bulkin BJ, Pearce EM (1979) J. Polym Sci, Polym Chem Ed 17:3121

7. Bulkin BJ, Chen CS, Pearce EM (1981) Polymer Prep, Polym Chem 22:301

8. Pearce EM, Bulkin BJ, Ng MY (1983) In: Craver CD (Ed) Adv. Chem Ser 203 Polymer Characterization, American Chemical Society, Washington DC, USA

9. Hummel DO, Univ of Cologne, personal communication

10. Liebman SA, Ahlstrom DH, Griffiths PR (1976) Appl Spectrosc 30:335

11. Lephardt JO, Fenner RA (1980) Appl Spectrosc 34:174

12. Lephardt JO (1982-3) Appl Spectrosc Rev 18:265

13. Lephardt JO (1984) In: Voorhees KJ (Ed) Analytical Pyrolysis, Butterworth

14. Jamieson A, McNeill IC (1974) J Polym Chem Ed 12:387

15. Davidson RG (1991) 2PPC, paper 27PZ01a

16. Ballistreri A, Fotis S, Maravigna P, Monfando G, Scamporrino E (1981) Polymer 22:131

17. Pidduck AJ (1985) J. Anal and Appl Pyrolysis 7:215

18. Kleps T, Piaskewicz M (1987) J. Thermal Anal 32:1785

19. Johnson PR (1976) Rubber Chem Technol 49:3

20. Reich L, Stivala SS (1971) Elements of Polymer Degradation.   McGraw-Hill,
     New York USA

21. Schnabel W (1981) Polymer Degradation: Principles and Practical Applications.
     Hanser International, Munich

22. Davidson RG (1987) J. Appl Polym Sci, 34:1631

23. Davidson RG (1991) J. Anal & Appl Pyrolysis, 21:181

24. Grassie N, Torrance BJD, (1968) J Polym Sci, Part A1 6:3315

25. Fenner RA, (1986) Fourier Transform Infrared Evolved Gas Analysis of the Thermal
     Decomposition Products of Polymers.   PhD Thesis, Virginia Commonwealth
     University, Richmond Virginia. UMI Dissentation Information Service 872 3590

26. Grayson MA, Wolf CJ (1984) J. Polym Sci, Polym Chem 22:1897

27. Ennis BC, Davidson RG, Pearce PJ, Morris CEM (1991) J Adhesion, in press

Interdomain Communicaton via Magnetic Spin Diffusion in a Microphase-Separated Polyurethane Elastomer

Laurence A. Belfiore
Department of Chemical Engineering
Polymer Physics and Engineering Laboratory
Colorado State University, Fort Collins, Colorado, USA 80523

Abstract

High-resolution carbon-13 solid state NMR spectroscopy is employed to monitor communication between the rigid and mobile domains in a polyether-polyurethane elastomer via proton magnetization transport. Results from thermal analysis and polarized optical microscopy indicate that the urethane-rich hard domains are semicrystalline and exhibit spherulitic growth that is somewhat disordered. Proton magnetization gradients are established via mobility differences between the two domains, and the ensuing spin-diffusion process is monitored via the carbon-13 nuclear manifold. Chemical structural differences between the two types of segments produce high-resolution ^{13}C NMR signals that allow one to track interdomain communication. Magnetic spin-temperature equilibration between the two domains occurs on the 10-100 ms time scale. This is consistent with a finely dispersed array of hard and soft microphases. The polyurethane proton-spin-diffusion results are similar to those obtained for a variety of SURLYN™ and KRATON™ commercial phase-separated copolymers.

Introduction

In this research contribution, high-resolution carbon-13 NMR spectroscopy is employed to probe the solid state morphology of a short-segmented polyurethane block copolymer. The particular experiment is designed to monitor proton spin diffusion between rigid and mobile domains. It was developed quite recently [1,2] as a slight modification of the Goldman-Shen pulse sequence [3]. It should be emphasized that the problem of domain size in polyurethane block copolymers is rather old and has been addressed by several research groups using a broad spectrum of experimental probes [4].

Y. Imanishi (Ed.)
Progress in Pacific Polymer Science 2
© Springer-Verlag Berlin Heidelberg 1992

Morphological studies of domain formation and microphase separation in urethane elastomers via solid state NMR spectroscopy have been limited primarily to wideline proton experiments. In these experiments, the free-induction decay of proton magnetization was monitored and related to domain mobility via the spin-spin relaxation time, T_2. This time constant governs the rate at which proton magnetization dephases in the rotating reference frame due to proton dipolar interactions [5,6]. Relaxation is most efficient and T_2 is subsequently shorter when the <u>static</u> contribution to the spectral density of a motional process is larger. Hence, rigid lattice dipolar couplings generate short spin-spin relaxation time constants. This is characteristic of the semicrystalline urethane hard segment, which has glass and melting transitions above ambient temperature. Considerable molecular motion in the soft segment partially averages homonuclear ^1H dipolar couplings. This mobility causes the soft-segment proton magnetization to dephase *slower* than the free-induction decay characteristic of the rigid urethane-rich hard-segment composite ^1H signal. Analysis of wideline proton NMR data has focused on the decomposition of transient free-induction decays into fast and slow components in the time domain. The data of Assink et al. [7-9] and Nierzwicki [10] provide evidence to support the claim that linear polyurethanes having either polyether or polyester soft segments exhibit proton free induction decays from pulsed wideline NMR which cannot be described by one exponential relaxation time constant (i.e., a single Lorentzian decay). In other words, the proton spin-spin relaxation time, T_2, characteristic of the polyether-rich or polyester-rich mobile domains is considerably *longer* than that in the rigid urethane-rich domains. This discrimination between domain mobility and, consequently, T_2 is exploited herein to (i) generate a spatial gradient in proton magnetization and (ii) monitor spin diffusion between domains as the proton nuclear manifolds approach spin-temperature equilibration in the presence of homonuclear dipolar couplings that persist across interdomain boundaries. These experiments have the potential to estimate (i) domain size, or (ii) the thickness of the interfacial region, based on the transient nature of the proton-spin-diffusion process. Morphological dimensions are obtained via order-of-magnitude estimates of the average proton-proton dipolar distance which employ a characteristic diffusion time constant as well as a diffusion coefficient for proton magnetization transport [11].

The slightly modified Goldman-Shen experiment employed herein is designed to detect and analyze carbon-13 NMR signals. In this respect, we rely on chemical structural differences as well as motional differences between the two types of segments to assign the ^{13}C spectrum and provide a fingerprint of each domain via resolution of isotropic chemicals shifts in the O$\underline{C}$H$_2$ region of the spectrum. To the best of our knowledge, only a few *high-resolution* solid state NMR studies of polyurethane microphase separation have appeared in the research literature [12-15,32]. Most notably, Eisenbach et al. [15] have addressed hydrogen bonding and domain morphology in a polyether-polyurethane, synthesized using selectively deuterated precursors such that the polydisperse hard segments were *partially* deuterated. Some of the results in that study [15] which focus on the concept of incomplete phase separation are somewhat ambiguous and difficult to interpret. This claim is based on the fact that an *intersegment* 1H-13C cross-polarization mechanism, which is required to prove that some of the rather short urethane segments are dispersed in the polyether-rich mobile domains [16], may not be the primary pathway by which 13C magnetization is generated in the hard segments. The ambiguity arises from the fact that some of the hard-segment proton sites were not deuterated--paving the way for *intrasegment* 1H-13C spin diffusion within the urethane segments during cross polarization.

The results discussed herein rely on proton polarization in the polyether-rich mobile domains to regenerate magnetization in the rigid urethane-rich domains via proton dipolar interactions which must be operative across domain boundaries. The data in Figure 5 suggest that 10 ms of proton spin diffusion in the presence of homonuclear dipolar couplings is sufficient to regenerate a "near-equilibrium" 13C spectrum of the hard segment. The time scale required for spin-temperature equilibration between hard- and soft-segment proton nuclear manifolds in the polyether-polyurethane elastomer is comparable to that observed in the SURLYN™ and KRATON™ commercial phase-separated copolymers [1,2,17]. In this respect, segmented polyurethane block copolymers represent the third class of microphase-separated systems whose morphological characteristic are accessible via interdomain proton magnetization transport and the modified Goldman-Shen experiment. However, detailed calculations of domain size and interfacial thickness based on the time-dependent spin-diffusion process require sophisticated numerical modelling developments before physically realistic morphological parameters can be obtained [11].

Experimental Considerations

Materials Investigated. The linear polyether-polyurethane investigated herein was obtained courtesy of Richard Oertel at the UpJohn Company in North Haven, CT. The *polydisperse* urethane hard segment is based on 4,4'-diphenylmethane di-isocyanate (MDI), chain extended with butanediol. Hydroxyl-equivalent data indicate that the mean hard-segment length consists of three-to-four urethane repeat units, which is sufficient for hard-segment crystallization to occur [18]. The *soft* segment is a 1000-molecular-weight poly(tetramethylene ether glycol), PTMEG, and the overall hard/soft segment weight fractions are 0.58/0.42. The chemical repeat formula of the polyurethane thermoplastic elastomer is provided below.

Differential Scanning Calorimetry. Thermal analysis was performed on a Perkin-Elmer DSC-7. Melting endotherms/crystallization exotherms were recorded under helium and nitrogen purges at a heating/cooling rate of 10°C/min. Differential power output was monitored via Perkin Elmer's TAC 7/DX thermal analysis controller in conjunction with the DSC7 multitasking software on a 386/33 MHz personal computer. The as-received polyurethane elastomer was heated initially to 240°C in a Carver Laboratory press. Both fast- and slow-cooled samples were then sealed in aluminum pans at ambient temperature prior to obtaining their subsequent heating and cooling curves in the calorimeter. It should be emphasized that the DSC traces reported herein represent the *second* heating and the *second* cooling cycles. In this respect, one should not overlook the possibility that high-temperature annealing could trigger urethane interchange reactions which might disrupt the statistical distribution of the initially conceived hard-segment lengths [18].

Polarized Optical Microscopy. Thermo-optical microscopy (TOM) was performed using a Nikon Optiphot-Pol fitted with cross polarizers, automatic-exposure-based 35-mm or 5"x4" Polaroid still frame cameras, and a high-resolution Javelin video camera. The video camera was interfaced to (i) a 310A video integrator manufactured by Colorado Video, Inc. (Boulder, CO), (ii) a Sony PVM-1343MD high-resolution colour monitor, and (iii) a Panasonic AG-6720A time-lapse video recorder. Temperature variation was accomplished via **convective** heating/cooling using filtered laboratory air in a hot stage manufactured by Fluid Inc. (Denver, CO.) The temperature of the heated stage was controlled to within 0.3°C by a Eurotherm 818P15 programmable microprocessor that receives feedback from a thermocouple placed directly on the cover slip above the sample, and has the capability of performing several ramp and soak thermal profiles. Video images were analyzed using a light-intensity (grey-scale) integrator manufactured by Colorado Video Inc. The output voltage signal from the integrator and the sample temperature relayed by the Eurotherm microprocessor were captured by a Strawberry Tree Inc.

data acquisition board, complete with graphical analysis software, interfaced to a Compaq Deskpro 386/20 MHz personal computer.

Carbon-13 Solid State NMR Spectroscopy. Proton-enhanced dipolar-decoupled carbon-13 solid state NMR spectra were obtained on a modified Nicolet NT-150 spectrometer at Colorado State University's solid state NMR center. The carbon frequency was 37.735 MHz and magic-angle spinning was performed at 3600 Hz. The spectrometer incorporates a double-resonance cross-polarization/magic-angle-spinning (CP/MAS) probe that was designed and constructed at the CSU NMR center. The spinner system is a modified version of Wind's [19] with a sample volume of 0.3 cm^3. A proton 90-degree pulse width of 5 µs was employed, corresponding to an r.f. field strength of 50 KHz. The r.f. field was maintained at 50 KHz during cross polarization and subsequent high-power ^{1}H decoupling. The ^{13}C free induction decay (FID) was accumulated via 2K time-domain data points using quadrature detection. Prior to Fourier transformation, the signal-averaged FID was zero-filled to 4K. The spectral width encompassed a ±10 KHz frequency range and 5 Hz of line broadening was employed. Following Stejskal and Schaefer [20], spin-temperature alternation in the rotating frame was used to suppress the build-up of artifacts which may occur in proton-enhanced spectra. The sample temperature was maintained at 15±2^{0}C by passing the spinner air through a copper cooling coil immersed in an ice bath. Carbon-13 chemical shifts were referenced externally to the methyl resonance of hexamethylbenzene, 17.355 ppm deshielded from tetramethylsilane [21]. Details of the modified Goldman-Shen experiment for measurement of proton spin diffusion between two microdomains of dissimilar mobility are provided in the Results and Discussion section.

Results and Discussion

Morphological Characterization via Thermal Analysis and Polarized Optical Microscopy: Support for Partially Crystallizable Urethane-Rich Hard Domains

Experimental results from differential scanning calorimetry are illustrated in Figure 1 to provide evidence that the urethane-rich hard domains are semicrystalline at ambient temperature. Thermograms A and C of the fast- and slow-cooled samples reveal a series of endothermic transitions during the *second* heating cycle, with a total caloric content of approximately 20-25 J/g in each case. For comparison, Hwang et al. [22] have calculated the heat of fusion to be 22 KJ/mole (65 J/g) for a hypothetical 100% crystalline hard segment containing three MDI units and

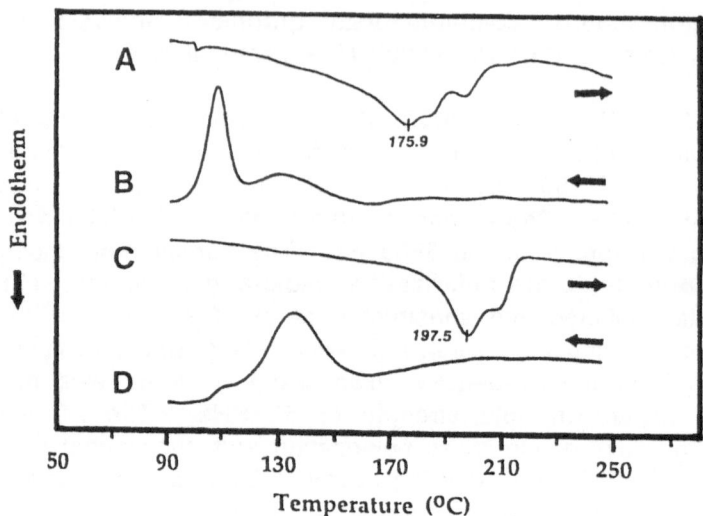

Figure 1 DSC thermograms of the polyether-polyurethane investigated in this study. The heating/cooling rate was 10°C/min
    a)   Second heating trace after <u>fast</u> cooling from 240°C in air
    b)   Subsequent cooling trace at 10°C/min
    c)   Second heating trace after very <u>slow</u> cooling from 240°C in a Carver Laboratory press
    d)   Subsequent cooling trace at 10°C/min

two butanediol units. In Figure 1, composite melting transitions are observed over the temperature range (i) 150-210°C for the sample that was cooled quickly in air (thermogram A), and (ii) 180-215°C for the sample that was cooled very slowly under slight pressure in the Carver Laboratory press (thermogram C) prior to obtaining the heat capacity-temperature data. Vallance et al. [23] have also measured complex multi-peaked endothermic behaviour in the vicinity of 170°C-230°C for an as-polymerized MDI-based polyether-polyurethane that was considered to be semicrystalline at room temperature by virtue of crystalline x-ray diffraction peaks superposed on a relatively strong amorphous scattering background. The semicrystalline thermoplastic polyurethane investigated by Vallance [23] was polymerized from a 5.8/4.6 molar ratio of MDI and butanediol, with a poly(tetramethylene oxide) soft-segment number-average molecular weight of 5430. Both the hard- and soft-segment lengths in the above-mentioned material [23] are larger than those for the commercial block copolymer studied herein.

Crystallization exotherms obtained at a cooling rate of 10°C/min are also illustrated in Figure 1 (thermograms B and D) during the second cooling cycle. Note that the crystallization kinetics during the *second* cooling cycle are *not* identical for two samples that were cooled at different rates during the first cooling cycle. For the sample that melts between 150°C and 210°C (thermogram A), the crystallization exotherm in Figure 1B exhibits a relatively sharp maximum in the vicinity of 110°C with a breadth of 8°C, as well as a rather broad but weaker exotherm in the 120-145°C range. This sample was cooled *quickly* in air during the first cooling cycle. In contrast, the crystallization exotherm in Figure 1D exhibits a maximum at ≈ 135°C with a breadth of 20°C for the sample that melts between 180°C and 215°C (thermogram C). This latter sample, which was cooled *very slowly* in the Carver Laboratory press during the first cooling cycle, also exhibits a low-temperature shoulder on the broad crystallization exotherm in the 110°C temperature range during the second cooling cycle--analogous to the primary crystallization behavior of the fast-cooled sample. A comparison of the *melting* endotherms during the second heating cycle in Figure 1A and 1C suggests that morphological differences between the two samples are prevalent as a result of the fast vs. slow first-cooling cycle [18]. Furthermore, comparison of the *crystallization* exotherms during the second cooling cycle in Figure 1B and 1D suggests that potential chemical modification of the urethane hard segment during the fast and slow first-cooling cycle produces distinguishable materials based on their crystallization kinetics [18].

Spherulite formation in this particular polyether-polyurethane was investigated via polarized optical microscopy at ambient temperature. The sample was moulded at 240°C under slight pressure, followed by very slow cooling to ambient temperature in the press. The micrograph in Figure 2 reveals spherulitic growth of the urethane lamellae with diffuse boundaries at a magnification of 370x. There are several dark non-birefringent regions that most likely represent amorphous material trapped, in some cases, within disordered spherulites. The average diameter of the spherulites is approximately 30-40 microns, which is much larger than the distances over which spin diffusion is operative on the 10 ms time scale between ^{1}H and ^{13}C dipolar-coupled nuclei.

The solid state NMR experiments described in the following sections focus on differences in chemical structure and molecular mobility between the hard and soft segments. In light of the results obtained from calorimetry and polarized optical microscopy, the urethane-rich hard domains are partially crystalline. Approximately two MDI-butanediol repeat units are required for hard-segment crystallization to occur, with rather low melting temperatures in the vicinity of 140°C compared to a melting point of ≈ 250°C for a high-molecular-weight homopolymer of MDI and butanediol [18]. The relatively rigid nature of the MDI-butanediol

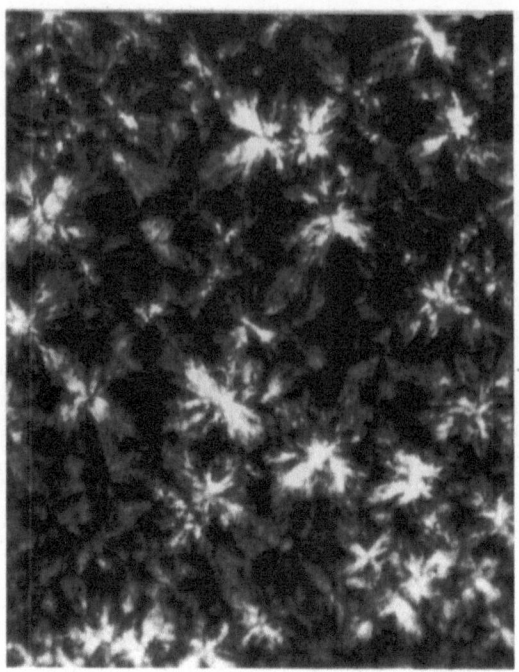

Figure 2 Polarized optical micrograph illustrating disordered spherulites at a magnification of 370x. The polyether-polyurethane sample was compression moulded at 240°C under slight pressure in a Carver Laboratory press, followed by very slow cooling to ambient temperature under pressure in the press prior to observing the birefringent pattern. The dark regions of the micrograph are representative of amorphous material that is unoriented and non-birefringent.

segments is verified by two types of NMR experiments which discriminate between rigid and mobile material. The 1000-molecular-weight polyether soft segments are predominantly mobile due to the fact that they are above their glass transition temperature, with no complicating effects due to soft-segment crystallization [16,23]. Soft-segment crystallization is also thwarted in the polyether-polyester segmented block copolymers produced commercially by DuPont under the tradename Hytrel™ [16]. In both the polyether-polyurethane investigated herein and Hytrel™, the polyether soft segment is poly(tetramethylene ether glycol) with number-average molecular weights between 1000 and 2000. Soft-segment mobility in the interfacial regions could be hindered as a consequence of incomplete phase separation and diffuse phase boundaries.

Carbon-13 Solid State NMR Spectral Assignments via ^{1}H-^{13}C Dipolar Dephasing

High-resolution solid state NMR spectra of the as-received polyether-polyurethane block copolymer are illustrated in Figure 3 at an ^{1}H-^{13}C cross-polarization thermal mixing (contact) time of 1 ms. The carbon-13 resonances characteristic of the polyether soft segment are highlighted in the upper spectrum of Figure 3. The internal methylene $\underline{C}H_2$ signals from the soft-segment polyol and the butanediol chain extender overlap extensively in the 25-27 ppm chemical shift region. However, the $O\underline{C}H_2$ resonances in the hard and soft segments are resolved in the vicinity of 70 ppm. The hard-segment $O\underline{C}H_2$ resonance of butanediol at 66 ppm is indicated via the dashed line in Figure 3. Evidence to support these spectral assignments is provided in the lower spectrum of Figure 3, which incorporates 50 µs of ^{1}H-^{13}C dipolar dephasing before data acquisition. Protonated carbon signals (except $\underline{C}H_3$) from relatively rigid domains are severely attenuated after 50 µs of "interrupted decoupling" due to near-

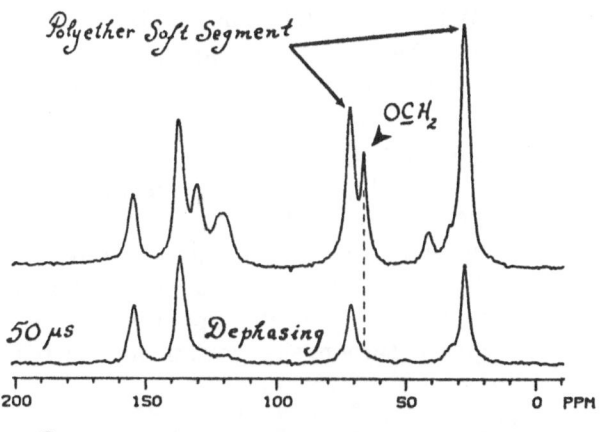

Carbon-13 Solid State Chemical Shift

Figure 3 High-resolution carbon-13 solid state NMR spectra of the MDI-based polyurethane elastomer via cross polarization.

Upper; the "equilibrium" spectrum based on 1 ms of cross-polarization contact and a 2-second pulse repetition delay. The $O\underline{C}H_2$ resonance of the butanediol chain extender is indicated by the dashed line, and the polyether soft-segment signals are identified.

Lower; the "interrupted-decoupling" spectrum of the same material. Only ^{13}C signals in the mobile polyether soft segment and the non-protonated ^{13}C resonances in the hard segment survive the 50 µs dipolar dephasing delay.

static dipolar couplings between directly bound proton and carbon nuclides. The OCH_2 resonance at $\approx$ 72 ppm, which *partially* survives the 50 μs dipolar dephasing delay in the lower spectrum of Figure 3, is unique to the polyether *soft* segment. As expected, the hard-segment OCH_2 resonance of butanediol at 66 ppm (dashed line) is severely attenuated after the dephasing delay. Attenuation of the soft-segment OCH_2 resonance at $\approx$ 72 ppm tentatively suggests that there are regions of restricted soft-segment mobility. The fortuitous resolution of OCH_2 resonances in the hard and soft segments of the polyether-polyurethane block copolymer is analogous to the NMR spectral discrimination observed in the Hytrel™ polyether-polyester block copolymers [16, 24] produced commercially by DuPont. In the following sub-section of this research contribution, the two oxygen-linked methylene ^{13}C resonances in the vicinity of 70 ppm are exploited to (i) fingerprint the hard and soft segments via isotropic chemical shift differences and (ii) monitor interdomain communication via proton magnetization transport.

Further chemical shift assignments for the *urethane hard segment* are possible from a comparison of the 110-160 ppm region of the spectra in Figure 3. The two signals on the left side of the lower spectrum, which almost completely survive the 50 μs dipolar dephasing delay, represent contributions from ^{13}C nuclei *without* directly bonded protons. Hence, the resonance envelope at $\approx$ 137 ppm in the lower spectrum is assigned to both non-protonated (substituted) carbons in the aromatic ring of MDI, whereas the two more shielded (i.e., upfield) resonances at approximately 120 ppm and 130 ppm in the upper spectrum encompass the four protonated aromatic carbons. The urethane carbonyl signal is furthest downfield at $\approx$ 154 ppm. Finally, the CH_2 resonance at $\approx$ 40 ppm characteristic of MDI is found only in the upper spectrum of Figure 3--as expected for a hard-segment carbon site having directly bonded protons.

Interdomain Communication Between the Urethane Hard Segment and the Polyether Soft Segment via Proton Spin Diffusion

Microdomain size in phase-separated copolymers plays a fundamental role in determining various macroscopic physical properties in the solid state. The difference in segmental mobility between the hard and soft domains governs the physical properties of microphase-separated polyurethane elastomers [7]. In this respect, the development of structure-property relations at the molecular level which relate directly to macroscopic behavior is the focus of this sub-section. One can exploit the well-documented difference between domain mobility [7-10] and the ^{13}C NMR chemical shift distinction between the OCH_2 resonances in the hard and soft segments to probe the microdomain morphology of polyether-polyurethane elastomers via proton magnetization transport. In proton-spin-diffusion studies from which domain size, interfacial thickness, and molecular proximity may in principle be determined, it is attractive to observe proton resonances directly via high-resolution ^{1}H NMR [25,26]. However, solid state ^{1}H spectra of polymers are often insufficiently resolved to fully separate signals from different domains or copolymer segments. In this respect, it is attractive to modify the Goldman-Shen experiment [1,3], as described below, such that ^{1}H spin-diffusion is monitored indirectly via carbon-13 sites directly bound to the protons of interest. The success of this morphological probe relies on the ability to resolve ^{13}C resonances in each microphase and/or copolymer segment. This was demonstrated in the previous subsection. Of critical importance is the establishment of a magnetization, or spin-temperature, gradient necessary to drive the spin-diffusion process. This is readily facilitated by the substantial difference in mobility between the hard (urethane-rich) and soft (polyether-rich) domains as documented by Assink et al. [7,9], Nierzwicki [10], and Figure 3 of this study. The fact that ^{1}H spin-spin relaxation rates are drastically different in the two domains of interest allows one to prepare the initial condition for proton-spin-diffusion studies.

The modified Goldman-Shen experiment is illustrated schematically in Figure 4. This experiment is a "second-generation" analog of the technique proposed by Goldman and Shen [3] almost two decades past. Initially, an ^{1}H 90°-pulse produces proton polarization in the transverse plane. After the time delay designated by τ_1 in Figure 4, proton magnetization unique to the urethane hard segment is severely attenuated by spin-spin relaxation processes. Dephasing of proton magnetization due to homonuclear interactions between dipolar-coupled protons is somewhat analogous to the dephasing of carbon-13 magnetization described in the previous subsection and illustrated in Figure 3. In both cases, strong dipolar interactions (homonuclear or heteronuclear) which persist in the urethane-rich hard domains result in short spin-spin relaxation times and, consequently, short lifetimes for transverse magnetization. Proton polarization in the polyether-rich mobile domains which survives the τ_1 time delay is subsequently returned to the $\pm$ z-axis by the second 90°-pulse. At the beginning of the τ_2-time interval, a gradient in proton magnetization between the rigid (urethane-rich) and mobile (polyether-rich) domains has been prepared. However, the modified Goldman-Shen experiment does not sample the proton signals directly and, hence, cannot detect spatial characteristics of the spin-temperature gradients in the proton spin system. If proton dipolar interactions are operative across domain boundaries as a consequence of relatively strong couplings in the interfacial regions, then the initial magnetization gradient will decrease with time, analogous to phenomenological transient diffusion-like

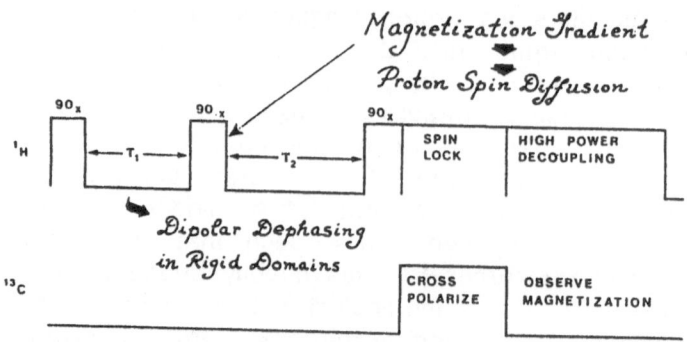

Figure 4 Schematic rf pulse representation of the slightly modified Goldman-Shen experiment to monitor proton spin diffusion between two domains of dissimilar mobility via high-resolution detection of carbon-13 signals. The dipolar dephasing delay (τ_1) was fixed at 15 µs, the spin-diffusion mixing period (τ_2) was either 0.01 µs or 10 ms, and the ^1H-^{13}C cross-polarization contact time was 100 µs.

processes [27,28], and proton polarization will be redistributed between the rigid and mobile microdomains until spin-temperature equilibration is achieved. By incorporating ^1H-^{13}C cross polarization [29,30] in the pulse sequence together with magic-angle sampling spinning and high-power ^1H-^{13}C dipolar decoupling (during data acquisition), one can map the effect of proton spin diffusion between microdomains of dissimilar mobility onto the carbon-13 nuclear spin system. If the thermal mixing time during cross polarization is relatively short (on the order of 50-100 µs), then one can effectively thwart long-range ^1H-^{13}C dipolar communication and obtain carbon intensities that are proportional to the magnetization of directly attached protons. Previous results for random ionic copolymers of ethylene and methacrylic acid, and a tri-block copolymer of styrene and butadiene [1,2,17], have indicated that a dipolar dephasing delay of approximately 10-20 µs is sufficient to nullify proton polarization in the rigid segments. Since homonuclear interactions between rigid dipolar-coupled protons are typically stronger than heteronuclear interactions between ^1H and ^{13}C, a 10-20 µs dephasing delay is sufficient to attenuate proton (OCH_2) magnetization whereas a delay of at least 50 µs is necessary to attenuate ^{13}C signals, as illustrated in the previous subsection. Proton polarization in the mobile domains is reduced to a much lesser degree during the 10-20 µs dephasing interval because ^1H homonuclear dipolar couplings in the polyether segments are partially averaged by molecular motion. Thus, a magnetization or spin-temperature gradient exists between the urethane-rich rigid domains and the polyether-rich mobile domains at the beginning of the mixing period, τ_2. Proton magnetization is subsequently redistributed between the two microdomains by spin diffusion during the mixing period. While the proton flip-flop terms in the dipolar Hamiltonian redistribute ^1H polarization between the hard- and soft-segment proton reservoirs during the mixing period [3], both reservoirs also approach spin-temperature equilibration with the lattice at a rate which is presumably much slower than the spin-diffusion rate. The effect (experimental artifact) of spin-lattice relaxation is nulled by storing proton magnetization alternately along the ±z-axis at the beginning of the mixing period and subtracting alternate signals in the time domain before Fourier transformation. It should be mentioned that the proposed experiment fails to monitor an undistorted proton-spin-diffusion phenomenon if the spin-diffusion rate is comparable to the spin-lattice relaxation rate.

Representative carbon-13 solid state NMR spectra from the modified Goldman-Shen experiment are illustrated in Figure 5 for the as-received polyether-polyurethane containing 58 wt.% hard segment (i.e. MDI and butanediol). The proton dipolar dephasing period was 15 µs, the mixing (spin-diffusion) time was 10 ms, the ^1H-^{13}C cross-polarization contact time was 100 µs, and proton magnetization was stored alternately along the ±z-axis at the start of the mixing period to suppress spin-lattice relaxation processes. It is evident from the lower spectrum in Figure 5 that 15 µs of proton dipolar dephasing is sufficient to null the rigid $O\underline{C}H_2$ signal (dashed line) characteristic of the butanediol hard-segment chain extender before proton spin diffusion is initiated. Furthermore, 100 µs of cross-polarization contact is short enough to suppress long-range ^1H-^{13}C dipolar communication and maintain the null in the hard-segment $O\underline{C}H_2$ signal intensity (lower spectrum in Figure 5). Redistribution of proton polarization between the hard and soft segments occurs (and most likely equilibrates) on the order of 10-100 ms. This claim is based on the appearance of the composite $O\underline{C}H_2$ signal in the upper spectrum of Figure 5 relative to the same resonance envelope in the upper spectrum of Figure 3. The polyurethane elastomer investigated herein represents the third kind of microphase-separated copolymer in which transient proton spin diffusion between rigid and mobile domains equilibrates on a time scale of 10-100 ms. Order of magnitude estimates based on a spin-diffusion time constant of 10-100 ms suggest that the average distance between dipolar-coupled protons is in the range of 25-70 angstroms. In these calculations [dipolar distance ≈ {(diffusion coefficient)(diffusion time constant)}$^{0.5}$], a spin-diffusion coefficient of 5×10^{-12} cm^2/sec has been employed, characteristic of proton magnetization transport in a material whose proton density and homonuclear proton dipolar couplings are somewhat smaller than those in polyethylene [11]. Other examples of proton spin diffusion focus on interdomain communication between (i)

126

crystalline and amorphous ethylenic segments in a zinc-neutralized random copolymer of ethylene and methacrylic acid, and (ii) rigid polystyrene occlusions and the mobile polybutadiene matrix, characteristic of SBS triblock copolymers [1,2,17].

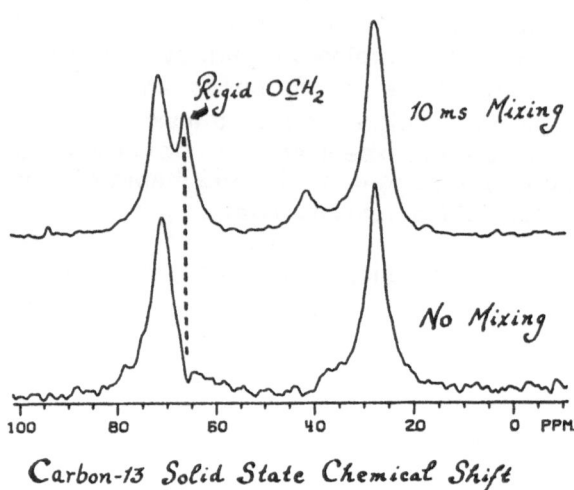

Carbon-13 Solid State Chemical Shift

Figure 5 High-resolution carbon-13 solid state NMR spectra of the MDI-based polyurethane elastomer from the slightly modified Goldman-Shen experiment.
Lower spectrum; the initial condition before proton spin diffusion is initiated. The two-strong signals at 27 ppm and 72 ppm which survive the 15 μs dipolar dephasing delay are unique to the polyether soft segment.
Upper spectrum; after 10 ms of proton spin diffusion, a "near-equilibrium" spectrum is observed based on 100 μs of cross-polarization contact and a 2-second pulse repetition delay. The dashed line and the arrow identify the hard-segment OCH₂ resonance of butanediol, generated via proton dipolar couplings across interfacial boundaries with the mobile polyether-rich microdomains. The internal methylene CH₂ signal of butanediol at ≈ 25 ppm resonates at a slightly different chemical shift than the CH₂ signal of the polyol soft segment at ≈ 27 ppm. Hence, the overall CH₂ resonance envelope in the upper spectrum is slightly broader than that in the lower spectrum because the hard-segment butanediol signals have been regenerated after 10 ms of proton spin diffusion.

Proton spin-temperature equilibration between the hard- and soft-segment-rich domains of the polyurethane elastomer on the order of 10-100 ms might be considered fast relative to a macroscopically phase-separated blend [26] or copolymer, but slow relative to a strongly interacting mixture [25]. This is reasonable for a microphase-separated material whose solid state morphology has been the subject of considerable theoretical and experimental research. Under fortuitous circumstances, intimate (near-neighbor) contact between dissimilar molecules in a mixture can be studied via direct measurement of proton spin diffusion in a two-dimensional application of the 1H-CRAMPS experiment (Combined Rotation And Multiple Pulse Spectroscopy). Belfiore et al. [17,25,31] have detected intermolecular dipolar communication in a hydrogen-bonded cocrystallized solid solution of poly(ethylene oxide) and resorcinol on the *100-μs* time scale, whereas Ernst and coworkers [26] report the absence of proton spin diffusion on the *100-ms* time scale for an immiscible blend of polystyrene and poly(vinyl methyl ether), cast from chloroform.

Conclusions

High-resolution carbon-13 solid state NMR spectroscopy has verified that both structural and motional differences between the hard and soft domains are characteristic of polyether-polyurethane thermoplastic elastomers. Implications about the molecular mobility of chemically distinguishable hard and soft segments, discussed herein, are consistent with previous solid state studies based on the transient free induction decay of proton magnetization in pulsed wideline NMR experiments. Interdomain communication via proton magnetization transport suggests that 1H-1H dipolar couplings persist in the interfacial regions as well as across domain boundaries. The transient spin-diffusion process equilibrates on a time scale of 10-100 ms. This indicates that (i) proton dipolar distances between the two domains in the vicinity of the interfacial regions lie within the range of 25-70 angstroms, and (ii) interfacial surface area is, most likely, substantial. Although the morphological conclusions are qualitative at best, the NMR results for polyurethanes and other microphase-separated copolymers, with proton spin diffusion time constants in the range of 10-100 ms, lie between those for intimately mixed molecular complexes ($\approx$100 μs) and macroscopically phase-separated polymer-polymer blends (>100 ms).

Acknowledgements

The author is grateful to the National Science Foundation (Grant #MSM-8811107), the Colorado Advanced Materials Institute, and Asahi Chemical Industry in Okayama, Japan for complete financial support of this research endeavor. The solid state NMR proton-spin-diffusion studies described herein were performed in collaboration with Dr. C.E. Bronnimann at the Colorado State University solid state NMR center, funded by NSF under Grant #ChE 8746548. Richard Oertel of The Upjohn Company in North Haven, CT is acknowledged for kindly supplying and chemically characterizing the polyurethane materials, and Hugh Graham is acknowleged for obtaining the polarized optical micrograph.

Literature References

1. L. A. Belfiore and A. A. Patwardhan, *ACE Proceedings, Div. Polym. Mater. Sci. Engr.,* **54**, 638 (1986).
2. L. A. Belfiore, R. J. Shah, and C. Cheng, in **"Contemporary Topics in Polymer Science, Multiphase Macromolecular Systems"**, Volume 6, B. M. Culbertson, editor, Plenum Press, **6**, 619 (1989).
3. M. Goldman and L. Shen, *Physical Review,* **144**(1), 321 (1966).
4. See, for example, **"Multiphase Polymers"**, S. L. Cooper and G. M. Estes, eds., *ACS Adv. Chem. Series,* **176** (1979).
5. T. C. Farrar and E. D. Becker, **"Pulse and Fourier Transform NMR"**, Ch. 2 and 4, Academic Press, New York (1971).
6. C. P. Slichter, **"Principles of Magnetic Resonance"**, 2nd edition, p. 42 and Ch. 3, Springer-Verlag, Heidelberg (1978).
7. R. A. Assink, *J. Polym. Sci., Polym. Phys. Ed.,* **15**, 59 (1977).
8. R. A. Assink and G. L. Wilkes, *Polym. Engr. Sci.,* **17**(8), 606 (1977).
9. R. A. Assink and G. L. Wilkes, *J. Appl. Polym. Sci.,* **26**, 3689 (1981).
10. W. Nierzwicki, *J. Appl. Polym. Sci.,* **29**, 1203 (1984).
11. J. R. Havens and D. L. VanderHart, *Macromolecules,* **18**, 1663 (1985).
12. T. Shiibashi, Y. Kitazawa, K. Arai, and E. Maekawa, *Kobunshi Ronbunshu,* **45**(2), 147 (1988).
13. J. J. Dumais, L. W. Jelinski, L. M. Leung, I. Gancarz, A. Galambos, and J. T. Koberstein, *Macromolecules,* **18**, 116 (1985).
14. A. Cholli, J. Koenig, T. Sun, and H. Zhou, *J. Appl. Polym. Sci.,* **28**, 3497 (1983).
15. C. D. Eisenbach and W. Gronski, *Makromol. Chem., Rapid Commun.,* **4**,707 (1983).
16. L. A. Belfiore, *Polymer,* **27**, 80 (1986).
17. L. A. Belfiore, T. J. Lutz, and C. Cheng, in **"Solid State NMR Characterization of Polymers"**, L. J. Mathias, editor, Plenum Press, p. 145 (1991).

18. C. D. Eisenbach, M. Baumgartner, and C. Gunter, in **"Advances in Elastomers and Rubber Elasticity"**, J. Lal and J. E. Mark, editors, Plenum (1987).
19. R. A. Wind, F. E. Anthonio, M. J. Duijvestijn, J. Smidt, J. Trommel, and G. M. C. DeVette, *J. Magn. Reson.*, **52**, 424 (1983).
20. E. O. Stejskal and J. Schaefer, *J. Magn. Reson.*, **18**, 560 (1975).
21. W. L. Earl and D. L. VanderHart, *J. Magn. Reson.*, **48**, 35 (1982).
22. K. K. S. Hwang, S. B. Lin, S. Y. Tsay, and S. L. Cooper, *Polymer*, **25**, 947 (1984).
23. M. A. Vallance, J. L. Castles, and S. L. Cooper, *Polymer*, **25**, 1734 (1984).
24. L. W. Jelinski, *Macromolecules*, **14**, 1341 (1981).
25. L. A. Belfiore, *Polymer Preprints*, **29**(1), 17 (1988).
26. P. Caravatti, P. Neuenschwander, and R. R. Ernst, *Macromolecules*, **18**, 119 (1985).
27. J. Crank, **"The Mathematics of Diffusion"**, Oxford University Press (1956).
28. H. S. Carslaw and J. C. Jaeger, **"Conduction of Heat in Solids"**, 2nd edition, Oxford University Press (1959).
29. S. R. Hartmann and E. L. Hahn, *Phys. Review*, **128**, 2042 (1962).
30. A. Pines, M. G. Gibby, and J. S.Waugh, *J. Chem. Phys.*, **59**, 569 (1973).
31. L. A. Belfiore, T. J. Lutz, C. Cheng, and C. E. Bronnimann, *Journal of Polymer Science; Polymer Physics Edition*, **8**, 1261 (1990).
32. M. D. Meadows, C. P. Christenson, W. L. Howard, M. A. Harthcock, R. E. Guerra, and R. B. Turner, *Macromolecules*, **23**, 2440 (1990).

Physical Characterization of Stereocomplexes

Robert E. Prud'homme and Anna M. Ritcey

Département de chimie, Centre de recherche en sciences et ingénierie de macromolécules,
Laval University, Sainte-Foy, Québec, Canada G1K 7P4

Abstract: Stereocomplexes have been formed by mixing the two isotactic poly(α-methyl-α-ethyl-β-propriolactones) (PMEPL) of opposite chirality. This leads to an insoluble complex exhibiting a melting transition which is 40°C above that of the initial isotactic components. Structural differences between these samples have been determined by nuclear magnetic resonance spectroscopy and electron diffraction. It was found by NMR that the stereocomplex crystallizes in a 2_1 helical conformation whereas the corresponding isotactic chains exhibit a helical or extended chain conformation depending upon the method of sample preparation. Electron diffraction confirms these measurements with the determination of a,b and c dimensions of the orthorombic unit cells.

INTRODUCTION

The polymerization of α,α-disubstituted β-propiolactones[1,2,3,4] leads to polyesters of the general formula $(CH_2CR_1R_2COO)_n$. Several polymers in this series have been found to be semi-crystalline despite their atactic nature and x-ray diffraction studies have revealed two crystal modifications[5,6,7,8], referred to as the α- and β-forms. In general, unoriented films exhibit the α-form and the transformation to the β-form is induced by stretching.

As a consequence of the α to β transition which accompanies orientation, fibre patterns of the α-form are difficult to obtain. Melt extruded fibres of poly(α,α-dimethyl-β-propiolactone), known as poly(pivalolactone) (PPL), do, however, crystallize in the α-form and retain their orientation upon high temperature annealing under tension[9]. Furthermore, oriented samples of poly(α-methyl-α-n-propyl-β-propiolactone) (PMPPL) have been prepared which show x-ray layer lines corresponding to the two phases[8]. The α-form of both PPL and PMPPL is characterized by a fibre repeat distance of about 6 Å, which has been identified as the periodicity of a 2_1 helix. Although oriented samples of poly(α-methyl-α-ethyl-β-propiolactone) (PMEPL) in the α-phase have never been reported, x-ray powder data[10] have been fitted with a monoclinic unit cell with c = 6.1 Å.

Y. Imanishi (Ed.)
Progress in Pacific Polymer Science 2
© Springer-Verlag Berlin Heidelberg 1992

Studies of the crystalline, thermal and mechanical properties of PPL[6] and PMPPL[8] indicate that the α-β transformation involves a change in conformation from a 2_1 helix to the fully extended chain. A fibre repeat distance of 4.75 Å, corresponding to the planar zigzag conformation, is evident in the x-ray diffraction patterns of the β-forms of PPL[6], PMPPL[8] and PMEPL[7].

A third polymorph, known as the γ-phase, has been found for PPL crystallized by rapid cooling from the melt[11,12]. From combined electron and x-ray diffraction data, the crystal structure of this phase was determined to be an orthorhombic lattice containing antiparallel 2_1 helices of the same chirality[13].

Conformational analysis[14] predicts that the general class of polyesters based on the substituted poly(β-propiolactone) backbone will have nearly identical crystalline conformations, that of a 2_1 helix with a periodicity of 6 Å. The planar zigzag conformation is also energetically allowed, with low energy barriers separating the two conformations[15].

To date, studies concerning the polymorphic behaviour of α,α-disubstituted poly(β-propiolactones) have been limited to atactic polymers. Polyesters which contain two different α-substituents, such as PMEPL and PMPPL, possess a chiral centre, and optically active polymers can therefore be prepared[16]. Although the melting temperature and enthalpy of fusion are found to increase with optical purity for samples of PMEPL and PMPPL of varying tacticity[17,10], only slight differences in unit cell dimensions were reported between the racemic and optically active products. In the case of PMEPL, however, a new and unusual crystal form was found in mixtures of the two isotactic polymers of opposite absolute configuration[18,19]. Equimolar blends crystallize with a morphology and crystal structure that are distinctly different from those of the individual isotactic components, indicating the formation of a stereocomplex. In nonequimolar blends, stereocomplex formation occurs preferentially over the crystallization of the isotactic polymers and controls the morphology over a wide concentration range. Perhaps most striking is the observation that the stereocomplex melts 40°C above the melting temperature of the individual isotactic components.

In the present paper, differences between isotactic PMEPL and the stereocomplex are examined by solid state nuclear magnetic resonance (NMR) spectroscopy. These studies reveal new polymorphic behaviour of the isotactic polymer and differences in crystal structure which depend on tacticity. Crystal structures of the various polymorphs were also determined by electron and x-ray diffraction studies.

EXPERIMENTAL

PMEPLs of varying degrees of optical purity were prepared and characterized by Grenier et al.[16] All samples were initially prepared as films cast from solution in hexafluoroisopropanol, the only known solvent for isotactic PMEPL. Melt crystallized samples were obtained by the subsequent heating of these films, under vacuum, to temperatures 20°C above the melting point, followed by cooling at 1-3°C/min. Films of the stereocomplex were prepared by casting from solutions containing equal concentrations of the two isotactic polymers of opposite absolute configuration.

Solid state CP/MAS NMR spectra were recorded with a 75 MHz Chemagnetics spectrometer. The standard pulse sequence for cross polarization with bilevel dipolar decoupling was employed with a contact time and recycle delay of 1 ms and 3 s, respectively. Chemical shifts were referenced to hexamethylbenzene at 17.4 ppm. Additional details concerning sample preparation and NMR studies can be found in a forthcoming paper[20].

Samples for electron microscopy were cast directly on carbon coated microscope grids (200 mesh) from single drops of 0.01% solutions in hexafluoroisopropanol. Melt crystallization was performed by heating the grids, either under vacuun or in a nitrogen atmosphere, to temperatures 20°C above the melting point of the polymer, and subsequently cooling them to room temperature at a cooling rate of 1°C/min.

Electron diffraction patterns were observed with a Phillips EM 420 electron microscope, operated at 100-120 kV. Gold, deposited on representative samples, served as a standard for the calibration of measured d-spacings.

RESULTS AND DISCUSSION

Solid state NMR spectra of isotactic PMEPL prepared by solution casting and melt crystallization are shown in Fig. 1. The peak assignments were confirmed by delayed decoupling spectra. Comparison of the two spectra of Fig. 1 indicates significant differences between the melt crystallized and the solution cast samples. In the melt crystallized spectrum, both C3 and C4 (defined in Fig. 2) appear as two well separated peaks rather than as single resonances, with the major components of these peak pairs located at 66.1 and 23.4 ppm, respectively. The NMR spectrum of the solution cast film contains only one peak for each of C3 and C4, located at 74.1 and 18.1 ppm, respectively. These chemical shifts correspond to those of the minor components of the two peak pairs

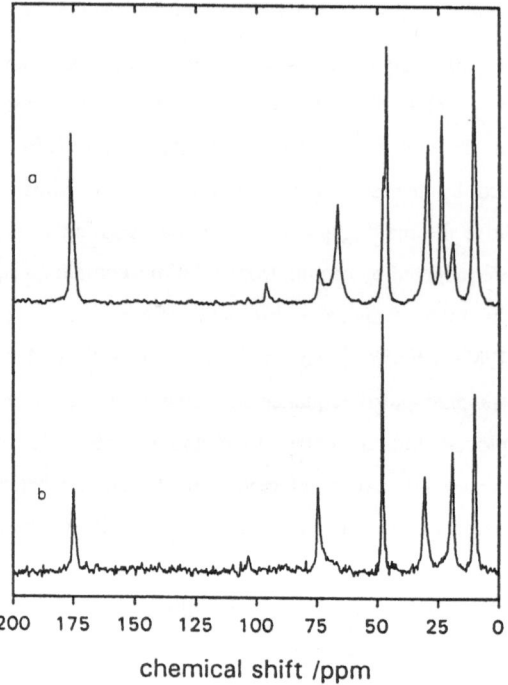

Figure 1. Solid state ^{13}C NMR spectra of (a) melt crystallized and (b) solution cast
isotactic PMEPL.

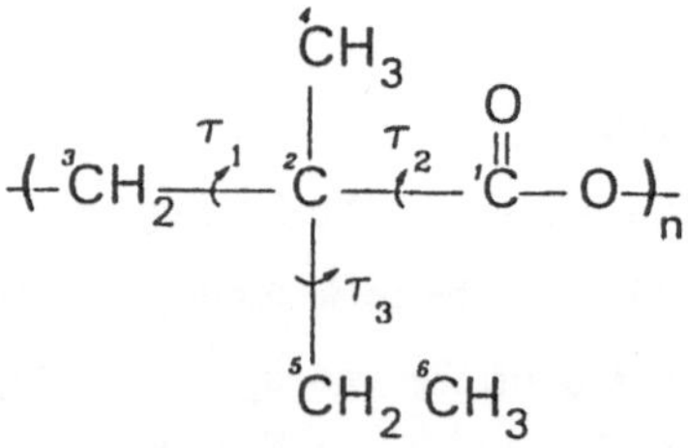

Figure 2. Chemical structure of PMEPL defining carbon atoms and torsion angles
referred to in the text.

in the melt crystallized sample. Peak splitting of a lesser magnitude is observed for the tertiary carbon, C2, in the melt crystallized sample, giving rise to peaks at 47.4 and 46.1 ppm. Once again, a single resonance is observed in the solution cast spectrum, occurring at the position of the minor component in the melt crystallized sample.

Spectra obtained for atactic PMEPL and the stereocomplex formed by equal molar mixtures of the two isotactic polymers of opposite absolute configuration are presented in Fig. 3. In contrast to the isotactic case, no dependence on thermal history is found. These spectra are unexpectedly similar, and resemble the solution cast isotactic spectrum of Fig. 1. NMR spectra were also recorded for polymers of intermediate tacticity. PMEPL of optical purity 75% shows the same dependence on sample preparation as is found for the isotactic polymer whereas samples of optical purity 25% behave as the atactic case. Similar observations were made by Grenier and Prud'homme[18] by the comparison of other properties, including x-ray patterns, solubility and morphology.

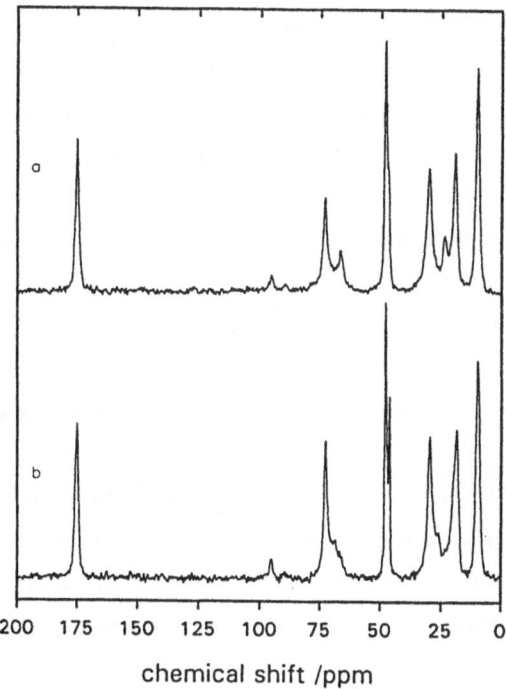

Figure 3. Solid state ^{13}C NMR spectra of (a) the stereocomplex and (b) atactic PMEPL.

The presence of two peaks for a single carbon atom in a given NMR spectrum indicates the existence of two different environments for that atom within the sample. The separation of the peak pairs assigned to C3 and C4 are too large (7.5 and 4.7 ppm, respectively) to be solely the result of variations in chain packing, and thus indicate the presence of two different polymer conformations. As outlined in the introduction, conformational calculations and structural studies of similar polyesters suggest that the probable conformations for PMEPL are the 2_1 helix and the planar zigzag. These two conformations differ in the torsion angles labelled τ_1 and τ_2 in Fig. 2.

Chemical shift displacements of the magnitude observed for C3 and C4 are frequently attributed to changes in the number of γ-substituents occupying gauche positions relative to the carbon atom of interest.[21] The two carbon atoms which show large shifts in peak position, i.e. C3 and C4, both occupy positions γ to the terminal methyl group of the α-ethyl sidechain, suggesting that the differences observed between the NMR spectra of the various PMEPL samples may be related to conformational changes involving this substituent. This interpretation does not, however, exclude the possibility of concurrent changes in backbone conformation. Elementary molecular models do indeed indicate that the conformation of the sidechains depend upon backbone conformation[20].

Thin films of isotactic PMEPL prepared by direct melt crystallization on a microscope grid show the single crystal-like electron diffraction pattern illustrated in Fig. 4. If it is assumed that this diffraction pattern, obtained with irradiation normal to the sample surface, is the hk0 reciprocal lattice, the unit cell parameters **a** and **b** can be explicitly evaluated as 9.10 Å and 7.44 Å, respectively, with $\gamma = 90°$. Unfortunately no higher order reflections were recorded from tilted samples, presumably because of the weak intensity of the hkl reflections accessible through such experiments. The periodicity along the c-direction was therefore evaluated from epitaxially crystallized samples. The c-axis dimension was found[22] to be 4.75 Å.

Thin films of PMEPL containing equal quantities of the two isotactic polymers of opposite absolution configuration give rise to the unique electron diffraction pattern shown in Fig. 5. The striking contrast between this pattern and that obtained for the pure isotactic components (Fig. 4) signifies that major differences in crystal structure exist between the stereocomplex and its individual isotactic components. If it is again assumed that this diffraction pattern is the hk0 reciprocal plane, the following unit cell parameters can be directly evaluated: $\mathbf{a} = 5.88$ Å, $\mathbf{b} = 34.2$ Å and $\gamma = 90°$. Upon tilting the sample[22], higher

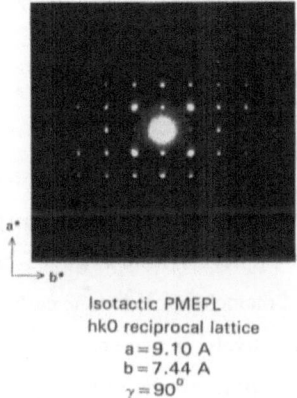

Isotactic PMEPL
hk0 reciprocal lattice
a = 9.10 A
b = 7.44 A
γ = 90°

Figure 4. Electron diffraction pattern observed for melt crystallized isotactic PMEPL with irradiation normal to the sample surface, identified as the hk0 reciprocal lattice. (a* vertical, b* horizontal)

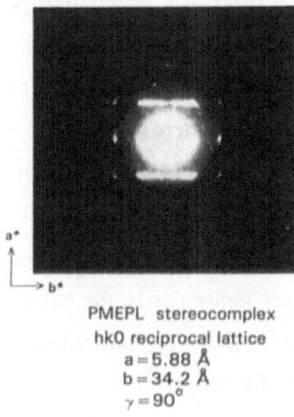

PMEPL stereocomplex
hk0 reciprocal lattice
a = 5.88 Å
b = 34.2 Å
γ = 90°

Figure 5. Electron diffraction pattern observed for the PMEPL stereocomplex with irradiation normal to the sample surface, identified as the hk0 reciprocal lattice. (a* vertical, b* horizontal)

order reflections are observed and can be indexed with $c = 7.09$ Å. This, however, requires doubling the a-dimension given above.

CONCLUSIONS

The solid state ^{13}C NMR spectrum of poly(α-methyl-α-ethyl-ß-propiolactone) shows high sensitivity to changes in polymer conformation. The spectra reveal that isotactic PMEPL crystallizes in two different forms which can be easily distinguished by the chemical shifts of the α-methyl and mainchain methylene carbon atoms. These resonances appear at 23.4 and 66.1 ppm, respectively, in the spectrum of melt crystallized isotactic PMEPL. By comparison to the spectrum of oriented samples, it is concluded that these peak positions are characteristic of the extended chain conformation. The second crystal form of isotactic PMEPL is obtained on solution casting and is identified by a NMR spectrum in which the α-methyl and mainchain methylene carbons are located at 18.1 and 74.1 ppm, respectively. Similar spectra are also obtained for atactic PMEPL and the stereocomplex and are attributed to the 2_1 helical conformation. The chemical shift differences observed between the planar zigzag and helical conformations can be explained satisfactorily in terms of rotational isomerism of the α-ethyl group.

Although NMR spectroscopy indicates that atactic PMEPL and the stereocomplex share similar conformations, electron and x-ray diffraction patterns reveal different crystal structures. The proposed unit cell parameters of the various samples are summarized in Table 1. Three polymorphs of PMEPL are identified. Melt crystallized isotactic PMEPL crystallizes in a monoclinic, pseudo-orthorhombic lattice. The unit cell parameters indicate that this form is equivalent to the previously reported ß-modification obtained by sample elongation and consists of polymer molecules in the extended conformation. The same polymorph is found for isotactic PMEPL crystallized from dilute solution at elevated temperatures. The stereocomplex formed by equimolar mixtures of isotactic PMEPL of opposite absolute configuration crystallizes in an orthorhombic lattice, with a periodicity along the chain direction equivalent to that of an ideal 2_1 helix, without deviation of ester linkage from planarity. The third polymorph of PMEPL is found for of atactic samples and solution cast isotactic films. X-ray diffraction data for these forms can be fitted satisfactorily with a unit cell analogous to the γ-phase of PPL, giving a value of c implying a helical conformation. No monoclinic phase corresponding to the α-phase of PPL was found for the samples examined.

Table 1. Summary of unit cell parameters for the various orthorhombic lattices found for PMEPL and PMPPL. Parameters for samples marked with an asterisk were evaluated from x-ray diffraction data only, indexed according to the γ-phase of PPL.

sample	a	b	c
PMEPL-isotactic -melt crystallized	9.10	7.44	4.75
PMEPL-isotactic* -solution cast	12.7	11.4	5.9
PMEPL-stereocomplex	11.8	34.2	7.1
PMEPL-atactic*	12.6	11.9	5.8

ACKNOWLEDGMENTS

The authors thank the Natural Sciences and Engineering Research Council of Canada and the Department of Education of the Province of Quebec (FCAR program) for financial support.

REFERENCES

1. Jérôme, R.; Teyssié, P. in *Comprehensive Polymer Science*; Eastmond, G.C.; Ledwith, A.; Russo, S.; Sigwalt, P. Ed., Pergamon Press, 1989; Vol.3, Chapter 34.

2. Thiebaut, R.; Fischer, N.; Etienne Y.; Coste, J. *Ind. Plat. Mod.* **1962**, 14, 1.

3. *Ring-Opening Polymerization*; Ivin, K.J.; Saegusa, T., Ed.; Elsevier, New York, 1984.

4. Hall, H.K. Jr. *Macromolecules* **1969**, 2, 488.

5. Knoblock, F.W.; Statton, W.O. *U.S. Patent* **1967**, 3, 299, 171.

6. Prud'homme, R.E.; Marchessault, R.H. *Macromolecules* **1974**, 7, 541.

7. Duchesne, D.; Prud'homme, R.E. *Polymer* **1979**, 20, 1199.

8. Cornibert, J.; Marchessault, R.H.; Allegrezza, E.A. Jr.; Lenz, R.W. *Macromolecules* **1973**, 6, 676.

9. Perego, G; Melis, A.; Cesari, M. *Makromol. Chem.* **1972**, 157, 269.

10. Grenier, D.; Prud'homme, R.E. *Macromolecules* **1983**, 16, 302.

11. Prud'homme, R.E.; Marchessault, R.H. *Macromolecules* **1974**, 175, 2701.

12. Meille, S.V.; Konishi, T.; Geil, P.H. *Polymer* **1984**, 25, 773.

13. Meille, S.V.; Brückner, S.; Lando, J.B. *Polymer* **1989**, 30, 786.

14. Cornibert, J.; Marchessault, R.H. *Macromolecules* **1975**, 8, 296.

15. Cornibert, J. *Ph.D. Thesis* **1972**, Université de Montréal.

16. Grenier, D.; Prud'homme, R.E.; Leborgne, A.; Spassky, N. *J. Polym. Sci.:Polym. Chem. Ed.* **1981**, 19, 1781.

17. Spassky, N.; Leborgne, A.; Reix, M.; Prud'homme, R.E.; Bigdeli, E.; Lenz, R.W. *Macromolecules* **1978**, 11, 716.

18. Grenier, D.; Prud'homme, R.E. *J. Polym. Sci.: Polym. Phys. Ed.*, **1984**, 22 577.

19. Lavallée, C.; Prud'homme, R.E. *Macromolecules* **1989**, 22, 2438.

20. Ritcey, A.M.; Prud'homme, R.E., Macromolecules, in press

21. Duddeck, H. in *Topics in Stereochemistry*, Eliel, E.; Wilen, S,; Allinger, N. Ed., Wiley, 1986, Vol. 16, p 219.

22. Ritcey, A.M.; Brisson, J.; Prud'homme, R.E. Macromolecules, submitted

Characterization of Polymer Systems by Real Time Pulsed NMR

T. Nishi and T. Hayashi

Department of Applied Physics, Faculty of Engineering.
The University of Tokyo, 7-3-1 Hongo, Bunkyo-ku, Tokyo,
113, Japan

Abstract: Application of real time pulsed NMR(RTPNMR) for the
characterization of polymer systems has been reviewed and
several new examples are given. They include RTPNMR of curing
process of epoxy resins and crosslinking process of polydiace-
tylenes. In both cases, there appear several relaxation times
for spin-spin relaxation time T_2 and spin-lattice relaxation
time in the rotating frame $T_{1\rho}^*$. However, spin-lattice relaxation time T_1 is single for epoxy resins and from these data,
heterogeneous nature of the system during curing and crosslinking are analysed. For the last example, orientation of
liquid crystalline polymers under magnetic field is detected by
RTPNMR at high temperatures. RTPNMR is proved to be a powerful
method for characterization of polymer systems in terms of
degree of molecular motion.

INTRODUCTION

Characterization of polymer systems is a very challenging problem both from scientific and engineering viewpoint. However,
the problem itself is very complex and there can be many ways
for the characterization. Table 1 shows an example for the
characterization of multicomponent polymer systems. It is
essential to combine several methods to get meaningful information for the characterization. In the table, E means an excellent method to get quantitative results and G means a good
method to get semi-quantitative results. F means a fair method
to get qualitative results. Blancks mean that there are not
enough reports to prove the usefulness of the methods.

So far, most characterization methods have been concerned with
polymer systems that are not changing with time. However, if
one take into account real polymer systems or industrial polymer
processes, it is very important to characterize polymer systems
that change with time. Typical examples are hardening processes
in polymer systems and they are classified into several cases as
shown in Figure 1.

Y. Imanishi (Ed.)
Progress in Pacific Polymer Science 2
© Springer-Verlag Berlin Heidelberg 1992

Table 1. Characterization of multicomponent polymer systems.

Category	Method	Intermolecular interaction	Molecular order miscibility	Surface & interface structure	Micro-phase separation	Phase separated structure	High order structure of crystals	Spherulite structure	IPN structure	Orientation
Scattering	SANS	E	E	G	G	F			G	G
	SAXS	E	E	G	E	G	E		G	G
	SALS		G	G	F	E		E		E
Microscopy	TEM		G	G	E	E	E	E	G	E
	SEM		F	G	G	E	G	E	F	G
	PCM		F	F	F	E	F	E		F
	PM		F	F		G	E	E		G
Relaxation	Pulsed NMR	G	E	G	G	E	F		G	F
	Dynamic mech.		G	F	G	G			G	G
	Dielc. relax.		G		G	G			G	F
Thermal analysis	DSC, DTA	F	E	F	G	G	G	F	G	G
	TOA		G		F	G	F		G	
	TSC		G	F	G	G			F	
Spectroscopy	FT-IR	E	F	G	F					G
	solid-HR-NMR	E	E	F						
	Fluorescence	G	E	F	G	G			G	G
	ESCA			E	F					
	SIMS		G	E		G				

E; Excellent, G; Good, F; Fair

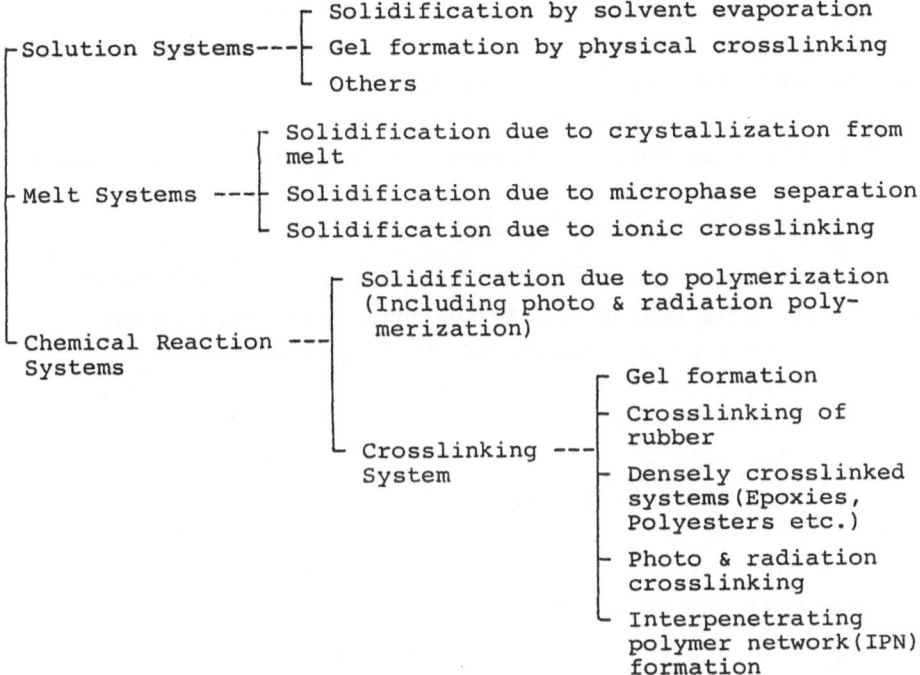

```
                           ┌─ Solidification by solvent evaporation
  ┌─Solution Systems───┤  Gel formation by physical crosslinking
  │                        └─ Others
  │
  │                      ┌ Solidification due to crystallization from
  │                      │ melt
  ├─Melt Systems ───┤ Solidification due to microphase separation
  │                      └ Solidification due to ionic crosslinking
  │
  │                        ┌─ Solidification due to polymerization
  │                        │  (Including photo & radiation poly-
  └─Chemical Reaction ───┤   merization)
    Systems              │                    ┌ Gel formation
                         │                    ├ Crosslinking of
                         │                    │ rubber
                         └─ Crosslinking ───┤ Densely crosslinked
                            System           │ systems(Epoxies,
                                             │ Polyesters etc.)
                                             ├ Photo & radiation
                                             │ crosslinking
                                             └ Interpenetrating
                                               polymer network(IPN)
                                               formation
```

Figure 1. Classification of hardening processes in polymer
          systems.

```
                              ┌─ Rheometer, Curemeter
  ┌─ Mechanical Measurement ───┤ Torsional braid analysis(TBA)
  │                            ├ Ultrasonic method(Sound velosity,
  │                            │ attenuation)
  │                            └ Others
  │
  ├─ Electrical Measurement ──── Dielectric measurement(Complex
  │                              dielectric constant), Conductivity
  │                              and so on.
  │
  ├─ Optical Measurement ──── Light scattering, Dynamic light
  │                           scattering, FT-IR etc.
  │
  └─ Other Method ──── X-ray diffraction, X-ray scattering (SOR),
                       Thermal analysis etc.
```

Figure 2. Characterization of hardening processes.

For the characterization of these processes there have been
developed several methods and they are shown in Figure 2. They
can be divided into two large groups. One is related to the
structure of polymer systems and the other is related to the
molecular motion in polymer systems. Information on the degree
of molecular motion is essential for the understanding of dynamical properties of polymer systems.

One of the best ways to characterize the degree of molecular
motion is to apply pulsed NMR to the system and measure spin-
spin relaxation time T_2, spin-lattice relaxation time T_1, spin-
lattice relaxation time in the rotating frame $T_{1\rho}$, and the
amount of their components since they are directly related to
the correlation time of molecular motion τ_c of the resonant
nuclei. There have been many experiments to study solid polymers by pulsed NMR and multiphase character has been revealed
for many systems including crystalline polymers, polymer blends,
filled polymers and so on(1).

Recently, we have developed a real time pulsed NMR(RTPNMR)
system(2) controlled by a microcomputer as shown in Figure 3 to
study dynamics of structure formation in polymer systems. It
has been successfully applied to study crystallization process
(3), gelation process(4) and so on. In this paper, application
of RTPNMR for characterization of curing process of epoxy resins
, polymerization & crosslinking process of diacetylenes, and
orientation process of liquid crystalline polymers(LCPs) under
magnetic field are described. Papers related to the detailed
analysis of these data will appear in the near future.

EXPERIMENTAL

Pulsed NMR equipment was manufactured by Bruker and the model
was PC-20(resonance frequency for proton being 20 MHz). Pulse
sequence used in T_2 measurement was the solid echo method($90^\circ_x \tau$
90°_y) to obtain a rapid free induction decay(FID) and the spin
echo(Carr-Purcell-Meiboom-Gill(CPMG)) method($90^\circ_x \tau (180^\circ_y \ 2\tau)_n$) to
obtain slow FID behavior to avoid the effect of inhomogeneity
of the static magnetic field. The solid echo train method($90^\circ_x \tau$
$(90_y \ 2\tau)_n$) was used to measure $T_{1\rho}^*$. The modified inversion
recovery method($180^\circ_x \tau 90^\circ_x \ \tau_0 90^\circ_y$) ($\tau_0$ = 5.0μs) was applied to
measure T_1. All the NMR measurements were carried out under the
control of microcomputer. The data acquisition is executed with

8 bits A/D converter unit(FIXX FSA100SV). Analysis of the data were accomplished by the non-linear least square method assuming Gaussian and/or exponential decay.

For the measurement of the orientation of LCP under magnetic field at high temperatures, special probe head as shown in the Figure 4 was prepared(5). It is usable up to 330°C and it was necessary to adjust the magnetic field by the reference just before the beginning of the measurement.

As an example for an epoxy resin, diglycidyl ether of bisphenol-A (Epon 828 by Yuka Shell Epoxy Co. Ltd.) was cured with 2-ethyl-4-methylimidazole or N,N-dimethylbenzylamine. The concentration of the curing agent was changed from 0.05 to 0.2 molar ratio. The curing temperature was changed from 40°C to 80°C.

As an example for a diacetylene, 1,6-diacrylate-2,4-hexadiyne(DAHD) was studied and the polymerization was initiated with a very small amount of azobisisobutyronitrile.

As an example for LCP, main chain type LCP which shows nematic phase between 230°C and 240°C (PB-7) was used and the structure is shown below.

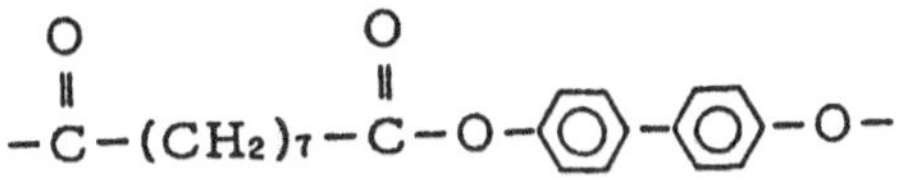

RESULTS AND DISCUSSION
Epoxy resins : Figure 5 shows an example of RTPNMR results on epoxy resin cured with 2-ethyl-4-methylimidazole(5%) at 50°C(6). (a) is the temporal change of T_2. At the initial stage there is only one T_2 indicating the homogeneous nature of the mixture. T_2 becomes shorter during the curing process due to the formation of polymer network and the increase in T_g of the system. However, there appears short T_2 component soon after the starting of cure. This means the non-uniform curing in the system. (b) shows the temporal change of the short T_2 component ϕ. This multi-relaxation time nature is also observed for the temporal change of $T_{1\rho}^*$ and it is shown in (c). On the other hand, there is observed only one T_1 during the curing process as shown in (d). In the T_1 measurement, spin diffusion effect is opera-

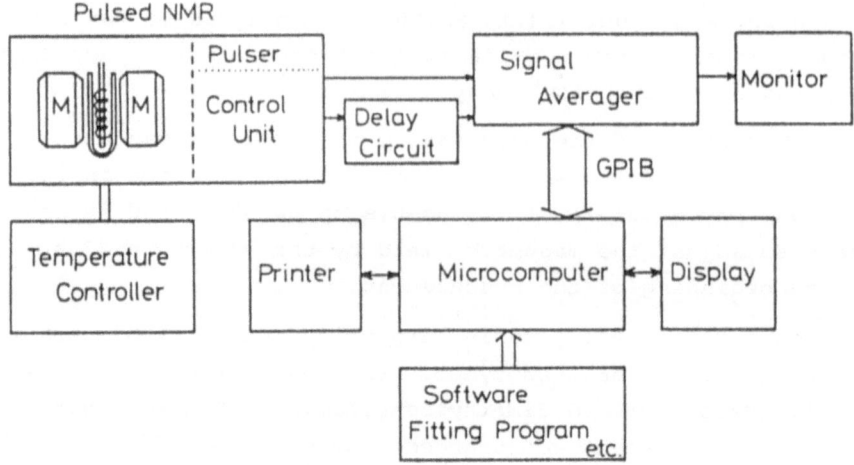

Figure 3. Block diagram of the real time pulsed NMR measurement system controlled by a microcomputer.

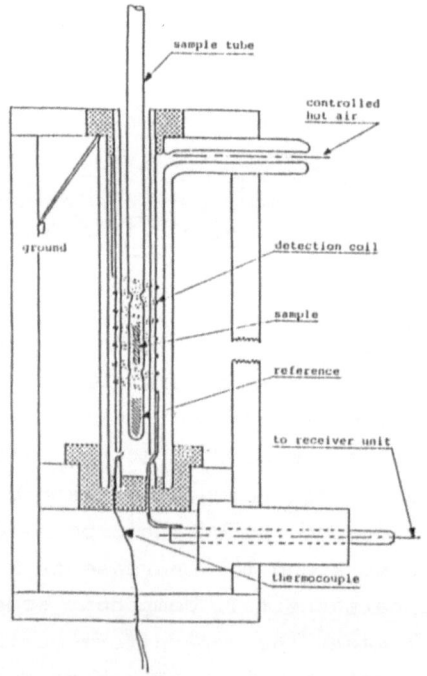

Figure 4. Probe head for the pulsed NMR measurement at high temperatures up to 330°C.

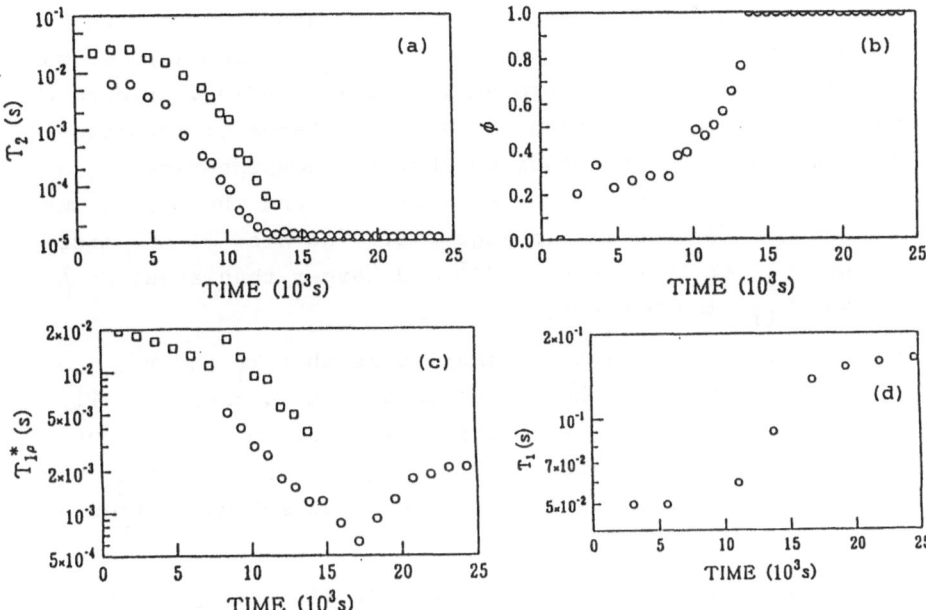

Figure 5. RTPNMR results on epoxy resin cured at 50°C. Temporal changes of T_2 (a), short T_2 component ϕ (b), $T_{1\rho}^*$ (c), and T_1 (d).

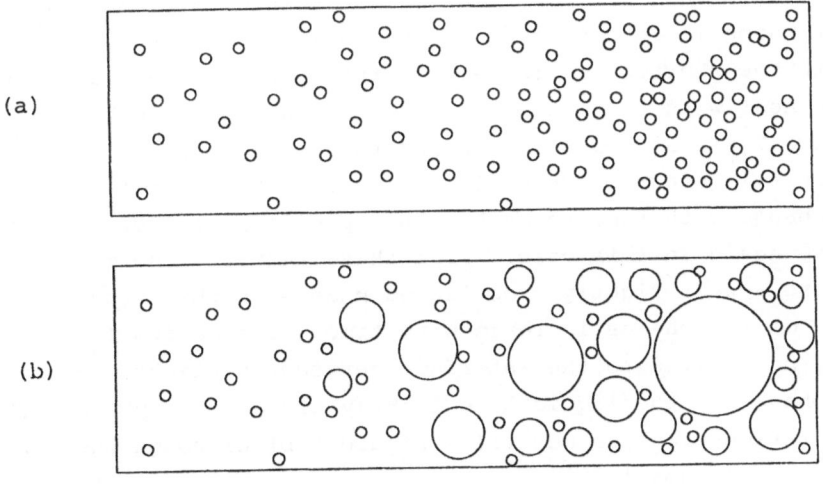

Figure 6. Schematic representation of temporal change of inhomogeneity during curing.

tive and the spin diffusion distance for T_1 and T_1 are usually about 100 Å and 10 Å, respectively(7). From (c) and (d), we can estimate the size of the heterogeneity during curing. Figure 6 shows schematic representation of temporal change of inhomogeneity during curing. The observed short T_2 component during curing may be due to the so-called micro-gel and the curing must be proceeding like type (a) in Figure 6. The size of the micro-gel may be smaller than about 100 Å and larger than about 10 Å from T_1 and $T_{1\rho}^*$ measurements.

The other interesting point in Figure 5 is that T_1 and $T_{1\rho}^*$ are still changing with time when there is almost no change in T_2. T_2 becomes insensitive after solidification of the system but T_1 and $T_{1\rho}^*$ are sensitive in the change of τ_c. This means that the curing is still proceeding after the apparent solidification of the epoxy resins.

The same type of behavior was observed for all the epoxy resins studied and the detailed analysis will be reported in the near future.

Polydiacetylenes(PDA): Figure 7 and 8 show RTPNMR results during the polymerization of DAHD at 50°C(8). The behavior is quite different from that of epoxy resins. In this case, there appear at least three T_2 components. In Figure 7(a), (b), the component with the shortest T_2 increases with time and the amount of the component with the medium T_2 is not changing so much. There appear two $T_{1\rho}^*$'s as shown in (c) and there are observed two T_1's at the initial stage as shown in Figure 8. The short $T_{1\rho}^*$ and T_1 do not change with time so much. The spin diffusion effect may be active again in this case. From these results, one may say that there appear tightly crosslinked phases in the system and the size of the phases increase with time. The medium T_2 component may be related to the interface between the tightly crosslinked phase and the liquid-like lightly crosslinked phase. The polymerization proceeds more like Figure 6(b) although the size of the phase is small.

Liquid crystalline polymers(LCP): Figure 9 shows temporal change of FID from PB-7 at 235°C under magnetic field(0.47T). The resonance frequency for the proton was 20 MHz. The polymer is in the nematic phase and the sample is jumped into the static magnetic

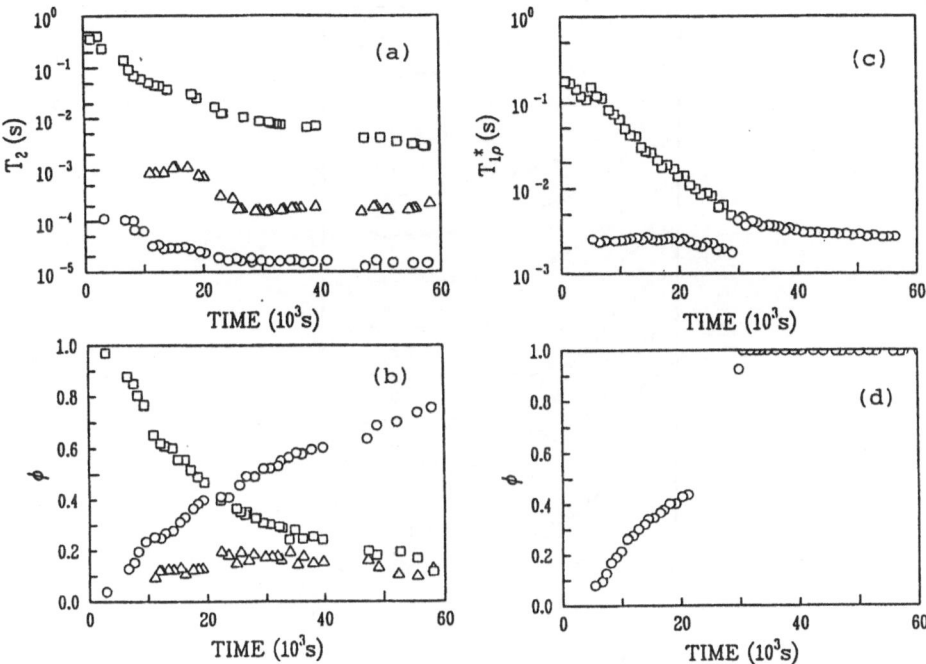

Figure 7. RTPNMR results for the polymerization of DAHD at 50°C.
Temporal change of T_2 (a) and its component (b).
Temporal change of $T_{1\rho}^*$ (c) and the short T_1^* component
(d).

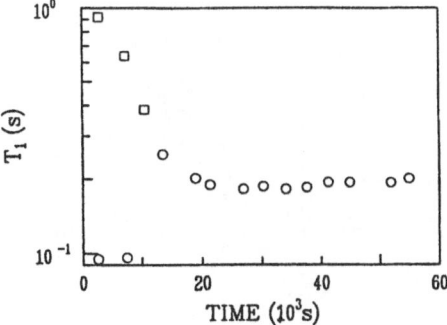

Figure 8. RTPNMR results for the polymerization of DAHD at 50°C.
Temporal change of T_1.

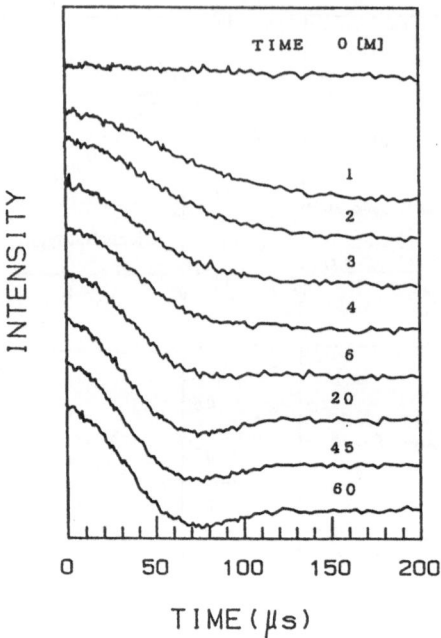

Figure 9. Typical FID of PB-7 in the nematic phase at 235°C
under magnetic field(0.47 T).

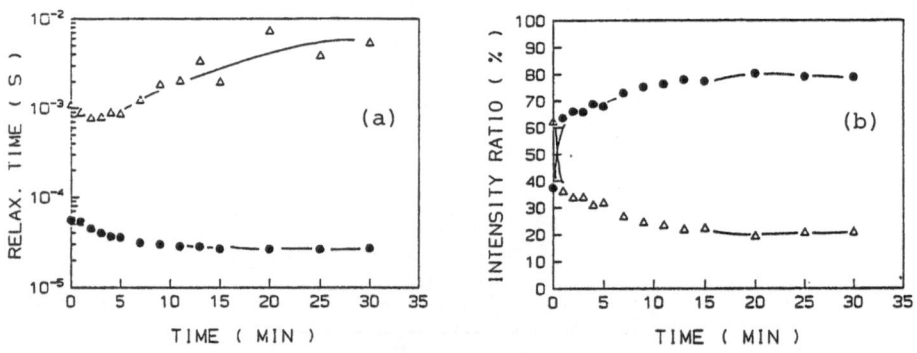

Figure 10. Temporal change of the T_2(a) and its component (b)
for PB-7 under magnetic field(0.47 T).

field. The FID keeps on changing for 20 to 40 min. due to the orientation of the LCP molecules. Figure 10 shows temporal change of the T_2 (a) and its component (b). The short T_2 component corresponds to the magnetically oriented fraction in LCP. The existence of long T_2 component and the T_2 increase with time suggest the degree of liquid crystallinity in LCP. The same kind of behavior was also observed for the other copolymer type main chain LCPs. The detailed analysis will be also given in the future.

CONCLUSION

As shown in the previous section, RTPNMR can reveal many aspects of polymer systems that are difficult to be detected by other methods and it is proved to be a powerful method for the characterization of polymer systems. However, it is necessary to apply this technique to many other systems and to compare the results with other methods to get more quantitative results.

The authors are grateful to Mr. Ryu Shioda for measuring LCPs and we are also grateful to Yuka Shell Epoxy Co. Ltd., Asahi Chemical Co. Ltd., and Mitsubishi Chemical Co. Ltd. for providing us epoxy resins, DAHD, and LCPs, respectively.

References

1 McBrierty VJ, Douglass DC (1981) J Polym Sci Macromol Rev 16: 295

2 Tanaka H, Nishi T (1986) J Appl Phys 60:1306

3 Tanaka H, Nishi T (1986) J Chem Phys 85:6197

4 Tanaka H, Fukumori K, Nishi T (1988) J Chem Phys 89:3363

5 Shioda R (1989) Master Thesis:Tokyo Univ

6 Hayashi T, Yamamura H, Nakamura N, Nishi T (1990) Rept Prog Polym Phys Jpn 33:495

7 Tanaka H, Nishi T (1986) Phys Rev B33:32

8 Hayashi T, Nishi T (1991) Ann Rept Eng Res Inst Fac Eng Univ Tokyo 50:87

Additivity Rule and Other Approaches to Non-Isothermal Crystallization

Z. H. Stachurski, J.R Griffiths and S. Chew

Department of Materials Engineering, Monash University
Melbourne, Victoria, Australia 3168

Abstract: A number of theories for predicting non-isothermal crystallization of polymers have been proposed (1-5). Some are based on Avrami solution (2,3), others were derived independently (1,4,5). All are based on an "additivity" principle which states that the rate of crystallization at a time, t, depends on the extent of crystallization at time t, but not on the previous history (i.e., it is path independent). In this paper we present experimental results which show that, within a certain regime of cooling rates, the additivity principle applies to isotactic polystyrene. In a previous publication we have shown the same to be true for polyethylene (6).

DEFINITIONS AND ASSUMPTIONS

Crystallization is a thermodynamic transformation from a liquid to a solid phase; it is a process of solidification. We define the degree of crystallization (or degree of solidification), $X(t,T)$, measured at a given time, t, and temperature, T, in terms of the mass fraction of the transformed material. Figure 1 and equations (1) given below illustrate the quantities involved in the definition.

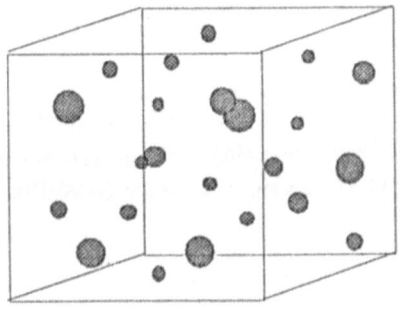

the total volume of molten polymer is V_o

n crystallizing particles; the volume of each particle is denoted V_i

Figure 1. When a liquid polymer is cooled to a temperature below its melting temperature then crystallizing particles (spherulites) appear at random in the liquid and grow with time. Notice impinging (overlapping) particles and the gradual exclusion of liquid volume for the appearance of new nuclei. The release rate of latent heat of crystallization must be less than the rate of heat removal from the polymer for the growth to continue.

Y. Imanishi (Ed.)
Progress in Pacific Polymer Science 2
© Springer-Verlag Berlin Heidelberg 1992

$$X(t,T) = \frac{mass\ of\ crystallized\ particels}{total\ mass\ of\ polymer} = \frac{\rho_c \sum [V(t)]_i}{\rho_l V_o}, \quad for\ i = 0\ to\ n$$

$$at\ t = 0 \qquad \sum [V]_i = 0, \qquad\qquad \therefore\ X = 0$$

$$at\ t = \infty \qquad \sum [V]_i = V_o \frac{\rho_l}{\rho_c}, \qquad \therefore\ X = 1$$

(1)

In equations (1) ρ_c and ρ_l are the crystalline and liquid densities respectively and n is the number of particles. In general, and in contrast to metals, the crystalline density of polymers depends on the temperature of crystallization, and may change with time and distance along the spherulite radius during crystallization (7). It is, therefore, necessary to distinguish between the degree of crystallisation, as defined above, and the degree of crystallinity, which is the more frequently measured quantity, and which may not have a one-to-one correspondence with the degree of crystallization.

Isothermal Kinetics

An isothermal crystallization is carried out by cooling a polymer liquid from above its melting temperature to the crystallization temperature, T_c, and holding it at that temperature until crystallization is completed. The study of isothermal crystallization is usually carried out under the following four assumptions:

(i) that the rate of change of temperature is slow compared to the thermal diffusion time through the sample; $dT/dt << DT/L^2$, where D is thermal diffusivity and L is sample size,

(ii) that the release rate of latent heat of crystallization is $<<$ than the heat transfer rate.

(iii) that the ratio of ρ_c/ρ_l is constant and independent of time and temperature.

(iv) that the process of crystallization, carried out in a finite volume, is described by the following differential rate equation:

$$dX(t,T)/dt = f(X,T) = k(T)g(X)t^{m-1}$$

(2)

where k(T) is a crystallization rate constant, g(X) is a function of X only, usually (1 - X), and $1 \le m \le 4$, is a constant related to the molecular mechanisms of nucleation and to the geometry of crystal growth. A mathematical solution of equation (1) is of the form shown below (1-4,8-10):

$$X(t,T) = 1 - \exp[-(t/\tau_X(T))^m]$$

(3)

where $\tau_X = (k/m)^{-m}$. It is generally observed that nucleation is a heterogeneous or "pseudo-homogeneous" process (11). Typically the number of nuclei initially increases with time to approach a constant value (saturation of sites), and to a first approximation can be described by the following equation (12):

$$n(t,T) = n_o(T)[1 - \exp(-t/\tau_n(T))]$$

(4)

where n_o and τ_n are constants characteristic of the polymer/nucleating agent system. Crystal growth occurs by the molecular mechanism of secondary nucleation (13). This process is described by the following equation, shown below:

$$\dot{G}(T) = G_o \exp[-\frac{E_D}{k_B T}] \exp[-\frac{\chi \, T_M}{k_B (T \, \Delta T)}] \tag{5}$$

where G_0 is a constant, E_D is activation energy for segmental diffusion, k_B is Boltzmann constant, and the parameter χ is related to crystal/liquid interfacial energy. In pure and monodispersed polymers the spherulite radius (at constant temperature) always increases in direct proportion to time. The volume of individual spherulite particle is, therefore, described by:

$$[V(t,T)]_i = \alpha_i [r(t,T)]_i^{\,m} = \alpha_i [\dot{G}(T)t]_i^{\,m} \tag{6}$$

where α is a geometrical constant, and $r(t,T)$ is the particle (spherulite) radius. The kinetics of nucleation and crystal growth, as described above, are contained in the value of the crystallization rate constant, $k(T)$, in equation (2).

Non-Isothermal Kinetics

Non-isothermal crystallization is usually carried out under continuous cooling conditions, with temperature and time uniquely related as follows:

$$T(t) = T_i - \xi(t) \tag{7}$$

where T_i is the initial temperature $(> T_M)$ and $\xi(t)$ is some function of time; in the simplest case of linear cooling $\xi(t) = Ct$, with $C = dT/dt$ as the cooling rate. It is assumed that the progress of crystallization under these conditions is also described by equation (2), with the crystallization rate constant, $k(T)$, now not constant, but varying with temperature and time through involvement in relationships (4) to (7).

Whereas studies of crystallization under isothermal conditions are invariably directed towards the understanding of the process through its analysis, non-isothermal crystallization studies are usually directed towards the development of a model for predicting the time and amount of the solidified polymer or degree of crystallinity (1,3,5). All non-isothermal transformation kinetic theories listed below rely implicitly on "additivity" principle. The analysis and testing of this principle has both practical and academic interest. In this paper we report the results of some studies of the additivity rule on a particular grade of isotactic polystyrene. During crystallization the temperature is varied in simple steps, and the time to achieve a certain amount of crystallization is measured and compared against that calculated from isothermal experiments. However the salient features of the non-isothermal theories are given first.

Transformation kinetics according to Nakamura and Ziabicki: Nakamura (3) extended Avrami theory to non-isothermal transformations and proposed the following equation:

$$X(t,T) = 1 - \exp\left\{ -\left[\int_0^t K(T(\tau))\, d\tau \right]^m \right\}$$

(8)

The earlier method, derived by Ziabicki (1), differs from the above only in that it assumed $m = 1$ in equation (2) and represented the integral by an approximated value of "kinetic crystallizability".

Transformation kinetics according to Ozawa: Ozawa (2) proposed a method to predict crystallization of polymers under constant cooling or heating rate by modifying the Avrami equation:

$$X(T) = 1 - \exp\left[-\frac{K'(T)}{C^m} \right]$$

(9)

where $K'(T)$ is the so-called cooling (or heating) function, C is the cooling (or heating) rate, and m is the Avrami exponent. The equation provides a simple way of predicting non-isothermal transformation kinetics for a known cooling rate from an experimentally derived master curve of $K'(T)$ vs T and the exponent m.

Integral method of transformation kinetics: Recently published results (14) for crystallization of a number of polymers under constant cooling rate which are well described by an equation of the following form:

$$X(T) = \frac{A}{C^m} \int_{T_i}^T n(\theta)\, \tilde{R}(\theta)\, \dot{G}(\theta)\, d\theta$$

$$\tilde{R}(\theta) = \int_{T_i}^T \dot{G}(\theta)\, d\theta$$

(10)

This approach relies on the availability of the experimental data for the number of nuclei, n(T), and crystal growth rate, G(T). These data are fundamental material properties, and can be measured independently of the measurement of crystallization rate.

ADDITIVITY PRINCIPLE

Additivity Rule According to Avrami and Christian

The concept of additivity was originally proposed to predict the incubation period required during continuous cooling conditions. Avrami (8) described the assumptions for additivity under so-called "isokinetic" conditions which require that the ratio, G(T)/N(T) is a constant. Later Cahn proposed a less restrictive regime, requiring that G(T) is a function of instantaneous temperature only, and that the number of nuclei is constant through early site saturation (15).

Christian (16) has demonstrated that additivity is best illustrated by considering the total time required to reach a specific amount of transformation, X_a, obtained by summing the fractions of time taken to reach this stage isothermally, until the sum reaches a value of one. This is equivalent to saying that an arbitrary cooling curve can be adequately approximated by a series of isothermal steps and that the sum of fractions transformed at each step adds up to X_a when:

$$\sum_{i=1}^{p} \frac{\Delta t_i}{[t_a(T)]_i} = 1 \tag{11}$$

where p is the number of isothermal steps, Δt_i is the time period of crystallization at i'th step, t_a is time of isothermal crystallization at i'th step required to reach X_a fraction of transformation. Then the sum, $\Sigma \Delta t_i$, is the time to achieve X_a fraction of transformation under continuous cooling conditions. In order to test the additivity rule one must carry out the following measurements:

(i) X versus time under isothermal conditions, over a range of temperatures, so that the values of X_a and t_a can be selected.

(ii) choose cooling rates, and approximate cooling curves by a series of isothermal steps.

(iii) carry out measurements in accordance with the schemes devised in (ii) above.

(iv) carry out calculations according to equation (12) .

Experimental Testing of Additivity

Material: The polymer studied was isotactic polystyrene (iPS) obtained from Polymer Laboratories, UK., of molecular weight, $M_W = 1570 \times 10^3$, $M_N = 245 \times 10^3$, and $T_M = 240°C$ and $T_g = 100°C$.

Method: Crystallization studies were carried out on an optical microscope, (Carl Zeiss), and a hot stage (Mettler FP-82). The temperature on the stage can be controlled to within 0.1°C and maximum cooling (or heating) rates of 20°C/min can be achieved. Glass slides and cover slips were cleaned in chromic acid and distiled water in ultrasonic bath. Polymer samples were placed between the glass slide and cover slip and pre-treated by melting to 280°C for 20 s. For crystallization studies the initial temperature, T_i, was set at 255 °C.

The thickness of the polymer film was approximately 10 μm, and the final size of spherulites was of the order of 50 μm. This corresponds to a disk-like two-dimensional growth geometry. Photographs were taken at intervals during the crystallization. The micrographs were analysed by digital image processing method to derive the quantities X = area covered by spherulites / total area, G = dr /dt, and n = number of spherulites per unit area as a function of time and temperature.

Results: The experimental results are presented in Figure 2 to 6 and Table 1. In view of the limited space the description of the results is confined to the captions only.

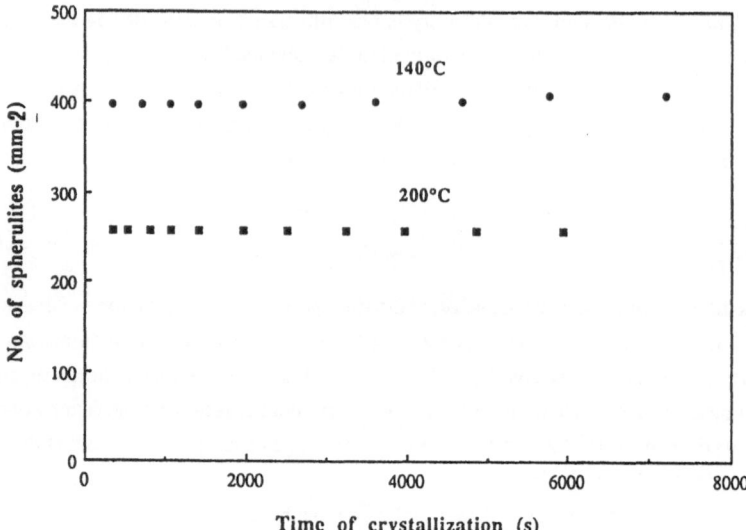

Figure 2. Variation of the number of visible spherulites as a function of the time of crystallization shown for two selected temperatures. Notice that the data can be fitted to equation (4) with $n_o(200°C) \approx 260$, $n_o(140°C) \approx 400$, and the time constant $\tau_n \approx 10$ s.

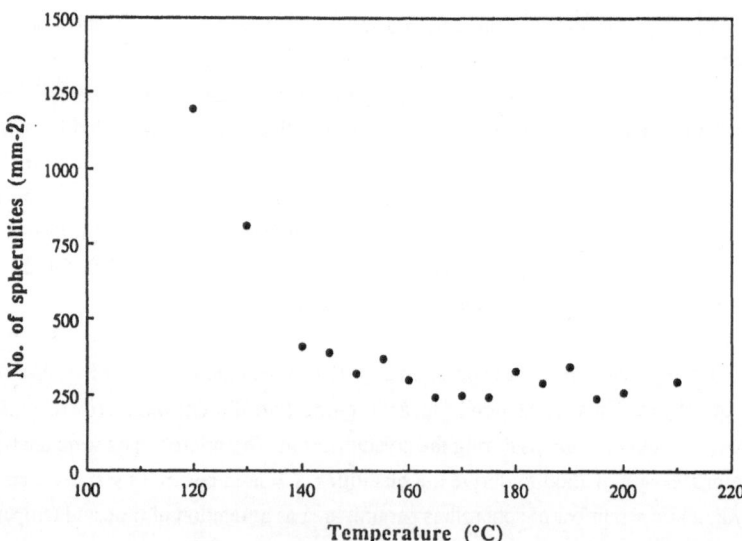

Figure 3. Variation of the final number of spherulites (the constant $n_o(T)$ in equation (4)) as a function of temperature of crystallization.

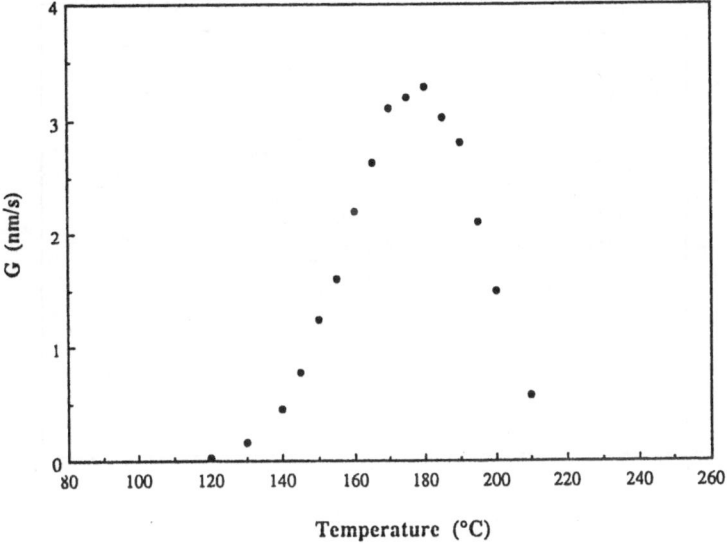

Figure 4. Variation of spherulite radius growth rate with temperature. Notice that the data can be fitted to equation (5) using the appropriate characteristic constants.

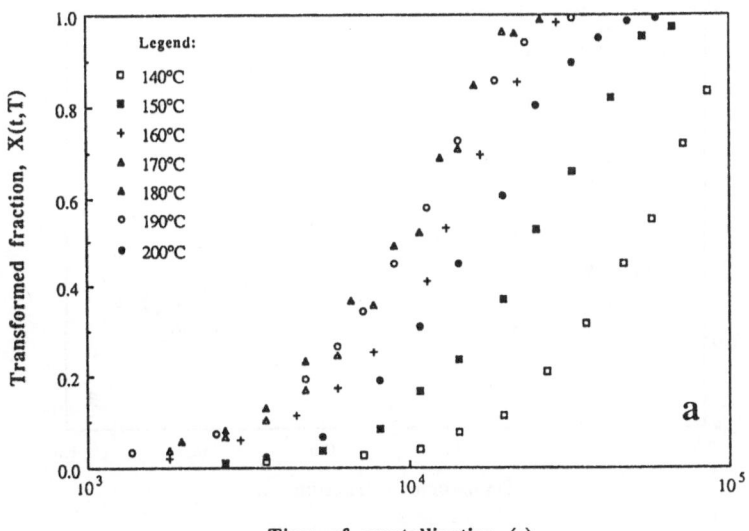

Figure 5 (a). Dependence of the crystallized fraction, X(t,T), on time of crystallization at a range of temperatures indicated on the diagram.

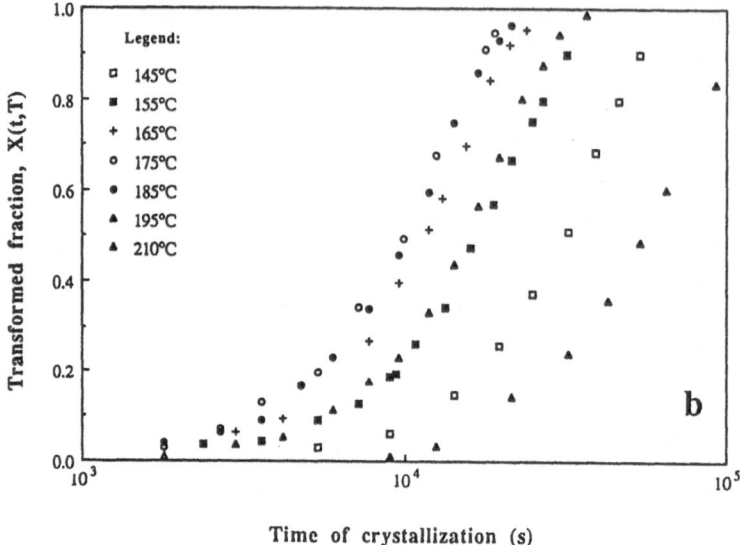

Figure 5 (b). Dependence of the crystallized fraction, X(t,T), on time of crystallization at a range of temperatures indicated on the diagram.

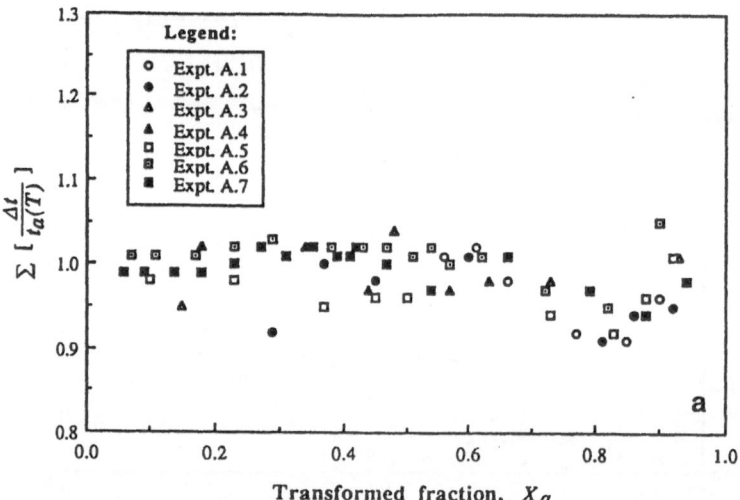

Figure 6. Results of calculations using equation (12) and experimental data from a number of experiments. These results show that the additivity rule is obeyed in iPS over the whole range of crystallized fractions.

Table 1. Results of selected seven non-isothermal experiments conducted to test the additivity rule for iPS crystallized from melt to $X_a = 0.5$. In A.1 cooling was approximated by two isothermal step experiment, whereas in A.7 cooling was approximated by 13 isothermal step experiment.

	Experiment						
	A.1	A.2	A.3	A.4	A.5	A.6	A.7
T_{c1}	200°C	200°C	200°C	200°C	200°C	200°C	200°C
t_1	3420s	3600s	3600s	3600s	3600s	3600s	3720s
T_{c2}	170°C	190°C	180°C	195°C	190°C	195°C	195°C
t_2	8460s	3600s	1920s	4500s	1800s	1500s	1200s
T_{c3}		180°C	160°C	190°C	180°C	190°C	190°C
t_3		4500s	7380s	3000s	1800s	900s	600s
T_{c4}			140°C	185°C	170°C	185°C	185°C
t_4			960s	1800s	1800s	1200s	900s
T_{c5}					160°C	180°C	180°C
t_5					1800s	900s	600s
T_{c6}					150°C	175°C	175°C
t_6					1800s	900s	900s
T_{c7}					140°C	170°C	170°C
t_7					1800s	1200s	600s
T_{c8}						165°C	165°C
t_8						900s	900s
T_{c9}						160°C	160°C
t_9						960s	600s
T_{c10}						155°C	155°C
t_{10}						600s	900s
T_{c11}							150°C
t_{11}							600s
T_{c12}							145°C
t_{12}							1200s
T_{c13}							140°C
t_{13}							6000s
$\sum \frac{\Delta t_i}{t_a(T)}$	1.02	1.01	1.00	0.96	0.96	1.00	1.01

DISCUSSION

Previous workers have made the observation that individual isotherms (shown in Figures 5 (a) and (b)) can be brought into coincidence simply by shifting each curve by a distance along the log time axis. The possibility of superimposing transformation isotherms for a variety of combinations of T_c and ΔT_c suggests that the additivity rule should be applicable when simulating a continuous cooling path with a series of decreasing isothermal crystallizations. Mathematically, the condition for additivity is that the function, $f(X,T)$, in equation (2) is a separable function of its arguments. A consequence of this is that all the isothermal transformation curves should fall on one master curve if plotted using a normalized time, $t/t_X(T)$, where $t_X(T)$ denotes the time taken to reach some specified fraction of X. The results shown in Figures 5(a) and (b) satisfy this condition.

The results presented in Table 1 and Figure 6 show clearly that additivity rule applies to iPS for non-isothermal crystallization within the limits of cooling rates between 0 to 10°C/min. In a previous publication we have shown that similar results were obtained for polyethylene (14). However separate studies on polyethylene according to Nakamura method (17) and Ozawa method (18) showed strong deviations from the theoretical predictions. This was interpreted in terms of a pronounced secondary crystallization, which is typical of polyethylene (19). In all polymers, for a given degree of crystallization, say $X = 0.5$, two experiments may result in different crystallinity, depending not only on the thermal history and pressure, but also on such other molecular parameters as nucleating conditions, molecular weight, presence of impurities or solvents, recrystallization, and the like. Therefore the kinetic measurements of the degree of crystallization and the degree of crystallinity rarely, if ever, show a one to one correspondence. In view of this the apparent disagreement between our and previous results can be reconciled by the fact that the optical microscopy method used here measures the true fraction of the liquid to solid phase transformation and not the degree of crystallinity as measured by DSC, X-ray diffraction and other methods. This is related to the third assumption listed on page 2. This assumption is not necessary, and not made in this work because the degree of crystallization, X, is given by:

$$X = \frac{\rho_c A_c h_c}{\rho_l A_l h_l} \tag{12}$$

where A_l is the total area of view under the microscope and h_l is the original thickness of the polymer film (when liquid). A_c is the area covered by spherulites at time t and h_c is the film thickness at time t. Applying the principle of conservation of mass, and noting that the film cannot shrink sideways, the following relationship must hold: $\rho_c/\rho_l = h_l/h_c$. Consequently $X = A_c/A_l$.

From the results shown in Figure 2 and 3 one can infer that the Avrami isokinetic condition is not satisfied, i.e., $N/G \neq$ constant. However, over the temperature range from 200 to 140 °C the number of nuclei is approximately constant, and Figure 2 shows that nuclei site saturation occurs very early in the process. This is in accordance with the conditions set out by Cahn (15). It should be noted that in the early days of the science of polymer crystallization it was assumed that polymers nucleate by a homogeneous, thermally activated process (20). However it is now generally accepted that, based on an overwhelming experimental evidence, most polymers show heterogeneous nucleation (11), with all potential nuclei developing into spherulites in the early stages of crystallization (X < 0.1). Another point to be made here, related to the specific nature of nucleation, is that additivity rule will not work in experiments in which the polymer is first quenched to below its glass transition temperature and then heated in a series of isothermal steps.

Finally the experimental conditions must be examined. Taking the density of iPS as 1.1 g/cm^3, heat capacity as 1.2 J/g.K and thermal conductivity as 0.1 W/m.K, we calculate the value of thermal diffusivity, $D = 0.076 \times 10^{-6}$ m^2/s. Thermal diffusivity of glass is an order of magnitude higher. Since our films were of the order of 10 μm thick, and heat conduction is primarily perpendicular to the glass slide, then the condition $dT/dt \ll DT/L^2$ is satisfied. Furthermore, since the specific heat of fusion is of the order of 0.1 kJ/cm^3 and the maximum crystallization rate at 180 °C is approximately 10^{-4} s^{-1}, then the second assumption from page 2 is also satisfied.

REFERENCES

1   Ziabicki A, (1967) Appl Polym Symposia 6:1

2   Ozawa T, (1971) Polym 12:150

3   Nakamura K, Watanabe T, Katayama K, Amano T, (1972) J Appl Polym Sci 16:1077

4   Malkin AY, Beghishev VP, Keapin IA, (1983) Polym 24:81

5   Friedl CF, McCaffrey NJ, (1991) SPE Antec 38:252

6   Chew S, Griffiths JR, Stachurski ZH, (1989) Polym 30:874

7   Calvert PD, Ryan TG, (1978) Polym 19:611

8   Avrami M, (1940) J Chem Phys 8:212

9   Evans UR, (1945) Trans Faraday Soc 41:365

10 Galenski A, Piorkowska E, (1983) Colloid & Polym Sci 261:1

11 Binsbergen FL, (1977) J Polym Sci: Polym Symposium 59:11

12 Aggarwal SL, Marker L, Kollar WL, Geroch R, (1966) J Polym Sci 4:715

13 Lauritzen JJJr, Hoffman JD, (1973) J Appl Phys 44:4340

14 Chew S, Stachurski ZH, Griffiths JR, Rose LFR, (1991) POLYMER 91 Conf Proc 272

15 Cahn JW, (1956) Acta Metall 4:572

16 Christian JW, (1965) Theory of Transformations in Metals and Alloys, Pergamon, New York

17 Kamal MR, Chu E, (1983) Polym Eng Sci 23:27

18 Eder M, Wlochowicz A, (1983) Polym 17:471

19 Hay JN, (1979) Brit Polym J 11:137

20 Mandelkern L, Quinn FA, Flory PJ, (1954) J Appl Phys 25:830

Use of Two-Dimensional NMR for Polymer Characterization

Kunio Hikichi

Department of Polymer Science, Faculty of Science,
Hokkaido University, Sapporo, 060 Japan

1. INTRODUCTION

It has been well recognized that the NMR spectroscopy is a quite powerful tool for characterizing synthetic polymers(1-3). The most important point of the NMR method for characterizing polymers is to assign the NMR spectra. Until recently, this has been done empirically by using model compounds of simple molecules and by assuming statistical model of polymerization mechanism.

Normally, proton and carbon-13 NMR spectra are widely used for analytical purpose. The proton NMR spectra have the advantage of high sensitivity, but have the disadvantage of low-resolution. The spectra are observed in a limited range of chemical shift of about 10ppm. In proton spectra, the two-bond and the three-bond J-coupling interactions occur. These J-coupling interactions make spectra too complex to be analyzed. The three-bond coupling, however, helps us to establish the connectivity of chemical bonds.

On the other hand, carbon-13 spectra suffer from the low sensitivity because of the low natural abundance, but enjoy the high resolution. The spectra are dispersed over a range of about 250ppm; the spectra are simple because of lack of the coupling interaction if one irradiates all proton resonances to decouple the proton--carbon-13 J-coupling. Carbon-13 spectra are, therefore, widely used for characterizing polymers.

In the past decade, the two-dimensional NMR method has been progressively developed(4,5). This method has been successfully used for the assignmental problem. In this paper, we would like to show the usefulness of the two-dimensional NMR method for characterizing synthetic polymers. We focus our attention to the characterization of the stereo-configuration of vinyl polymer, in particular, poly(vinyl alcohol)(PVA)(6,7).

2. EXPERIMENTAL

Sample

The PVA sample was a commercial product obtained from Nippon Gosei Kagaku Co. Ltd. The molecular weight is about 4,400.

NMR Experiments

The NMR experiments were performed on a JEOL JNM-GX500 spectrometer operating a proton resonance frequenciy of 500MHz. The NMR spectra were observed in D2O solution at 353K. The concentrations of samples for proton and carbon-13 measurements were 15%(w/v) and 40%(w/v), respectively. Various types of 2D experiments including the proto J-resolved spectroscopy, proton F1-axis broad-band decoupled correlation spectroscopy(BDCOSY), the F1-axis(proton) broad-band decoupled carbon-13-proton chemical shift correlation spectroscopy(CHCOSY), heteronuclear multiple quantum coherence spectroscopy(HMQC), and INADEQUATE 2D experiment were performed.

Y. Imanishi (Ed.)
Progress in Pacific Polymer Science 2
© Springer-Verlag Berlin Heidelberg 1992

3. RESULTS AND DISCUSSION

Figure 1 shows the resolution-enhanced 500MHz proton spectrum of PVA in D2O. The spectrum consists of two parts, the low-field part comes from methine protons and high-field part from methylene protons. The spectrum is very complicated due to the dispersion of chemical shifts of different stereo-configurations as well as the splitting by J-coupling between different protons. It is quite difficult to analyze this spectrum.

Figure 2 shows the proton broad-band decoupled carbon-13 spectrum of PVA in D2O. The spectrum is dispersed over 30ppm wide. The spectrum consists of just different chemical shifts but does not contain the J-coupling. The spectrum, therefore, just reflects different stereo-configuration of PVA. This is the reason why the carbon-13 spectra are widely used for polymer characterization. It can be seen that the spectrum is resolved at the pentad-hexad level. What we have to do is to assign this carbon-13 spectrum in an absolute manner using various 2D-NMR techniques.

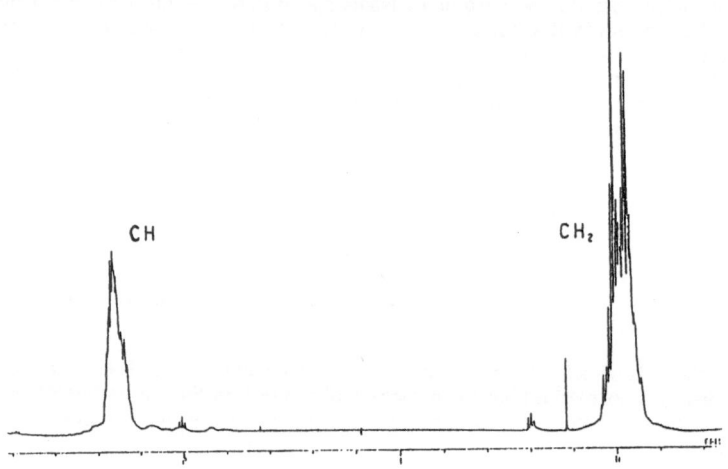

Figure 1. Resolution-enhanced 500MHz proton NMR spectrum of PVA in D2O at 353K.

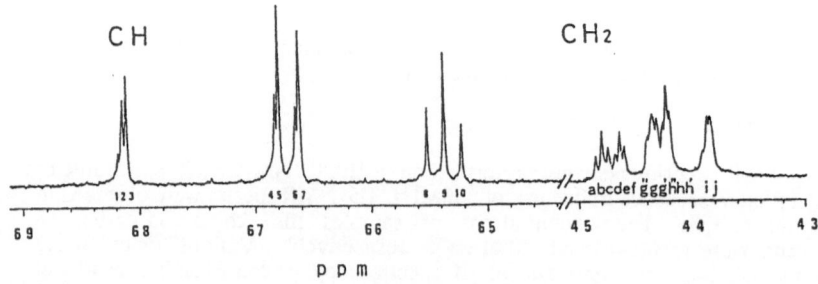

p p m

Figure 2. 125MHz proton noise-decoupled carbon-13 NMR spectrum of PVA in D2O at 353K.

Firstly, we assign proton spectra. In vinyl polymers, we have two different types of methylene protons, one is non-equivalent proton and the other is equivalent.

For example, two protons of a methylene in the mmm configuration are in different environments. These two protons are non-equivalent, show different chemical shifts, and split due to the geminal J-coupling between the two protons. There are also vicinal couplings between protons three bonds apart, therefore the spectra are very complex. The vicinal coupling helps us to connect a methylene and the neighboring methine.

On the other hand, two protons of a methylene in the rrr or the mrm configurations encounter the same environment. The geminal coupling does not, therefore, splits the resonance; the spectrum is simple.

Figure 3 shows the J-resolved 2D spectrum of PVA. The horizontal axis represents the normal chemical-shift axis and the vertical axis just the J-splitting. In the upper part the normal resolution-enhanced spectrum is shown, and in the lower part the projection along the line of slope 1. The projected spectrum contains only chemical shift free from the coupling interaction. Thus, the projected spectrum reflects different stereo-configurations. It is found that three peaks are in the methine region and 9 peaks in the methylene region. The three methine peaks and nine methylene peaks suggest that methine and methylene proton resonances of PVA are resolved at the triad-tetrad level. Peaks 3, 7, and 8 do not show the large splitting due to the geminal coupling(-15Hz). Two of the three are, therefore, equivalent methylenes, i.e., rrr and mrm methylenes.

At the tetrad level, we expect 10 different methylene resonances, becasue the mmm methylene protons are non-equivalent, and give two resonances; mmr methylene

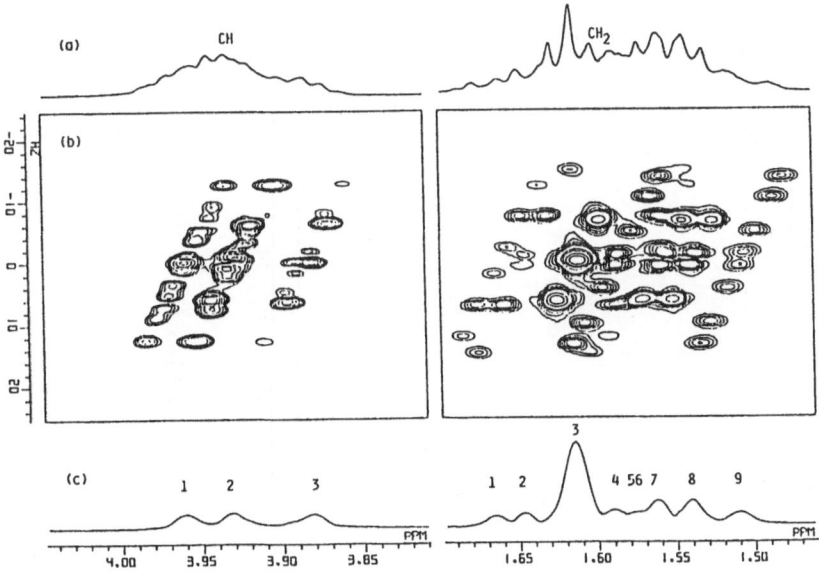

Figure 3. Proton J-resolved 2D NMR spectrum of PVA.

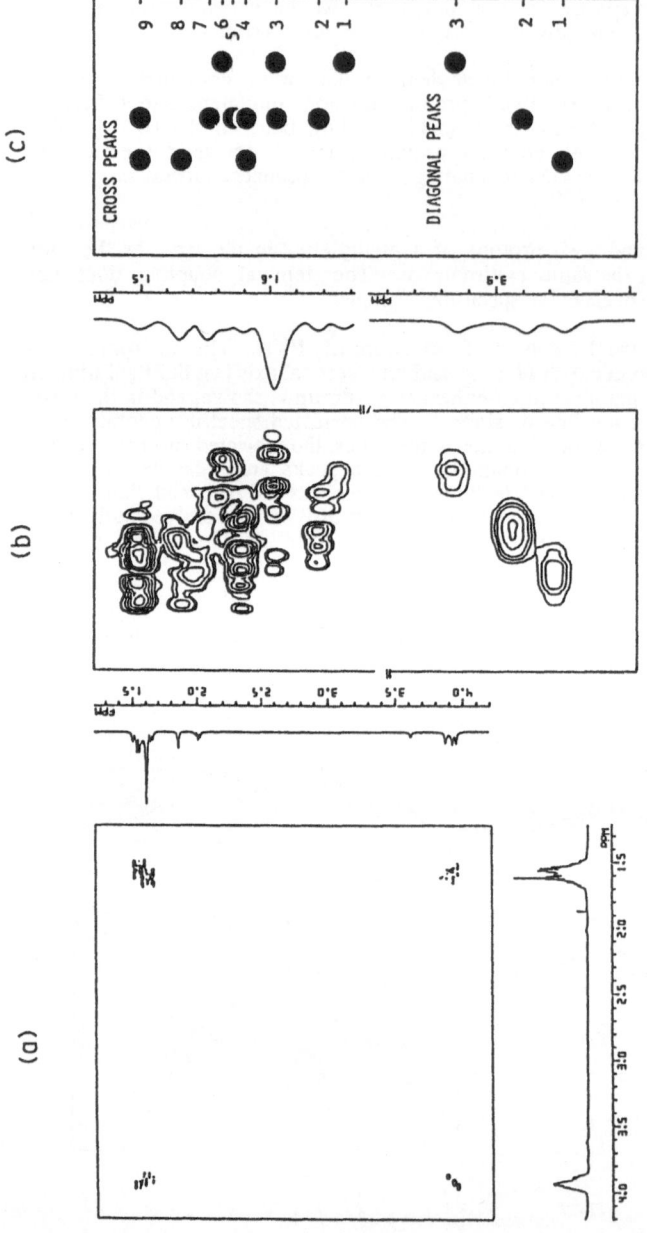

Figure 4. F1-axis broad-band decoupled COSY spectrum of PVA.

protons are also non-equivalent, and give two resonances; mrm methylene protons are equivalent, and give just one resonance, and so on. We expect total 10 resonances. We observed here, however, only 9 resonances. One of non-equivalent methylenes is accidentally equivalent.

There is a rule for the connectivity. For example, a methylene neighboring to a mm methine is mmm or mnιr methylene. A methine in the mr configuration can be conncected to mmr, mrm, mrr, and rmr methylenes, and a methine in the rr configuration conncected to mrr and rrr methylenes.

On the left-hand side of Figure 4, the BDCOSY spectrum of PVA is shown. On the diagonal line the normal 1D spectrum appears, and in the off-diagonal region cross peaks between methine and methylenes appear. In the middle of Figure 4, the enlarged off-diagonal region is shown where the cross peaks between methine and methylenes and the diagonal peaks of methines appear. On the right-hand side of Figure 4, the schematic diagram of the middle spectrum is shown.

From the connectivity rule, we have the following rules for cross-peaks of COSY spectra. Firstly, a methine connecting maximum 3 methylene peaks is rr methine; secondly, a methine connecting maximum 4 peaks is mm methine; and thirdly, a methine connecting maximum 7 methylene peaks is mr methine. Using these rules, we can easily assign mr methine, because this methine is connected to 6 methylene peaks. We can not, however, determine mm and rr methine yet.

There are additional rules for assigning methylenes. Firstly, a methylene connecting only one methine peak is either one of mmm, rmr, mrm, and rrr; secondly, methylenes showing no cross-peak in the methylene-methylene corss-peak region is mrm or rrr; and thirdly, a methylene connecting two methines is mmr or mrr.

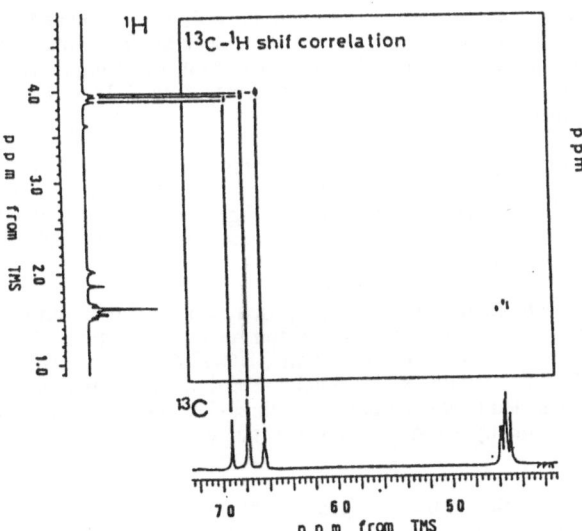

Figure 5. Proton broad-band decoupled CHCOSY of PVA.

Table I. Assignments of the ^{1}H spectra of PVA

	Peak No.	Configuration	Chemical shift/ppm
CH	1	*rr*	3.957
	2	*mr*	3.930
	3	*mm*	3.879
CH$_2$	1	*mmmm*	1.660
	2	*rmr*	1.645
	3	*mmr*	1.614
	4	*mrr*	1.586
	5	*rmr*	1.574
	6	*mmm*	1.566
	7	*mrm*	1.558
	8	*rrr*	1.539
	9	*mrr*	1.508

As mentioned before, two of the three peaks 3, 7, and 8 are equivalent methylene, rrr and mrm, and rrr methylene is connected to only rr methine and mrm is connected to only mr methine. Because the central methine peak has been found to be mr methine, peak 8 methylene can be assigned to rrr and 7 to mrm. Using all rules we can assign proton spectra of PVA in non-empirical manner. Table I lists the results of assignment of proton spectra of PVA in D2O.

The next step is to assign carbon-13 spectra using the results of proton assignment. This can be done by CHCOSY. This method correlates carbon resonance with proton which is directly attached to the carbon. Figure 5 shows the CHCOSY spectrum of PVA. Since we have assigned the proton spectrum, we can easily assign carbon-13 spectra at the triad-tetra level.

More recently, HMQC method has been proposed to be useful for correlating proton--carbon-13 spectra. This method observe proton resonance instead of carbon resonance. HMQC is, therefore, more sensitive than CHCOSY. Figure 6 shows the HMQC spectrum of PVA. This method is extremely useful to detect methyl carbon, because methyl carbon is detected by three protons, and three-fold sensitivity is obtained. In this spectrum we can see a methyl signal of residual acetate group.

As mentioned before, the carbon-13 spectrum of PVA is resolved at the pentad-hexad level. In order to assign the carbon-13 spectrum at this level, we apply INADEQUATE 2D method, which detects the carbon-13--carbon-13 connectivity by the carbon-13 J-coupling. This is extremely insensitive, because the probability of the carbon-13--carbon-13 connectivity in natural abundance is only 0.01%. Neverthless, if

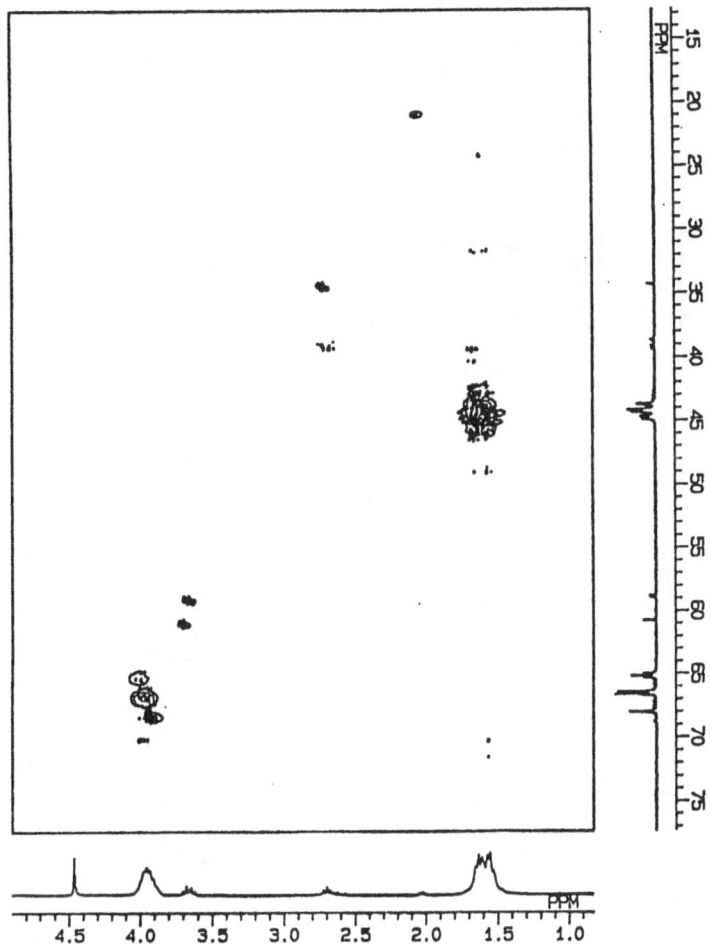

Figure 6. HMQC spectrum of PVA.

this method is successful, it is quite powerful. Figure 7 shows the INADEQUATE spectrum of PVA. The circled regions are cross peaks of ππ centered methine carbons and methylene carbons.

Figure 8 shows the schematic diagram of the connectivity between ππ centered methine peaks 8, 9, and 10 and methylene peaks a, b, c, g, and g". Figure 9 shows the theoretical connectivity between rr-centered pentad methine and hexad methylene. If we compare this with the experimental results, we can assign carbon-13 spectrum at the pentad-hexad level. Table II lists the results of carbon-13 resonance assignment of PVA.

Using this assignment, we calculate the fraction of each configuration. Table III shows the fractional area of methine peaks. The best fits of Bernoulli trial, the 1st-Markov trial, and the 2nd Markov trial are also listed. The data at the triad level are also shown here. The data at the traid level are all in agreement with any statistics. However, the data at more higher level are not agreement with any statistics. The

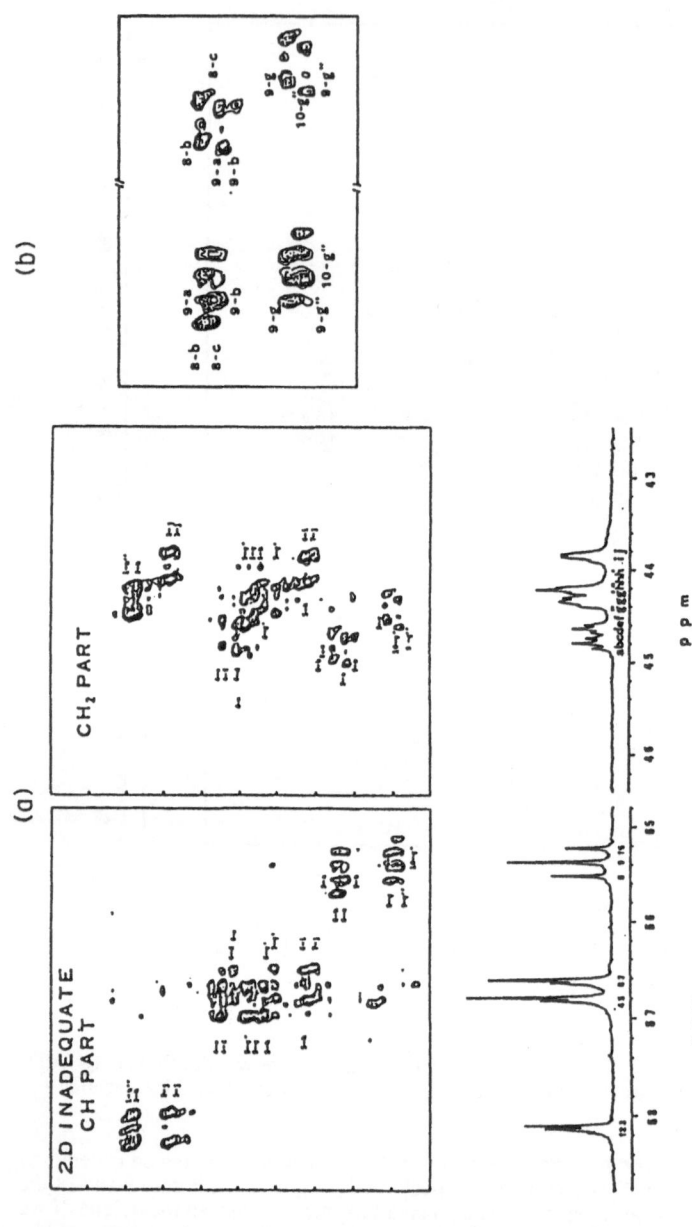

Figure 7. INADEQUATE 2D spectrum of PVA.

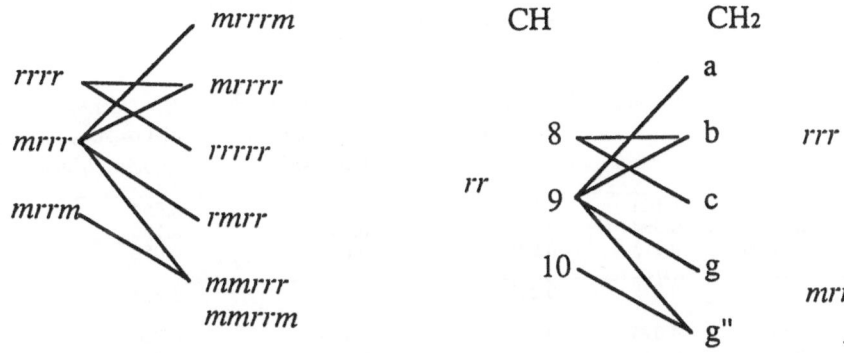

Figure 8. Observed connectivity of CH and CH₂.

Figure 9. Theoretical connectivity of CH and CH₂.

Table II. Assignments of the carbon-13 spectra of PVA in D₂O

	Peak No.	Configuration	Chemical shift/ppm
CH	1	*rmmr*	68.18
	2	*mmmr*	68.14
	3	*mmmm*	68.10
	4	*rmrr*	66.83
	5	*mmrr*	66.79
	6	*rmrm*	66.65
	7	*mmrm*	66.61
	8	*rrrr*	65.53
	9	*mrrr*	65.37
	10	*mrrm*	65.22
CH₂	a	*mrrrm*	44.85
	b	*mrrrr*	44.79
	c	*rrrrr*	44.74
	d	*mrmrm*	44.68
	e	*mrmrr*	44.64
	f	*rrmrr*	44.59
	g″,g,g′	*mrr*	44.38, 44.34, 44.30
	h″,h,h′	*mmr*	44.24, 44.22, 44.19
	i, j	*mmm,mrm*	43.86

Table III. Fractional areas of methine peaks of PVA

Peak No.	pentad	fraction	Bernoulli trail Pm=0.469	1st-order Markov trial Pm/r=.563, Pr/m=.498	2nd- order Markov trial α=.496, β=.442, γ=0.441, δ=.446
1	rmmr	0.021	0.062	0.047	0.054
2	mmmr	0.093	0.110	0.104	0.106
3	mmmm	0.115	0.048	0.059	0.052
	(mm)	0.229	0.220	0.210	0.212
4	rmrr	0.075	0.140	0.100	0.054
5	mmrr	0.174	0.124	0.128	0.124
6	rmrm	0.070	0.124	0.164	0.131
7	mmrm	0.160	0.110	0.127	0.093
	(mr)	0.479	0.498	0.519	0.402
8	rrr	0.082	0.080	0.074	0.086
9	mrrr	0.150	0.140	0.147	0.150
10	mrrm	0.060	0.062	0.073	0.065
	(rr)	0.292	0.282	0.284	0.301

failure of the applicability of the statistics prevents us to use the empirical statistical method for spectral assignment. We do not know why the statistics can not apply to PVA.

Acknowledgments: This work was supported by the Ministry of Education, Scinece, and Culture of Japan. The author wishes to thank Mr. Yasuda for assistance.

References:

1 Bovey F A (1972) High Resolution NMR of Macromolecules, Academic Press, New York
2 Randall J C (1977) Polymer Sequence Determination. Carbon- 13 NMR Method, Academic Press, New York
3 Bovey F A (1988) Nuclear Magnetic Resonance Spectroscopy, Academic Press, New York
4 Bax A (1982) Two-dimensional Nuclear Magnetic Resonancein Liquids, Delft Univeristy Press, Delft
5 Ernst R R, Bodenhausen G, Wokaun A (1987) Principles of Nuclear Magnetic Resonance in One and Two Dimensions, Clarendon Press, Oxford
6 Gippert G P, Brown L R (1984) Polym Bull 11:585
7 Hikichi K, Yasuda M (1987) Polym J 19:1003

Self Assembly of Polymer Blends at Phase Transition-Morphology Control by Pinning of Domain Growth

Takeji Hashimoto

Department of Polymer Chemistry, Faculty of Engineering,
Kyoto University, Kyoto 606, Japan

Abstract : Spinodal decomposition of polymer blends with critical and off-critical compositions were studied with a particular emphasis on their morphology control. This control utilizes processes which lead to pinning down of the domain growth in the mixtures. The pinning processes discussed are spontaneous pinning, physical pinning due to crystallization or vitrification, and chemical pinning. The spontaneous pinning is applicable only to the off-critical blends but the latter two pinnings are applicable to both off-critical and critical mixtures.

I. INTRODUCTION

We quench polymer blends with critical or off-critical compositions inside the spinodal phase boundary, and we investigate their self-assembling (ordering) processes, patterns (morphology), and dynamics via spinodal decomposition (SD) (1, 2). Basic information obtained in the studies of self-assembly will eventually lead us to control the patterns, functionalities and properties of polymer blends. As a methodology for controlling the patterns, we shall discuss various processes which pin down the pattern growth.

Time-evolution of the patterns via SD may be classified into three stages at least: (a) early stage, (b) intermediate stage and (c) late stage, as shown in Fig. 1 (3-5) which gives the spatial concentration variation $\phi_A(\vec{r};t)$ of a constituent polymer A in A/B binary mixtures at time t. In the early stage SD, the wavelength Λ of the dominant mode of the fluctuation stays nearly constant with time, but the concentration (or composition) difference $\Delta\phi(t)$ of polymer in A-rich and B-rich domains increases with t (Fig. 1 (a)). In the intermediate stage, both Λ and $\Delta\phi(t)$ grow with time (Fig. 1(b)). The growth of Λ reflects the nonlinear nature of the time-evolution equation of the pattern and hence the mode-mode coupling effect (6). The growth of $\Delta\phi(t)$ slows down in this stage, slower than the exponential growth in the

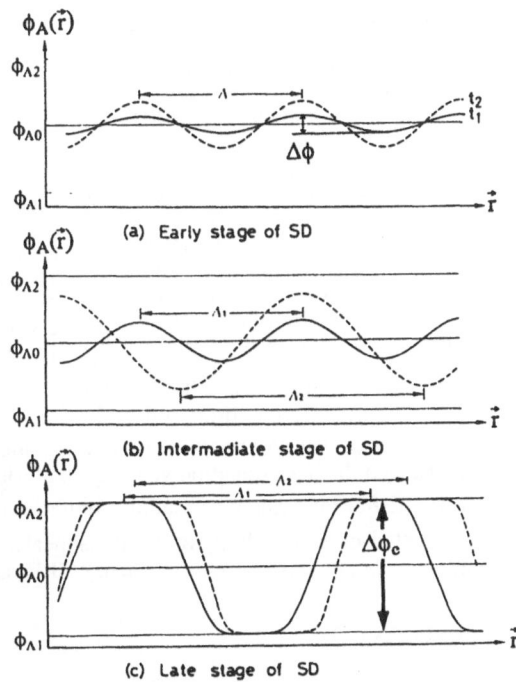

(a) Early stage of SD

(b) Intermadiate stage of SD

(c) Late stage of SD

Fig.1. Sketches of time-evolution of the composition (concentration) fluctuations in the (a) early, (b) intermediate, and (c) late stages of spinodal decomposition. ϕ_{A1} and ϕ_{A2}, respectively, are the equilibrium compositions of B-rich and A-rich domains at the phase separation temperature T_x. From T. Hashimoto, M. Itakura and H. Hasegawa (1986).

Y. Imanishi (Ed.)
Progress in Pacific Polymer Science 2
© Springer-Verlag Berlin Heidelberg 1992

early stage. In the late stage $\Delta\phi(t)$ reaches an equilibrium value $\Delta\phi_e$ (7) determined by the coexistence curve and temperature of the phase separation (Fig. 1(c)), $\Delta\phi_e = \phi_{A2} - \phi_{A1}$ in which ϕ_{A2} and ϕ_{A1} are, respectively, the equilibrium composition of A in A-rich and B-rich domains. In this stage the interface is well developed, and the wavelength Λ grows with time from Λ_1 to Λ_2. For various polymer mixtures (8,9), as well as for mixtures of small molecules (2), a global characteristic of the pattern growth (i.e., characteristic of the pattern growth at the length scale $r\approx$)$\Lambda(t)$) was found to scale with the time-dependent characteristic length parameter $\Lambda(t)$, so that

$$\phi_A(\vec{r}/\Lambda_1, t_1) = \phi_A(\vec{r}/\Lambda_2, t_2) \tag{1}$$

When this "dynamical scaling hypothesis" (10,11) is valid, the pattern grows with a dynamical self-similarity. The principle of dynamical scaling or dynamical self-similarity obviously breaks down when one more precisely focuses on time-evolution of the local structure, i.e., the structure at $r = t_I(T) \ll \Lambda(t)$ where $t_I(t)$ is the characteristic interfacial thickness (4,12). There is a time regime in which $\Lambda(t)$ increases but $t_I(t)$ decreases with t, which will lead to the four-stage model of SD (4,5).

Figure 2 shows a light micrograph demonstrating the dynamically self-similar growth of the pattern in the late stage SD. The patterns show a phase separation into anisotropic liquid domains composed of X-7G and isotropic liquid domains of poly(ethylene terephthalate) under crossed polarizers (13). As time elapses, the patterns grow as shown in the image change from (a) to (b). However when the magnification for the image (b) is lowered, the image changes as shown in part (c). The image (c) at 60 sec is seen to be statistically identical to the image (a) at 6 sec after the onset of the phase separation at 270°C. This observation demonstrates that only the

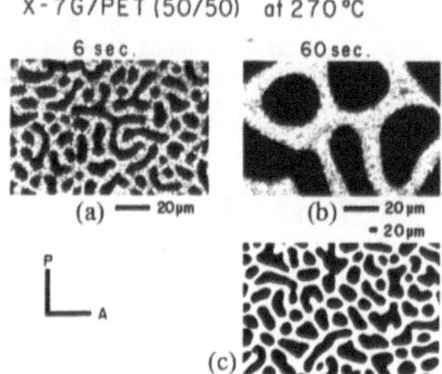

Fig. 2. Polarized light micrographs showing pattern growth with dynamical self-similarity. Polarizer and analyzer axes are set in the vertical and horizontal directions, respectively. The pattern (a) is obtained 6 s and the patterns (b) and (c) 60 s after onset of unmixing at 270°C for X-7G/PET (50/50 wt/wt) mixture. The patterns (b) and (c) are different only in magnification.

characteristic length scale Λ changes with t but the morphological characteristics of the pattern remain unchanged. In fact a quantitative image analysis showed that the contrast variation of the image $\rho(\vec{r},t)$ satisfies the dynamical scaling hypothesis of eq.1 when when $\phi_A(\vec{r},t)$ is replaced by $\rho(\vec{r},t)$. We shall discuss the validity of this dynamical scaling hypothesis for near critical and off-critical mixtures of poly(styrene-ran-butadiene) (SBR) and polybutadiene (PB).

II. EXPERIMENTALS

II-1. Specimens

The characteristics of the samples used are summarized in Table 1 where SBR, PB and PI designate, respectively, a random copolymer of poly(styrene-random-butadiene), polybutadiene, and polyisoprene. Binary mixtures of PP/EPR (14), X-7G/PET (15), both having (50/50 wt/wt composition), and PS/SB (16) (35/65 wt/wt) were occasionally used, where PP, EPR, X-7G, PET, PS and SB, designate, respectively, isotactic poly(propylene), a random copolymer of poly(ethylene-ran-propylene), X-7G® thermotropic liquid crystalline copolyester (14), poly(ethylene terephthalate), polystyrene, and poly(styrene-block-butadiene).

II-2. Film Preparation

Film specimens of SBR1/PB19, SBR1/PB55 and SBR2/PI55 were obtained by a solvent-cast method, using homogeneous solutions of the mixtures with toluene (17,18). Those with PP/EPR were prepared as follows (14). The mixtures were first dissolved in hot xylene. A dilute homogeneous solution containing ca.1 wt% of PP and EPR was poured into ice water. The precipitates were collected and cold-pressed into thin films of ca.50 to 100 μm thickness between two glass plates. Those with X-7G/PET were prepared from a dilute solution of ca. 1 wt% of X-7G and PET with orthochlorophenol, by casting the solution onto a glass plate and subsequently evaporating rapidly the solvent in a vacuum oven at 60°C (13,15). The thin glassy films of 1-10 μm thickness thus obtained were transparent, optically isotropic, and homogeneous under optical microscopic observation.

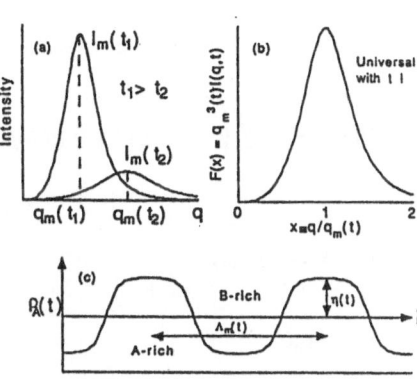

Fig. 3. (a) Time-change of the scattering profile during the later stage SD, (b) scaled structure factor $F(x)$ universal with time t, and (c) time-dependent spatial contrast variation $\rho_A(r)$.

II-3. Methods

We can analyze the SD process by scattering methods, particularly by light scattering method in this work. In this method the growth of the domains in the SD process is analyzed in terms of (a) their characteristic size $\Lambda_m(t)$ and (b) mean square fluctuations $<\eta^2(t)>$ of the scattering contrast, related to square of the composition difference $\Delta\phi^2(t)$ as shown in Fig.3 (part c). The characteristic parameters $\Lambda_m(t)$ and $<\eta^2(t)>$ which characterize the growing pattern are determined experimentally from the time evolution of the scattering profile, e.g., from the peak scattering vector $q_m(t)$ and the peak intensity $I_m(t)$ (Fig. 3(a)),

$$\Lambda_m(t) = 2\pi/q_m(t) \sim t^\alpha \tag{2}$$

$$I_m(t) = <\eta^2(t)>q_m^{-3}(t)\overline{F}(x=1) \sim t^\beta. \tag{3}$$

Table 1 Characterizations of Samples Uesd

Specimen	$M_w\times10^{-4}$	M_w/M_n	wt%PS[a]	wt%PP[b]
SBR1	11.8	1.18	20	-
SBR2	19.0	1.21	15	-
PB19	19.0	1.16	-	-
PI55	54.6	1.02	-	-
PP	23.5	4.1	-	100
EPR	14.9	2.5	-	27
PS	21.3	1.41	-	-
SB	8.5	1.7	40	-

a) wt% of styrene monomer , b) wt% of propylene monomer

The evolution of the parameters can be studied on the basis of power laws given by eqs. 2 and 3 in which α and β are the respective exponents characterizing the time-evolution of the pattern. In eq.3 $\overline{F}(x)$ is the scaled structure factor defined by

$$\overline{F}(x) = F(x) \bigg/ \int_0^\infty q^2 I(q) dq \sim F(x)/<\eta^2(t)> \tag{4}$$

with

$$F(x) \equiv q_m^3(t) I(x) \tag{5}$$

and

$$x \equiv q/q_m(t) \tag{6}$$

The scaled structure factor $\overline{F}(x)$ or $F(x)$ characterizes morphological aspects of the growing pattern such as those shown in Fig. 2, while the size of the pattern at t is characterized by $q_m(t)$ or $\Lambda_m(t)$ (eq.2). If the pattern grows with the dynamical self-similarity at a given T and $<\eta^2(t)>$ is invariant with t, $F(x)$ is universal with t as shown schematically in Fig.3(b). Obtaining this universal scaling function $F(x)$ confirms the validity of the dynamical scaling hypothesis. The dynamical scaling hypothesis also yields the following relationship between α and β,

$$\beta = 3\alpha \tag{7}$$

because in this criterion $<\eta^2(t)>$ and $\overline{F}(x=1)$ are time invariant in eq.3. Eq.7 provides another simplified way to confirm the dynamical scaling hypothesis. When the scaled structure factor $F(x)$ is to be discussed as a function of T, one has to take into account the T dependence of the equilibrium mean square fluctuation $<\eta^2>_e$.

For this purpose $\overline{F}(x)$ is more convenient than $F(x)$. The universal scaled structure factor with respect to both t and T was recently reported for the critical mixture of SBR/PB(12) and PI/PB(4, 19).

III. GROWTH OF PATTERNS

III-1. Critical Mixtures

The time changes of $I_m(t)$ and $q_m(t)$ during pattern formation via SD were measured at various T by time-resolved light scattering experiments for SBR1/PB19 mixtures having near critical composition (58 wt/42 wt) (17). The results obtained were subjected to the reduced plot as shown in Fig.4 (17), in which Q_m, $\tilde{I}_m$ and τ are the reduced characteristic wavenumber, reduced intensity and reduced time, respectively, which are defined by

$$Q_m = q_m(t;T)/q_m(0;T) \tag{8}$$

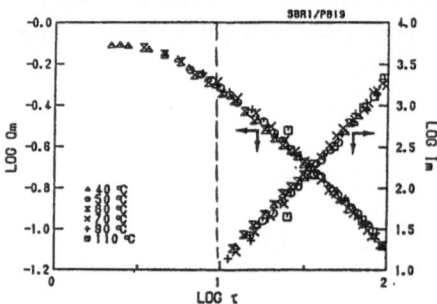

Fig. 4. Reduced Q_m and $\tilde{I}_m$ as a function of the reduced time τ for the SBR1/PB19 58 wt%/42 wt% mixture. The vertical broken line indicates the crossover between the intermediate and late stages. From T. Izumitani, M. Takenaka and T. Hashimoto (1990)

$$\bar{I}_m = I_m(t;T)q_m^3(0;T)\bigg/\!\!\int_0^\infty q^2 I(q;t;T)dq \tag{9}$$

$$\tau = t/t_c(T) \tag{10}$$

and

$$t_c(T) = [q_m^2(0;T)D_{app}(T)]^{-1} \tag{11}$$

In these equations, $q_m(0;T)$ and $D_{app}(T)$ are the quantities experimentally determined quantities, $q_m(0;T)$ is the characteristic wavenumber of the dominant mode of the fluctuation in the early stage SD given by

$$q_m(0;T) = 2\pi/\Lambda = [\sqrt{2}\xi(T)]^{-1}$$

(12)

where $\xi(T)$ is the thermal correlation length of the mixture, and $D_{app}(T)$ is the mutual (collective) diffusivity in the early stage SD.

The data $q_m(t;T)$ and $I_m(t;T)$ obtained at different T are shown to smoothly fall onto the respective master curves, $Q_m(\tau)$ and $\tilde{I}_m(\tau)$, indicating that the pattern grows according to the same coarsening mechanism over the temperature range covered in the experiments and that temperature affects only the time scale (through $t_c(T)$) and spatial scale (through $q_m(0;T)$) of the self-assembling patterns. This polymer system is found to obey a universal law found for small molecule systems, satisfying the Chou-Goldburg scaling postulate (20). The vertical broken line in the figure shows the crossover reduced time from the intermediate to late stage SD, $\tau_{cr} \cong 10$.

Now that the time-change of the size of the self-assembling pattern has been characterized by $Q_m(\tau)$ in Fig. 4, the time change of the morphology of the pattern is desired to be characterized. Fig.5 shows a result of such a characterization which gives the time-change of the scaled structure factor for the SBR1/PB19 58 wt/42 wt mixture at 40°C. The time progresses in the order of part c (from t = 64.86 to 220.91 m or τ = 1.9 to 6.6, the early to intermediate

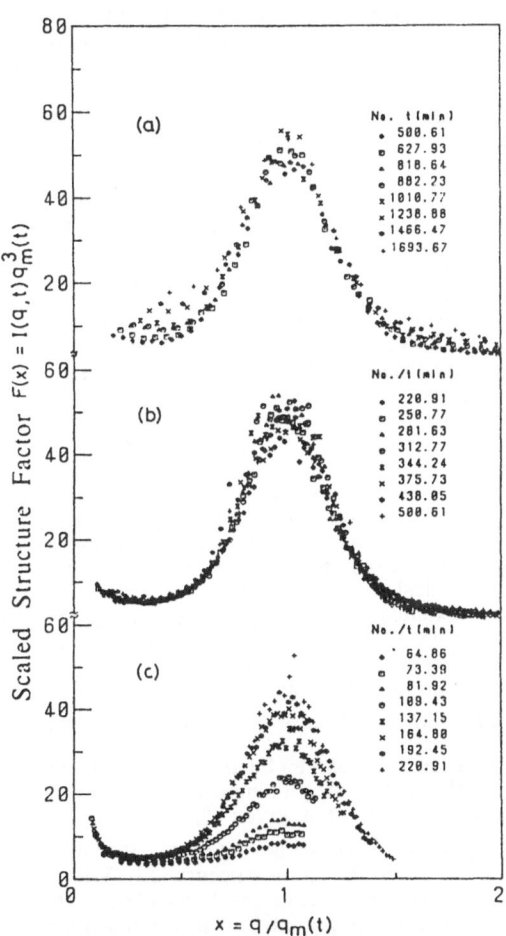

Fig. 5. Scaled structure factors F(x) obtained at 40°C for SBR1/PB19 with 58/42 wt%/wt% mixture where the time elapses in the order of parts c to a. In part c, F(x) monotonically increases with time but in parts a and b it is essentially independent of time.

The reduced time τ is obtained by dividing the real time by t_c = 2000 s. From M. Takenaka, T. Izumitani and T. Hashimoto (1990).

stage), part b (from t = 220.91 to 500.61 m or τ = 6.6 to 15, the intermediate to late stage) and part a (from t = 500.61 m to 1693.67 m or τ = 15.0 to 50.8, the late stage SD) (12). In the early-to-intermediate stage SD, F(x) is not universal with t but rather increasing with t (part c), which is primarily due to the increase of $<\eta^2(t)>$. F(x) is seen to become sharp with t, which reflects an ordering of the self-assembling pattern. However

in the late stage τ > 10, the structure factor is universal with t (parts a and b). Thus this polymer system obeys the universal law found for small molecules systems, satisfying the dynamical scaling hypothesis (10,11). The scaled structure factor was also found to be relevant to a bicontinuous periodic self-assembling pattern (11,21).

III-2. Off-Critical Mixtures

So far we described the characteristic features for self-assembling patterns in the later-stage SD (intermediate stage and late stage) for near critical mixtures. Here we discuss those for off-critical mixtures, i.e., the mixtures with off-critical compositions.

Figure 6 shows the coarsening in the later stage for SBR1/PI55 mixtures at 60°C, as observed by the time-change of the characteristic wavenumber $q_m(t)$ toward smaller values (part a) and by that of $I_m(t)$ to higher values (part b). The values $q_m(t)$ and $I_m(t)$ continue to change with t and hence the pattern continues to grow for the near critical mixture (the 50/50 wt/wt mixture). However the changes of $q_m(t)$ and $I_m(t)$ are pinned down, and hence the pattern growth also is pinned down at a level depending on the composition of the off-critical mixtures (the 20/80 and 30/70 wt/wt mixtures). The pinning occurs at a time t > t_p or at a level $q_m < q_{mp}$, in which t_p and q_{mp} are the pinning time and wavenumber, respectively.

The detailed studies clarified that, at a given composition ϕ of SBR1, the higher the

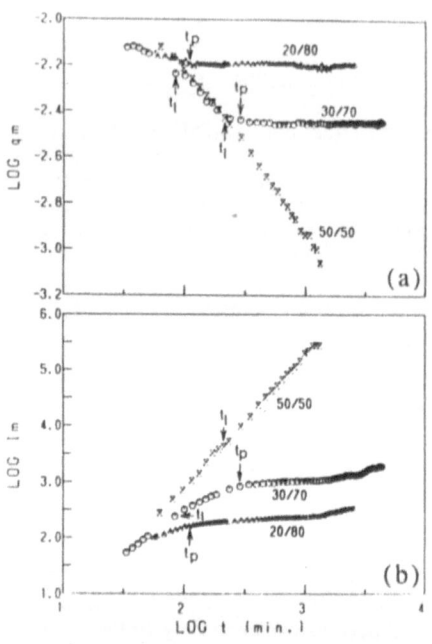

Fig. 6. Time-changes of (a) q_m and (b) I_m for the critical (50/50 wt/wt) and off-critical mixtures (20/80 wt/wt and 30/70 wt/wt) of SBR1/PI55 at 60°C. t_p and t_l refer, respectively, to the time at which the pinning and the crossover between the intermediate and late stages occur. From M. Takenaka, K. Tanaka and T. Hashimoto (1989).

temperature T, the shorter the time t_p, or the higher the level q_{mp}, i.e., the pinning occurs at an earlier stage,, and that, at a given T, the more the composition ϕ is biased toward 0 or 1, the earlier the stage where the pinning occurs (22). We can interpret these experimental results as follows. This mixture has seemingly a UCST phase diagram as shown in Fig.7(18). Inside the spinodal phase boundary, there may be a region where the percolating domain structure is maintained until very late stage SD (region B) and a region where the percolating domain structure exists initially but is later broken up into clusters of spheres via "dynamical percolation-to-cluster transition" (region A). The unique pinning phenomena observed for the off-critical mixtures with high molecular weight polymers is proposed

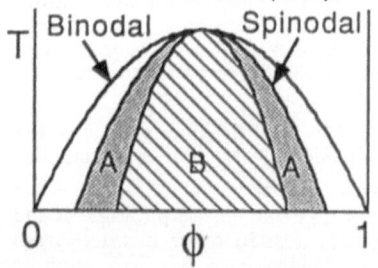

Fig. 7. Two regions A and B in the spinodal phase boundary. In region A, the pattern is initially percolated, but is later transformed into clusters of spheres, due to the dynamical percolation-to-cluster transition. In region B, the percolating pattern is maintained until the very late stage.

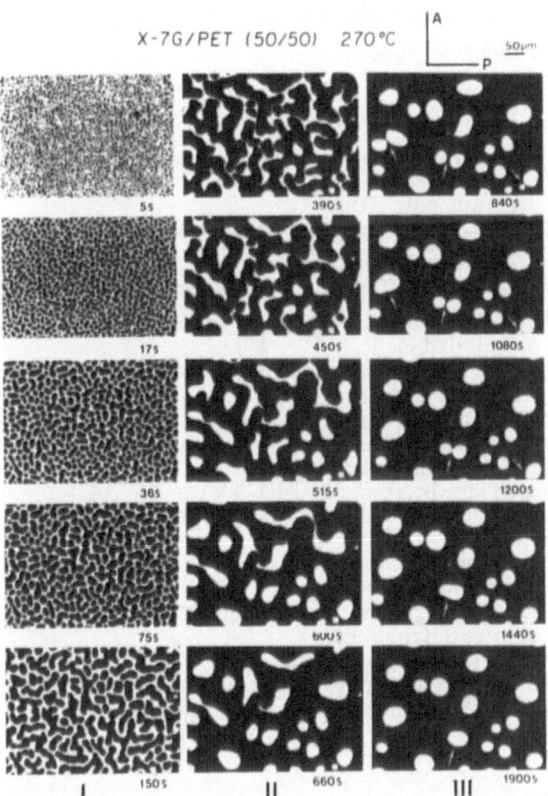

Fig. 8. Time evolution of the unmixing structure showing the dynamical percolation-to-cluster transition. In regime I the percolating structure grows with dynamical self-similarity. The transition takes place in regime II and clusters of spheres are found in regime III. From H. Hasegawa, T. Shiwaku, A. Nakai, and T. Hashimoto (1988).

to originate from the dynamical percolation-to-cluster transition occuring during the coarsening process in SD (23).

The phase separation via SD develops a bicontinuous, periodic domain structure with a well defined interface at the late stage. The pattern initially has the macroscopic percolation, the coexisting two domains being continuous in the whole sample space. However as the phase separation proceeds and the pattern grows, the domains rich in minority component may not be able to maintain the macroscopic percolation, resulting in a break-up into fragments with local percolation only. The fragments eventually degenerate into clusters of spherical droplets, which is driven by interfacial tension. The distribution of the interdroplet distance is sharp enough to give a scattering maximum, because the droplets originated from the periodic, percolating structure. Although we do not have direct experimental evidence obtained by real-space analysis to support this idea for the particular mixture of SBR1/PI55, we have it for another mixture of X-7G/PET, as shown in Fig. 8 (13).

At $t < t_p$ in which the percolating pattern exists, the pattern grows via motion of the interface which is continuous. The thermal motion causes instability and breaking up of the network into a larger mesh size and a thicker percolating network. At $t > t_p$, the percolating pattern may be transformed into the cluster pattern for which the growth occurs as a consequence of the diffusion-coalescence of the droplets. The diffusion of droplets requires either bulk diffusion of molecules or surface diffusion of molecules. In bulk diffusion, the diffusion of the droplets occurs through such a process that the molecules forming droplets diffuse from the droplets to the matrix and condense into different portion of the same droplets or different droplets. The bulk diffusion process has to overcome a kinetic barrier associated with the

182

enthalpy of mixing ΔH_{mix} of polymers having a net repulsive interaction as shown schematically in Fig.9. Now ΔH_{mix} per chain is given by

$$\Delta H_{mix} \sim \chi N k_B T \tag{13}$$

where N is the degree of polymerization (DP), the DP's for the two polymers are assumed to be identical to simplify the arguments. Thus when χN is very large as is the case of SBR1/PI55, the kinetic barrier is too large to be overcome, and the rate of the diffusion proportional to $\exp(-\chi N)$ is very small (23). Thus the bulk diffusion process is kinetically frozen-in, though there is a thermodynamical driving force for the domain to grow as the state II has a free energy lower than the state I (Fig.9). The heavily suppressed diffusion process may cause the pinning of the pattern growth, and hence the pinning may be a unique feature in the self-assembly for the off-critical and high molecular weight mixture.

It is obvious that the dynamical scaling hypothesis is not valid through the dynamical percolation-to-cluster transition. This is because the pattern changes from the bicontinuous periodic structure to the cluster of spheres. In fact the scaled structure factor was found to substantially broaden before and after the transition.

For mixtures having smaller molecular weight molecules, $\Delta H_{mix} < k_B T$, so that the growth of the droplets can always take place. The same argument may be easily extended to the surface diffusion where ΔH_{mix} may be

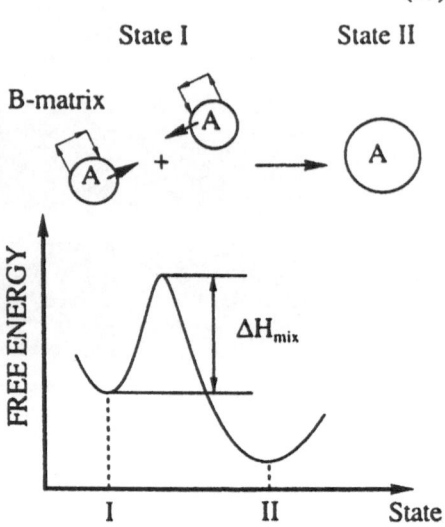

Fig. 9. Schematic illustration for the kinetic energy barrier for the diffusion-coalescence process. Since the state I has a higher free energy than the stage II, a transformation from the state I to II is thermodynamically favored. However it must overcome the barrier associated with the heat of mixing of unlike chains in the diffusion-coalescence process.

replaced by an excess free energy $\Delta S\gamma$ associated with the change of the interfacial area ΔS involved in the surface diffusion process. Here γ is the interfacial tension.

IV. PINNING OF PATTERN GROWTH AND MORPHOLOGY CONTROL

The various features found for the pattern growth of critical and off-critical mixtures, as discussed in sec. III, can be used as guiding principles to control morphology of binary polymer mixtures. In this section we shall discuss the processes which pin down the pattern growth with an aim to control the morphology. The processes to be discussed are physical pinning involving crystallization and involving vitrification, chemical pinning and spontaneous pinning, all of which commonly involve freezing-in the translational diffusion of centers of mass of one constituent polymers at least. These processes can be applied to pin down the growing domain morphology at a certain stage of the phase separation, which leads to control over some desired patterns. The physical and chemical pinning can be applied both for critical and off-critical mixtures but the spontaneous pinning can be applied only for the off-critical mixture with high molecular weights.

IV-1. Physical Pinning Due to Crystallization

An example of the physical pinning due to crystallization is seen in the solidification process of PP/EPR. This method involves first the domain growth by SD up to a certain stage and then a subsequent pinning of further growth of the domain by crystallization of PP.

For this purpose, cold-pressed thin films of the 50/50 wt/wt mixture of PP/EPR were subjected to a temperature T_x above the melting temperature Tm of PP, at which the phase separation occurs according to SD and a pattern having the desired Λ_m and $\Delta\phi$ is developed. After the pattern formation, the films were quenched to a crystallization temperature T_c below T_m at which a **rapid crystallization (diffusion-limited crystallization)** takes place by nucleation in a region rich in PP. The crystallization progresses and the crystallization front advances following region rich in PP without a significant perturbation of the pattern formed at T_x. The patterns formed at T_x were found to be essentially conserved during the crystallization at T_c and to be locked-in in the specimens solidified by the crystallization (14)

Figures 10 and 11 show two extreme morphologies developed for PP/EPR by the isothermal unmixing at T_x and the subsequent diffusion-limited crystallization (14). The morphology shown in Fig.10 was developed by unmixing at 200°C for 5 min. and subsequent crystallization at 125°C for 5 min., followed by an eventual rapid cooling below the vitrification temperature of the mixture. On the other hand the morphology shown in Fig.11 was developed by unmixing at 200°C for 20 min. and subsequent crystallization by quenching the mixture in an ice-water bath.

Figure 10a shows the view under polarized light microscopy where the polarizer and analyzer axes are vertical and horizontal. The volume-filling spherulites with an

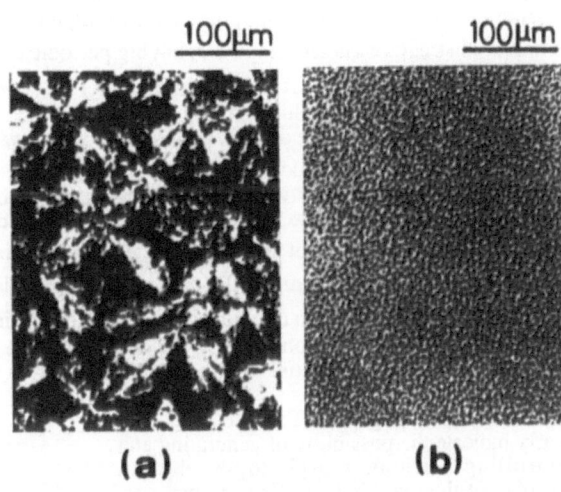

(a) **(b)**

Fig. 10. (a) Polarized light micrograph (POM) and (b) phase-contrast light micrograph (PCM) for PP/EPR mixture (50/50 wt/wt) unmixed at 200°C for 5 min. and subsequently crystallized at 125°C for 5 min. The POM which was obtained by setting polarizer and analyzer in vertical and horizontal directions clearly shows the volume-filling spherulites, while the PCM clearly shows the fine structure developed by SD. The two pictures were obtained on the same field of the same sample. From N. Inaba, T. Yamada, S. Suzuki, and T. Hashimoto (1988).

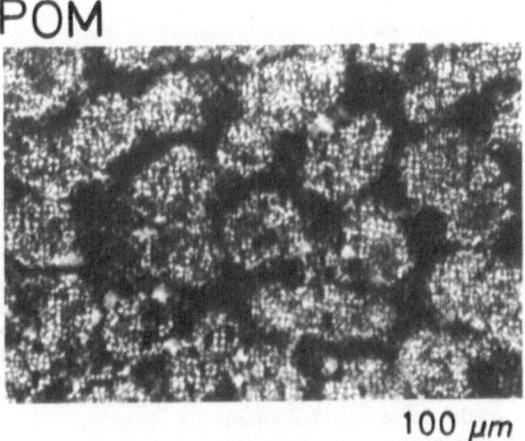

Fig. 11. Polarized light micrograph showing the bicontinuous percolaing structure developed by SD for the PP/EPR mixture (50/50 wt/wt) unmixed at 200°C for 20 min. and subsequently crystallized by quenching the mixture in an ice-water bath (polarizer and analyzer being set in vertical and horizontal directions). Note that the percolating PP-richdomains contain the volume-filling spherulites and have high optical anisotropy. From N. Inaba, T. Yamada, S. Suzuki, and T. Hashimoto (1988).

average diameter $D_s = 140$ μm are seen in the micrograph. On the other hand Fig. 10b shows the same field as in Fig. 10a under phase contrast light microscopy where the pattern is typical of the unmixed structure via late stage SD, having periodic concentration fluctuation with $\Lambda_m = 4.4$ μm as measured by the scattering method (14). These two pictures in Fig. 10 clearly show a unique morphology of the volume-filling spherulites which contain, as their internal structure, the percolating network structure of EPR and PP with a narrow distribution of the mesh size centered at Λ_m resulting from the spinodal decomposition above T_m.

The picture in Fig.11 was observed under the polarized light microscopy, polarizer and analyzer being again set in the vertical and horizontal directions. The picture shows the pattern, at a large distance scale of about 100 μm, typical of the unmixed structure at the late stage SD, i.e., a bicontinuous network structure of PP-rich and EPR-rich domains with characteristic spacing $\Lambda_m = 60$ μm. The bright PP-rich domains contain volume-filling spherulites having $D_s = 10$ μm and hence have high optical anisotropy. On the other hand, the dark EPR-rich domains have small or no optical anisotropy because of no or low crystallinity.

The two figures (Figs.10 and 11) clearly indicate the possibility of generating and controlling various morphologies by an interplay of the two kinds of phase transitions. The patterns developed contain dual morphological entities, i.e., of the spherulites with Ds and the bicontinuous, periodic domains with Λ_m. Their size and spatial organization may be controlled at will by controlling the phase transitions.

IV-2. Physical Pinning Due to Vitrification

A straightforward application of the physical pinning due to vitrification for the purpose of morphology control involves a quenching of the system undergoing the pattern growth below its glass transition temperature T_g. The pinning due to vitrification may occur also during self-assembly via SD when the mixtures contain polymer of high T_g (e.g. polymer A) and that of low Tg (e.g. polymer B). As the fluctuation grows, the composition difference $\Delta\phi$ of polymer A in A-rich and A-poor domains increases, leading to increasing T_{g1} for A-rich domains and decreasing T_{g2} for A-poor domains. As $\Delta\phi$ further increases, T_{g1} could cross the phase separation temperature T_x. At this stage A-rich domains are vitrified, and hence the self-assembling pattern is spontaneously pinned (**spontaneous pinning due to vitrification**).

An intriguing pattern formation involving the pinning due to vitrification was observed for the films of PS/PB, PS/SB, and PS/PB/SB prepared by evaporating homogeneous toluene solutions containing about 5 wt% polymers in total. The mixtures studied have an unequal fraction of each polymer. Figure 12 shows an example of

S/SB-65 Solvent Cast Film

(a)

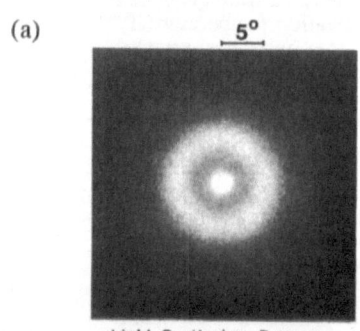

Light Scattering Pattern

(b)

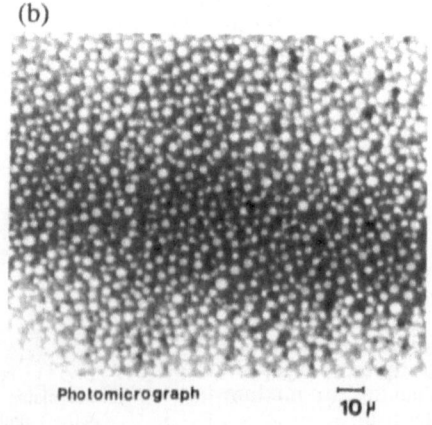

Photomicrograph   10 μ

Fig. 12. Light scattering pattern (a) and photomicrograph (b) of the solvent-cast film of S/SB-65 stained by O_sO_4. Toluene is used as a solvent. From T. Hashimoto, K. Sasaki, and H. Kawai (1984).

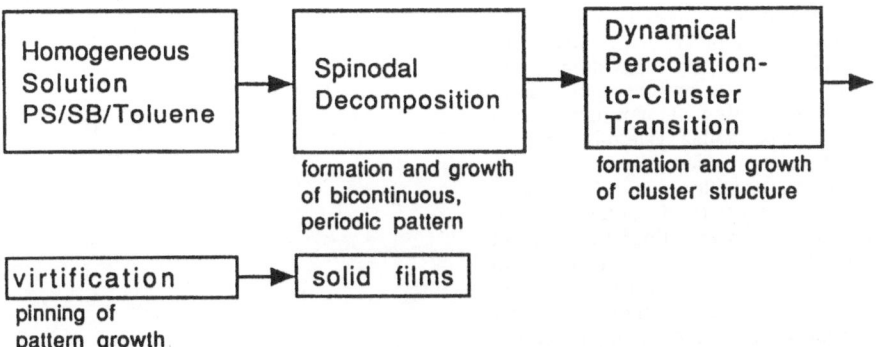

Fig, 13. Various processes encountered in the pattern formation of polymer mixtures by the solvent-casting method. The concentration of the polymers increases from the top left to the bottom right during the solvent evaporation.

typical light scattering pattern (a) and a light micrograph for the solvent-cast films of the PS/SB 35 wt/65 wt mixtures (16). The bright spheres correspond to the PS domains and the dark matrix corresponds to the SB domains stained by osmium tetraoxide. The SB domains have periodic microdomains composed of PS and PB microdomains which are unstained and stained by osmium tetraoxide, respectively. The microdomain structure, with a characteristic length scale of order of 100 Å, however, cannot be resolved under optical-microscopic observation, and hence the SB matrix appears to be a continuous, featureless, dark phase.

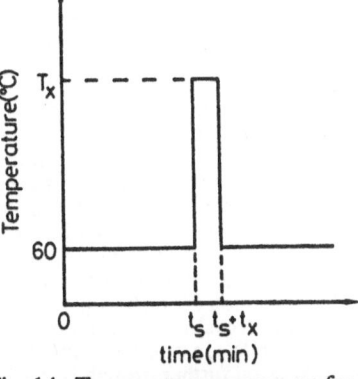

Detailed analyses of the results indicated that the scattering maximum reflects regularity of the intersphere distance and the distance estimated from the corresponding scattering angle is well correlated to that estimated from the micrograph (16). Time-resolved light scattering studies in the early stage phase separation of the PS/SB/toluene system indicated that the phase separation into domains of PS-rich solution and SB-rich solution occurs according to the SD mechanism (24). These two

Fig. 14. Temperature programme for the chemical crosslinking of the mixture during spinodal decomposition. T_x is crosslinking temperature, t_x is time period for the crosslinking reaction. From T. Hashimoto, M. Takenaka and H. Jinnai (1989).

results suggest that the solvent cast process involves a complex pattern formation process. Figure 13 shows a proposed series of the events encountered in the process.

During the solvent evaporation process, the concentration c of the total polymers increases. When c is greater than c*, the overlap concentration for polymer coils, phase separation starts to occur according to the SD mechanism, generating the domains rich in polystyrene and those rich in SB block copolymers which are periodic and bicontinuous. As the domains grow, minority domains (PS domains) may not be able to maintain the spatial continuity, which may then result in the dynamical percolation-to-cluster transition as discussed in sec. III-2, generating clusters of the PS spheres as shown in Figure 12(b) in the matrix of SB solution. The growth of the domains due to the diffusion-coalescence mechanism as discussed in sec. III-2 may not be necessarily pinned down in this case, because the effective energy barrier (proportional to ΔH_{mix}) for the mutual diffusion is significantly reduced by presence of the neutral solvent toluene, but the growth is slowed down as c increases. The growth is eventually pinned down

186

when c reaches such a level that the glass transition temperature of PS macrospheres as seen in Fig.12(b) or the PS microdomains in the SB matrix reaches the ambient temperature for the solvent-casting. The microdomain formation for this particular system is expected to occur at a concentration higher than that where the pattern formation due to SD occurs. The pattern is essentially frozen-in by the vitrification. Further evaporation of the solvent causes a shrinkage of the pattern, resulting in the pattern observed in Fig. 12(b) in the solid films. The intersphere distance is regular because the cluster pattern contains a memory of the periodicity in the bicontinuous patterns.

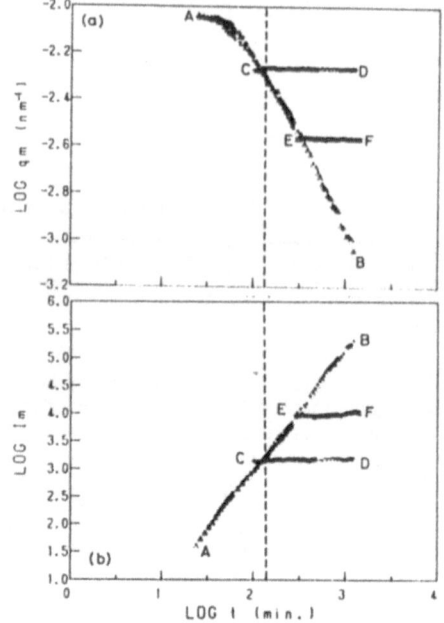

IV-3. Chemical Pinning

For mixtures such as PB/SBR and PI/SBR which contain unsaturated double bonds, one can pin down the pattern growth by crosslinking with peroxide. In order to test this idea, we added a peroxide by 1.5 wt % to the 50/50 wt/wt mixture of SBR2/PI55. The mixture was subjected to SD at a low temperature, e.g., 60°C at which the crosslinking reaction is inactive, and the pattern was grown to a desired stage for a certain period of time t_s as shown in Fig. 14 (25). Then we rapidly raised the temperature of the mixture to T_x at which we induce the crosslinking reaction for a short period of time t_x. After the crosslinking we dropped the temperature to 60°C in order to study the effect of crosslinking on further growth of the domain structure (see Fig.14).

Fig. 15. (a) The change of the wavenumber qm and (b) the peak intensity Im of the mixture during the coarsening of the unmixed structure at the later stage SD (the curves AB) and effects of the chemical pinning on q_m and I_m (the curves ACD for the pinning at the intermediate stage SD and the curves AEF for the pinning at the late stage SD).

Figure 15 shows the time change of the pattern as observed by those of q_m and I_m in such experiments with $T_x = 150°C$ and $t_x = 5$ min. (much longer than the half time for the radical generation of the particular peroxide used in this experiment (36 s) at 150°C). Without peroxide the pattern for the near critical mixture continues to grow as shown in the curve AB. However the crosslinking reaction is observed to be quite effective on the pinning of the pattern growth as seen by the cessation of the time changes of I_m and q_m at the point C or E where the crosslinking reaction was activated. The vertical line indicates the crossover time from the intermediate to late stage SD. Hence the crosslinking at the point C generates a pattern chemically pinned in the intermediate stage of SD and that at point E a pattern pinned in the late stage.

In this way we can produce interesting heterogeneous elastomers composed of two

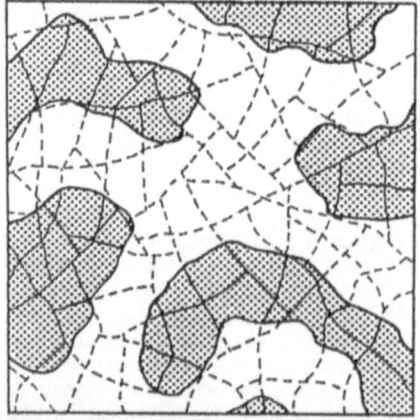

Fig. 16. Schematic illustration of bicontinuous and periodic domains of rubber-like polymer blends formed by SD. Note that both domains are chemically crosslinked.

component rubber-like polymers which are crosslinked. The two types of elastomers are phase-separated into bicontinuous and periodic domains having crosslinked networks as sketched in

Fig. 16. The size of the domain $\Lambda_m(t)$ and the level of demixing of the unlike segments or $\Delta\phi(t)$ can be controlled in principle.

IV-4. Spontaneous Pinning

The spontaneous pinning can be utilized to control the cluster structure of the spheres for the off-critical mixtures. For a given polymer pair, the cluster structure is controlled by changing temperature and composition as discussed in sec. III-2.

ACKNOWLEDGEMENTS

This work was supported in part by a Grant-in-Aid for Scientific Research in Priority Area "New Functionality Materials, Design, Preparation and Control", Ministry of Education, Science and Culture, Japan.

References

(1) Cahn J W (1965) J Chem Phys 42:93
(2) Gunton J S, San Miguel M, Sahn P S (1983) In: Domb C, Lebowitz JL (Eds) Phase Transition and Critical Phenomena, Vol 8, Academic Press London
(3) Hashimoto T, Itakura M, Hasegawa H (1986) J Chem Phys 85:6118
(4) Hashimoto T, Takenaka M (1991) J Appl Cryst 24:457
(5) Bates F S, Wiltzius P (1989) J Chem Phys 91:3258
(6) Langer J S, Bar-on M, Miller H D (1975) Phys Rev A 11:1417
(7) Jinnai H, Hasegawa H, Hashimoto T, Han C C (1991) Macormolecules 24:282
(8) Hashimoto T (1988) Phase Transitions 12:47
(9) Hashimoto T (1991) In: Cahn R W, Haasen P, Kramer E J (Eds) Materials Science and Technology, vol 12. Structure and Properties of Polymers (Chapter 6) VCE, Weinheim, in press
(10) Binder K, Stauffer D (1974) Phys Rev Lett 33:1006
(11) Binder K (1977) Phys Rev B15:4425
(12) Hashimoto T, Takenaka M, Izumitani T, (1989) Polymer Commun 30:45, Takenaka M, Izumitani T, Hashimoto T (1990) J Chem Phys 92:4566
(13) Hasegawa H, Shiwaku T, Nakai A, Hashimoto T (1988) In: Komura S, Furukawa H (Eds) Dynamics of Ordering Processes in Condensed Matter, Plenum NY
(14) Inaba N, Yamada T, Suzuki S, Hashimoto T (1988) Macromolecules 21:407
(15) Nakai A, Shiwaku T, Hasegawa H, Hashimoto T (1986) Macromolecules, 19:3010
(16) Hashimoto T, Sasaki K, Kawai H (1984) Macromolecules 17:2812
(17) Izumitani T, Takenaka M, Hashimoto T (1990) J Chem Phys 92:3213
(18) Izumitani T, Hashimoto T (1985) J Chem Phys 83: 3694
(19) Takenaka M, Hashimoto T (1991) J Chem phys, in press
(20) Chou YC, Goldburg WI (1979) Phys Rev A 20:2105
(21) Furukawa H (1986) Phys Rev B, Rapid Commun 33:638
(22) Hashimoto T, Takenaka M, Izumitani T, in preparation
(23) Takenaka M, Tanaka K, Hashimoto T(1989) In: Culbertson BM (Ed), Multiphase Macromolecular Systems, Plenum NY
(24) Sasaki K, Hashimoto T (1984) Macromolecules 17:2818
(25) Hashimoto T, Takenaka M, Jinnai H (1989) Polymer Commun, 30:177

Third-Ordered Susceptibilities of Nematic and Isotropic Solutions of a Rodlike Polymer (PBT)

Hedi Mattoussi and Guy C. Berry

Carnegie Mellon University, Pittsburgh, PA 15213, USA

Abstract: The anisotropic third-order nonlinear optical susceptibility $\chi^{(3)}$ of monodomain samples of nematic solutions of poly[1,4-phenylene-2,6-benzobisthiazole], PBT, a rodlike polymer, have been studied by third harmonic generation. Corresponding measurements have also been made for isotropic solutions of PBT. A substantial off-diagonal component of $\chi^{(3)}(-3\omega;\omega,\omega,\omega)$ is measured, resulting in the generation of an elliptically polarized third harmonic signal for a fundamental linearly polarized along the director.

INTRODUCTION

Nonlinear optical properties of organic materials have been of intense interest for the past three decades [1-8]. In the dipole approximation, the macroscopic polarization $P(\omega)$ of a material subject to electric fields $E(\omega_1)$, $E(\omega_2)$, etc., propagating with frequencies ω_1, ω_2, etc., may be expressed in Cartesian coordinates as [1-8]:

$$P_i(\omega) = \varepsilon_0[\chi_{ij}^{(1)} E_j(\omega) + \chi_{ijk}^{(2)} E_j(\omega_1)E_k(\omega_2) + \chi_{ijkl}^{(3)} E_j(\omega_1)E_k(\omega_2)E_l(\omega_3) + ...] \quad (1)$$

where $\omega = \Sigma\omega_v$; the component notation, in which the dependence of the $\chi^{(n)}(-\omega;\omega_1,\omega_2,...,\omega_n)$ on frequency is suppressed for convenience, will be used below. The susceptibilities $\chi^{(n)}$ are macroscopic properties; their relation to molecular characteristics is discussed below. In this study, we report measurements on $\chi^{(3)}(-3\omega;\omega,\omega,\omega)$ by third-harmonic generation on nematic monodomains made of solutions of the rodlike polymer poly[1,4-phenylene-2,6-benzobisthiazole], PBT:

PBT

Y. Imanishi (Ed.)
Progress in Pacific Polymer Science 2
© Springer-Verlag Berlin Heidelberg 1992

Measurements of $\chi^{(3)}(-3\omega;\omega,\omega,\omega)$ on isotropic solutions of PBT are also reported for comparison. In both cases, the polymer is dissolved in a strong protic acid and is protonated [9]. Related studies on PBT include measurements of $\chi^{(3)}(-3\omega;\omega,\omega,\omega)$ for isotropic solutions [10] and PBT dispersed in an organic matrix [11], along with on $\chi^{(3)}(-3\omega;\omega,\omega,\omega)$ and $\chi^{(3)}(-\omega;\omega,-\omega,\omega)$ for undiluted PBT [12] ($\chi^{(3)}(-\omega;\omega,-\omega,\omega)$ being derived from degenerate four wave mixing). The good alignment of the nematic solutions studied here, not available in the PBT samples previously studied, affords the opportunity to elucidate certain of the components $|\chi^{(3)}|_{ijkl}$ of the susceptibility tensor.

Previous studies with the nematic PBT monodomains studied here have shown them to exhibit marked birefringence Δn, with Δn increasing markedly with the the polymer volume fraction φ [13]. These measurements also showed that both the extraordinary and the ordinary refractive indices, n_E and n_O, respectively, depend on the wavelength λ:

$$\frac{(n_X^2 - 1)}{(n_X^2 + 2)} = K\frac{\lambda^2}{\lambda^2 - \lambda_X^2} \tag{2}$$

where λ_E and λ_O were 250 and 200 nm, respectively, and K was 0.27 for both n_E and n_O [13]. Analysis of the experimental results reported here requires these refractive indices and their dispersions.

EXPERIMENTAL

Materials: The PBT (M_w = 34,000), provided by SRI International, was dried in vacuum and dissolved in distilled methane sulfonic acid (MSA) using procedures discussed elsewhere [9]. Nematic solutions of PBT are formed when the polymer volume fraction φ exceeds a critical value φ_{NI} [14], where φ_{NI} depends on the chain aspect ratio and the temperature; $\varphi_{NI} \approx 0.03$ for the PBT used. The alignment procedure consists of a surface alignment preparation using a suitable flow in a rectangular channel, followed by exposure to an external magnetic field (5 to 7 Tesla) to speed the bulk alignment [10,15]. This procedure is found to provide a bulk alignment stable for an indefinite duration.

Nonlinear Optical Measurements: The third-order susceptibility of PBT in nematic and isotropic solutions has been determined using a third harmonic generation (THG) technique [1,2,10]. The Maker Fringe Pattern (MFP) [2,3,7] generated on rotation of a plane parallel sample or translation of a wedge-shaped cell. Fundamental input at the wavelengths of 1542 or 1907 nm was used to study the third-order susceptibility. The THG apparatus utilizes a Raman cell filled with methane or hydrogen gas to provide a fundamental intensity at $\lambda = 1542$ nm and $\lambda = 1907$ nm, respectively, when pumped at $\lambda = 1064$ nm by a pulsed Nd:YAG laser, for third harmonic signals $I_{3\omega}$ at $\lambda = 514$ nm and 636 nm, respectively [10]. A portion of the incident beam is directed to a reference to account for fluctuations in the incident intensity. A weakly focused beam generated a sufficiently intense signal for accurate detection with the samples of interest here. A Fresnel Rhomb in the incident beam permits control of the input polarization, and an analyzer in the THG beam permits selection of the polarization of the output signal.

Experiments used flat and wedge-shaped cells, provided by Hellma Cells Inc. In each case, the cell walls were specified to $\lambda/4$ surface flatness; 375 μm sample thickness for the parallel cell, and a wedge angle $\beta = 0.95°$ allowing a thickness scan between 200 to 800 μm over the length of the sample for the wedge cell. The reduced intensity $R(\theta,L)$ is given by

$$R(\theta,L) = \frac{I_{3\omega}(\theta,L)_{SAMP}}{I_{3\omega}(0,L)_{REFR}} \tag{3}$$

where SAMP and REFR stand for the sample and reference, respectively, and θ is the incidence angle and L the thickness; θ is varied by rotation of the sample (fixed L) in the use of a parallel slab, and L is varied by translation of the sample at normal incidence in the use of a wedge-shaped sample. A fused silica slab (L = 1.00 mm) served as the reference.

A general expression for $R(\theta,L)$ for an isotropic material gives [1-4]:

$$R(\theta,L) = K_R \left(\frac{[Q(\theta)W(\theta,L)]_{SAMP}}{[Q(0)W(0,L)]_{REFR}} \right)^2 \tag{4}$$

$$W(q,L) = \frac{|\chi^{(3)}|_{EFF}}{|n_{3\omega}{}^2 - n_\omega{}^2|} \Omega(\theta,L) \tag{5}$$

where $|\chi^{(3)}|_{EFF}$ is the effective component of $\chi^{(3)}$, e.g., see Eqn (1), $Q(\theta)$ is a weakly decreasing function of increasing angle, related in part to Fresnel factors ($Q(0)$ is unity), K_R is the ratio of the incident beam directed to the sample to that directed to the reference, and n_ω and $n_{3\omega}$ are the refractive indices at the incident frequency ω and the THG frequency 3ω, respectively [1-4]. The MFP exhibits symmetry about $\theta = 0$ for the rotating parallel slab cell, or oscillations for the wedge-shaped cell.

The function $\Omega(\theta,L)$ accounts for the interference effects giving rise to the MFP. For the case with weak absorption characterized by absorbances α_ω and $\alpha_{3\omega}$ for the incident and THG frequencies, respectively,

$$\Omega^2(\theta,L) = 2b\left(\frac{a-b}{2b} + \sin^2[\psi(\theta,L)]\right) \tag{6a}$$

$$\psi(\theta,L) = \pi L/2L_c(\theta) \tag{6b}$$

$$4a = \exp(-\alpha_{3\omega}) + \exp(-2\alpha_\omega) \tag{6c}$$

$$2b = \exp(-\alpha_{3\omega}/2)\exp(-\alpha_\omega)\exp(-\eta^2) \tag{6d}$$

where $\alpha = \mu L/\cos\theta'$, with μ the extinction; the last term in Eq. 6d accounts for the finite beam waist in the sample, and η is zero for a parallel-slab cell geometry or $(\pi d_0/4L_c(0))\tan\beta$ for a wedge-shaped cell, d_0 being the beam waist in the sample; the coherence length $L_c(\theta)$ may be expressed as

$$L_c(\theta) = \lambda/6|n_{3\omega}\cos(\theta'_{3\omega})-n_\omega\cos(\theta'_\omega)| \tag{7}$$

where the primes denote propagation angles in the sample, related to angles in air through Snell's law, using the appropriate refractive index, see below [2]. For use with a wedge-shaped cell, all angles are essentially zero, and the sample thickness is given by

$$L = L_R + (\Delta l)\tan\beta \tag{8}$$

where Δl is the translation of the cell from some reference position for which $L = L_R$. Under the conditions used, $\exp(-\eta^2) \approx 0.37$. In practice, the coefficient $|\chi^{(3)}|_{EFF}$ is evaluated by determination of the response $R_{STD}(0)$ for a standard with known $|\chi^{(3)}|_{EFF}$, $n_{3\omega}$, n_ω, and thickness L (a BK7 plate, $L = 3.00$ mm, was used for this purpose). It should be noted that $R(0,L)$ may take on any value from zero to its value R_{MAX} for $\psi \approx 0$, as $\psi(\theta,L)$ is not usually an integral multiple of π at $\theta = 0$. However, $\sin\psi$ may be taken as unity for the successive maxima in the fringe pattern, providing a method to estimate of $|\chi^{(3)}|_{EFF}/|n_{3\omega}^2 - n_\omega^2|$ from data on $R(\theta,L)$.

For an anisotropic material, $|\chi^{(3)}|_{EFF}$ becomes the modulus of an appropriate component of the tensor $\chi^{(3)}$, depending on the polarization of the light and the orientation of the director, see below. In addition, as discussed below, the birefringence expected with an anisotropic material may modify the MFP.

RESULTS

Isotropic Solutions: Maker fringe patterns obtained for an isotropic solution of PBT (volume fraction $\varphi \approx 0.021$) using the plane parallel cell with incident wavelengths of 1542 and 1907 nm are given in Figure 1. Similarly, the MFP obtained with the wedge-shaped cell are given in Figure 2. These data were obtained using parallel polars for the incident and THG beams. The same patterns resulted with any orientation of the polars with respect to the rotation axis for the parallel slab, or the wedge direction for the wedge-shaped cell. In addition, the THG signal was essentially nil for experiments with crossed polars ($R(\theta,L) < 0.01$). Separate experiments with cells filled with MSA provided negligible $R(\theta,L)$ at all θ and L, permitting a simplified analysis of the $R(\theta,L)$ as due only to THG from the polymer solute. Since the signal was nil for λ in the range 532 to 700 nm for THG at 514 nm, any fluorescence contribution to $R(\theta,L)$ measured at 514 nm caused by absorption at $\lambda_{3\omega}$ was negligible [15]; it follows that the fluorescence contribution for THG at 636 nm given the even smaller absorption at 636 nm. Data could not be obtained with an incident wavelength of 1064 nm (center YAG line), owing to the appreciable absorbance at wavelengths below 480 nm. In that case, however, appreciable fluorescence was observed, presumably from absorption of the THG signal at 355 nm. The MFP from the parallel-slab cells are symmetric, as expected. The intensity at successive minima are essentially zero for incident wavelength of 1542 nm. The weak dependence of $Q(\theta)$ on θ is evident in the slow variation of the successive maxima with θ for the MFP with the rotating slab at 1542 nm incident wavelength. Moreover, $R(0,L)$ is constant at the maxima for the data with the wedge-shaped cell, demonstrating that absorption effects are unimportant for input at 1542 nm. By contrast, the results with 1907 nm

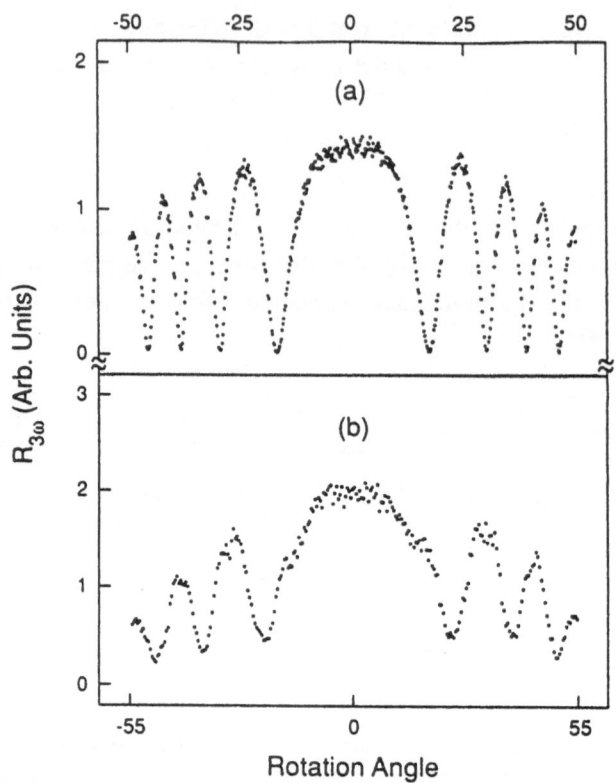

FIGURE 1: Maker Fringe Patterns from a parallel-slab cell (L = 375 μm) for an isotropic solution of PBT (φ = 0.021), with parallel polars for two incident wavelengths: (a) 1542 nm and (b) 1907 nm.

incident wavelength do not exhibit zero intensity at the successive minima. In addition, the R(0,L) at the maxima decrease and the minima increase markedly with increasing L for the data in the wedge-shaped cell for input at 1907 nm. These characteristics indicate that absorption at the incident wavelength, λ_ω = 1907 nm (the absorption is negligible at 636 nm) contributes substantially to R(q,L) for this case (see Eqn 4-6).

The data for incident wavelength of 1542 nm may be analyzed to give $|\chi^{(3)}|_{EFF} = |\chi^{(3)}|_{iso}$ using $n_{3\omega}$ and n_ω discussed in the preceding. In this case, the average refractive indices needed are calculated as $(n_E + 2n_O)/3$ for each frequency. Values of $|\chi^{(3)}|_{iso}/\varphi$ so determined are given in Table 1.

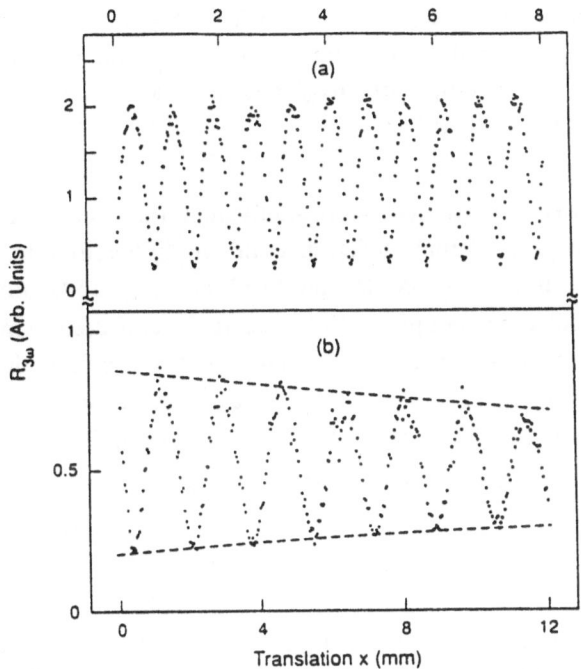

<u>FIGURE 2:</u>   Maker Fringe Patterns from a wedge-shaped cell (β = 0.95°) for an isotropic solution of PBT (φ = 0.021), with parallel polars for two incident wavelengths: (a) 1542 nm and (b) 1907 nm.

<u>Nematic Solutions:</u> In this case, the ratio $R(\theta,L)$ depends on the polarizations of the incident and THG signals, and the orientation of these relative to the sample director.   Several geometric arrangements are of interest, depending on the relative orientations of the polarizations of the incident and THG fields, E_{INC} and E_{THG}, respectively, and the optic axis of the monodomain n with respect to the axis of rotation; the latter is kept vertical in all the arrangements used.   As only the rotating plane parallel cell was used with the nematic monodomains, the dependence of $R(\theta,L)$ on L is suppressed hereafter.   All data were obtained using 1542 nm as input wavelength.   The large birefringence of the PBT monodomain complicates the analysis in some cases, as a relevant refractive index varies with θ for certain arrangements [15], see below.   In the following, the notation gives the orientations of these fields in the order $E_{THG}:n:E_{INC}$,

e.g., V-h-H designates an arrangement for which E_{THG} is vertical (parallel to the rotation axis) and both E_{INC} and n are horizontal. As may be seen in the preceding, knowledge of $n_{3\omega}$ and n_ω is required in order to determine $|\chi^{(3)}|_{EFF}$ from the MFP.

The MFP for $R_{VhV}(\theta)$ and $R_{HvH}(\theta)$ were equivalent for a nematic solution of PBT ($\varphi=0.055$), except for small difference in amplitude due to the Fresnel factors, and similar in appearance to the MFP at $\lambda_\omega = 1542$ nm shown in Figure 1 for the isotropic sample. This result is expected, since all rays propagate with refractive index n_O. The analysis using the measured refractive indices gave $|\chi^{(3)}|_{VhV}/\varphi = |\chi^{(3)}|_{HvH}/\varphi$, with results given in Table 1. The result is smaller than $|\chi^{(3)}|_{ISO}/\varphi$, as expected if the components to $\chi^{(3)}$ along the chain axis exceed those orthogonal to that axis. Both $|\chi^{(3)}|_{VhV}$ and $|\chi^{(3)}|_{HvH}$ correspond to $|\chi^{(3)}|_{yyyy}$, with the x and z Cartesian coordinates along the director and orthogonal to the sample plane, respectively.

The depolarized signals $R_{HhV}(\theta)$ and $R_{VvH}(\theta)$ were both nil ($R(\theta) < 0.01$), showing that the signals $R_{HvH}(\theta)$ and $R_{VhV}(\theta)$ discussed above correspond to plane-polarized THG waves. In the component notation used above, $|\chi^{(3)}|_{HhV}$ and $|\chi^{(3)}|_{VvH}$ both correspond to $|\chi^{(3)}|_{xyyy}$.

As shown in Figures 3 and 4, the behavior is more complex for data on $R_{VvV}(\theta)$, $R_{HhH}(\theta)$, $R_{HvV}(\theta)$ and $R_{VhH}(\theta)$: substantial nonzero minima are observed in all cases, and only one broad peak is observed for $R_{HvV}(\theta)$. Since no fluorescence signal is observed for λ from 532 to 800 nm, fluorescence contribution at $\lambda_{3\omega} = 514$ nm can not be the cause of the nonzero minima in Figure 3. Although the nonzero minima are reminiscent of the behavior attributed to absorbance effects for the isotropic sample with incident wavelength of 1907 nm, calculations of the relevant $\alpha_{3\omega}$ and α_ω show that absorption is not the source of the behavior observed here [15]. The substantial $R_{HvV}(\theta)$ proves that $\chi^{(3)}$ must have a nonzero $|\chi^{(3)}|_{yxxx}$ component, but as discussed below, the evaluation of that component may be compromised by the birefringent nature of the material. Part of the behavior giving the nonzero minima is clearly related to the refractive index for the propagating beams in the birefringent material. Thus, for $R_{VvV}(\theta)$, the incident beam propagates with refractive index n_E for all θ, however for $R_{HhH}(\theta)$, the incident beam propagates with refractive index $n_e(\theta)$, where

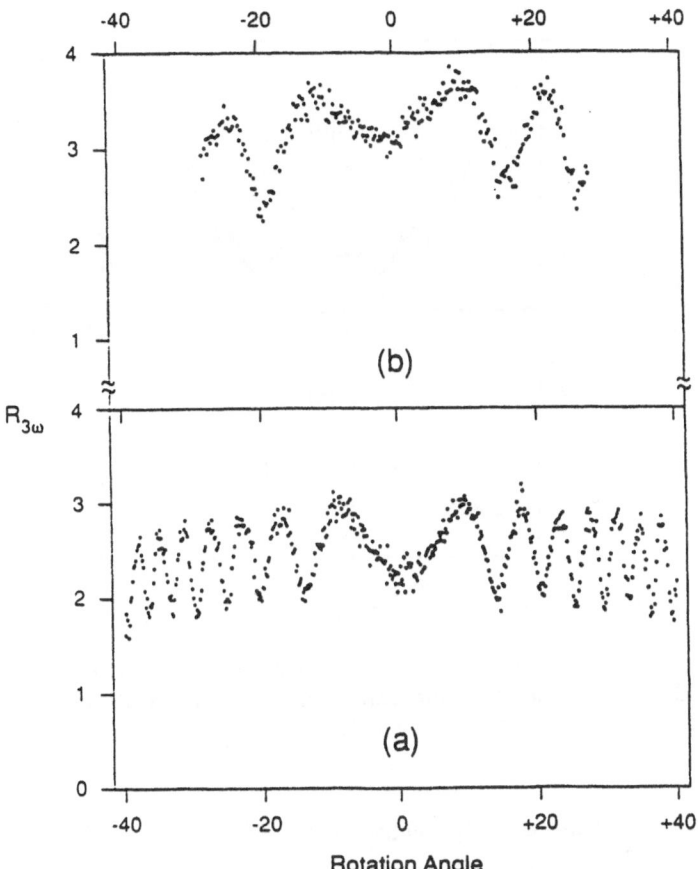

FIGURE 3: Maker Fringe Patterns for a nematic solution of PBT ($\varphi=0.055$), with parallel polars and the incident polarization parallel to the director arrangement: (a) V-v-V and (b) H-h-H.

$$n_e(\theta) = \left\{ \left(\frac{\cos\theta}{n_E}\right)^2 + \left(\frac{\sin\theta}{n_O}\right)^2 \right\}^{-1/2} \qquad (9)$$

In addition, the appearance of only slowly varying curves for $R_{HvV}(\theta)$ and $R_{vhH}(\theta)$ is due in part to the materials high birefringence. For example, consideration of the values of $n_{E,\omega}$ and $n_{O,3\omega}$ [13], shows that these are so close that $\psi(\theta) < \pi$ for the available range in θ, resulting in the broad peak observed for $R_{HvV}(\theta)$ in Figure 4. The increasing intensity noted for $R_{vhH}(\theta)$ with increasing θ in Figure 4b is due to the increasing difference between $n_{e,\omega}(\theta)$ and $n_{O,3\omega}$ when the incidence angle is varied.

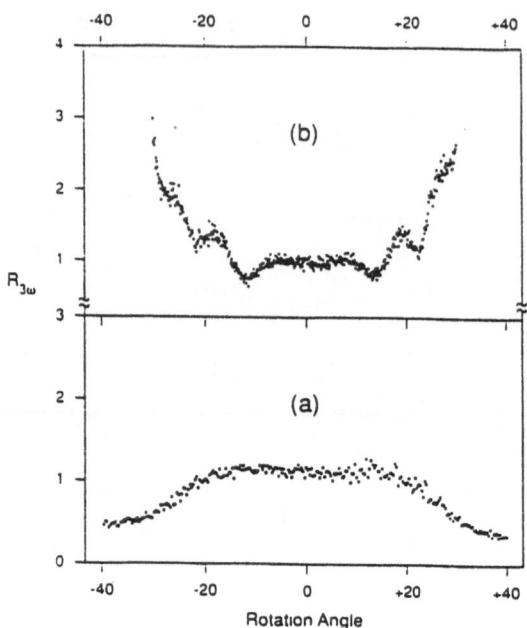

FIGURE 4: Maker Fringe Patterns for a nematic solution of PBT ($\varphi=0.055$), with crossed polars and the incident polarization parallel to the director arrangement: (a) H-v-V and (b) V-h-H.

TABLE I Third-order susceptibilities for PBT

P_i	$E_{j(kl)}$	COMPONENT	$|\chi^{(3)}|_{ijjj}/\varphi$ 10^{14}esu
ISOTROPIC SAMPLE;			$\varphi = 0.021$
x(y)	x(y)	Polarized	1120
x(y)	y(x)	Depolarized	< 0.1
NEMATIC SAMPLE;			$\varphi = 0.055$
x	x	xxxx	3640
y	y	yyyy	560
y	x	yxxx	19
x	y	xyyy	< 0.1

x-axis along the cell axis (the director for nematic sample); φ is the PBT volume fraction

In addition to these considerations, a THG signal created with polarization neither parallel nor orthogonal to the director will propagate as two waves, one each polarized in these directions in the birefringent sample, and propagating with distinct refractive indices that may depend on θ, producing elliptically polarized beam for the observed THG signal. This results in the nonzero minima noted in Figures 3 and 4.

Apparent values of $|\chi^{(3)}|_{vvv}$, $|\chi^{(3)}|_{HhH}$, $|\chi^{(3)}|_{vhH}$, and $|\chi^{(3)}|_{Hvv}$ deduced from the analysis of $R_{vhH}(0)$, $R_{HhH}(0)$, $R_{vhH}(0)$, and $R_{Hvv}(0)$, respectively, using the relevant refractive indices are listed in Table 1. With neglect of the complications from the birefringence, both $|\chi^{(3)}|_{vvv}$ and $|\chi^{(3)}|_{HhH}$ would correspond to $|\chi^{(3)}|_{xxxx}$, whereas both $|\chi^{(3)}|_{vhH}$ and $|\chi^{(3)}|_{Hvv}$ would correspond to $|\chi^{(3)}|_{yxxx}$. Owing to the birefringent nature of the material, the numerical values reported in Table 1 for these parameters may be suspect, but it seems unlikely that their general ranking will be altered if the birefringence is taken into account.

DISCUSSION

Summarizing the results, in component notation, $|\chi^{(3)}|_{xxxx} \gg |\chi^{(3)}|_{yyyy} > |\chi^{(3)}|_{yxxx} > |\chi^{(3)}|_{xyyy}$. For an isotropic sample, $|\chi^{(3)}|_{iso} = \Sigma |\chi^{(3)}|_{ijij}/5$. The value of $|\chi^{(3)}|_{ISO}/\varphi$ for the isotropic sample is seen to be larger than $\Sigma |\chi^{(3)}|_{ijij}/5\varphi = 960$ for the components given in Table 1, showing that terms such as $|\chi^{(3)}|_{xyxy}$, etc., do not vanish (it is assumed that $|\chi^{(3)}|_{zzzz} = |\chi^{(3)}|_{yyyy}$, etc.).

Of course, the principal interest here is not only in the $|\chi^{(3)}|_{ijkl}$ but in the corresponding components of the second hyperpolarizability tensor γ. In general,

$$\chi^{(3)}(-\omega; \omega_1,\omega_2,\omega_3) = f(\omega)f(\omega_1)f(\omega_2)f(\omega_3)N\gamma(-\omega; \omega_1,\omega_2,\omega_3) \tag{10}$$

where the $f(\omega_i)$ are local field factors and N is the number of molecules per unit volume. An eventual goal is to reliably estimate the parameter $\gamma(-\omega; \omega_1,\omega_2,\omega_3)$ from information on the molecular structure, and to be able to maximize the desired response by manipulation of the latter. The behavior reported here indicates that the maximum component to γ is off the molecular axis, contrary to what one might have anticipated for conjugated chain, for which π–electrons are assumed to be at the origin of the nonlinear coefficient [6,16]. This behavior requires that a relevant electronic transition in the NLO response of the solvated PBT be at an appreciable angle to the chain axis, in contrast with the result of calculations on conjugated cis and trans polyenes [6,16]. For PBT molecules, a possible candidate for a vector associated with an electronic transition of the rodlike molecular axis is the vector along the sulfur-sulfur atoms in the repeating unit of the chain (see above). This possibility opens the question of whether γ is influenced by the polymeric rodlike character of the PBT molecule, or it would be similar for a short oligomeric PBT (perhaps even a model of the repeat unit, i.e., the repeat unit terminated by protons).

Acknowledgements: This study was supported in part by a grant from the Air Force Office of Scientific Research. Discussions with Professor G. D. Patterson and Dr. P. G. Kaatz are appreciatively acknowledged.

REFERENCES

1 Nonlinear Optical Properties of Organic Molecules and Crystals, Ed. by S. Chemla, J. Zyss, Academic Press, Vol. 1 and 2, 1987.
2 a) Y. R. Shen, The Principle of Nonlinear Optics, J. Wiley, New York, 1984;
  b) S. K. Kurtz, in Quantum Electronics, Ed. by H. Rabin and C. L. Tang Vol. 1, Part A, Academic Press, N.Y., 1975, Chapt. 3, and references therein;
  c) F. Kajzar, J. Messier, Phys. Rev. A, 32, 2353 (1985).
3 A Yariv, Quantum Electronics, J. Wiley, New York, 1989.
4 Nonlinear Optical Properties of Polymers, Ed. by A. J. Heeger, J. Orenstein, D. R. Ulrich, Matl. Res. Soc. Symp. Proceed. Vol. 109, 1988.
5 Optical and Electronic Properties of Polymers, Ed. by J. M. Torkelson and J. A. Emerson, Matl. Res. Soc Symp. Proceed. Vol. 214, 1991.
6 P. N. Prasad, D. J. Williams, Introduction to Nonlinear Optical Effects in Molecules and Polymers, J. Wiley, New York, 1991.
7 J. Zyss, D. S. Chemla, Chapt. II-1 in ref. 1.
8 F. Kajzar, J. Messier, Chapt. III-2 in ref. 1.
9 a) G. C. Berry, P. R. Eisaman, J. Polym. Sci., Polym. Phys. Ed. 12:2253 (1974);
  b) C. C. Lee, S. G. Chu, G. C. Berry, J. Polym. Sci., Polym. Phys. Ed. 21:1573 (1983).
10 a) H. Mattoussi, P. G. Kaatz, G. D. Patterson, G. C. Berry, in ref. 5, p. 11.
   b) H. Mattoussi, G. C. Berry, ACS Polym. Preprints, ACS Meeting, 32(3): 690 (1991).
11 H. Vanherzeele, J. S. Meth, S. A. Jenekhe, M. F. Roberts, Appl. Phys. Lett. 58: 663 (1991).
12 D. N. Rao, J. Swiatkiewicz, P Chopra, S, K. Choshal, P. N. Prasad, Appl. Phys. Lett. 48: 1187 (1986).
13 a) H. Mattoussi, M. Srinivasarao, P. G. Kaatz, G. C. Berry, in ref. 5, p. 157;
   b) H. Mattoussi, M. Srinivasarao, P. G. Kaatz, G. C. Berry, Macromolecules, in press.
14 G. C. Berry, Disc. Faraday Soc. No. 79, 141 (1985).
15 H. Mattoussi, G. C. Berry, to be submitted.
16 A. F. Garito, C. C. Teng, Proc. Soc. Photo-Opt. Instrum. Eng. No. 613: 146 (1986).

Viscoelastic Properties of Semidilute Polyelectrolyte Solutions

I. Noda

Department of Applied Chemistry, Nagoya University
Furo-cho, Chikusa-ku, Nagoya 464-01, Japan

ABSTRACT: Zero-shear viscosity η^0 and steady-state compliance J_e of poly(N-methyl-2-vinylpyridinium chloride) solutions were measured in the absence and presence of added-salt over a wide range of polymer concentration. The polymer concentration dependence of η^0 in semidilute regions increases with increasing added-salt concentrations C_s and almost agrees with that for non-ionic polymers in good solvents at high added-salt concentrations. J_e in dilute regions depends on molecular weight M, polymer concentration C and C_s more strongly than that of non-ionic polymer solutions, as given by $J_e \propto M^{-2} C^{-3} C_s$. In entangled regions, on the other hand, J_e depends on polymer concentration only in the same manner as that of non-ionic polymer solutions, but its concentration dependence is weaker than that of non-ionic polymer solutions, as given by $J_e \propto C^{-1.3}$. Moreover, the polymer concentration dependence of the weight-average relaxation time τ_w increases with increasing added-salt concentration in the semidilute region for η^0 and the entangled region for J_e. These viscoelastic properties of entangled-polyelectrolyte solutions can be well explained by the reptation model assuming that the correlation lengths related to entanglements are determined by the electrostatic interactions evaluated from the Donnan equilibrium.

INTRODUCTION

The linear viscoelastic properties of polymers in the terminal regions can be discussed in terms of two parameters representing energy dissipation and storage, such as zero-shear viscosity η^0 and steady-state compliance J_e, respectively. η^0 of non-ionic polymer solutions can be understood by classifying the solutions into at least three regions, i.e., dilute, semidilute and concentrated regions, depending on concentration C and degree of coil-overlapping C/C^*, where C^* is defined by[1,2].

$$C^* = 3M/(4\pi \langle S^2 \rangle^{3/2} N_A) \qquad (1)$$

where M and $\langle S^2 \rangle$ are the molecular weight and the mean square radius of gyration of polymer at infinite dilution, respectively, and N_A is Avogadro's number.

Y. Imanishi (Ed.)
Progress in Pacific Polymer Science 2
© Springer-Verlag Berlin Heidelberg 1992

In dilute regions, the reduced zero-shear viscosity η^0_R defined by $\eta^0_{sp}/C[\eta]$ is expressed in expansion forms as

$$\eta^0_R = 1 + k'C[\eta] + \cdots \qquad (2a)$$
$$= 1 + k(C/C^*) + \cdots \qquad (2b)$$

where $\eta^0_{sp} = (\eta^0 - \eta_s)/\eta_s$, η_s is the solvent viscosity, k' is the Huggins' constant, $k = 3k'\phi'/4N_A$ and $[\eta]$ is the intrinsic viscosity. Here, the second equation is obtained by using the Flory-Fox equation, $[\eta] = \phi'\langle S^2 \rangle^{3/2}/M$, where ϕ' is the Flory viscosity factor.

In semidilute regions, η^0_R is given by the following scaling law by assuming that $\eta^0_{sp} \propto M^{3.4}$ in entangled systems[2-5].

$$\eta^0_R \propto (C/C^*)^{(4.4-3\nu)/(3\nu-1)} \qquad (3a)$$
$$\propto (C[\eta])^{(4.4-3\nu)/(3\nu-1)} \qquad (3b)$$

where ν is the exponent in the relationship between $\langle S^2 \rangle$ and M, $\langle S^2 \rangle \propto M^{2\nu}$. Thus, η^0_R can be expressed as universal functions of C/C^* or $C[\eta]$, eqs 2 and 3, from dilute through semidilute regions, if the molecular weights of samples are high enough[5].

In concentrated regions, the concentration dependence of η^0_{sp} becomes stronger than eq 3, and η^0_R deviates from the scaling law.

On the other hand, J_e of non-ionic polymer solutions is understood by classifying the solutions into two regions, *i.e.*, dilute (not-entangled) and entangled regions[2,5]. In dilute regions, a polymer chain is isolated so that J_e and the reduced steady-state compliance J_{eR} defined by $(J_e CRT/M)[\eta^0/(\eta^0 - \eta_s)]^2$ are given by

$$J_e \propto M \cdot C^{-1} \qquad (4a)$$
$$J_{eR} \doteq constant \qquad (4b)$$

In entangled regions[5], polymer chains form approximately a uniform network, regardless of solvent power, so that J_e is given by

$$J_e \propto M^0 \cdot C^{-2} \qquad (5a)$$
$$J_{eR} \propto (CM)^{-1} \qquad (5b)$$

Thus, J_{eR} is expressed as a universal function of CM. Moreover, it was reported that the cross-over concentration from dilute to entangled regions for J_e is about 5 times higher than that from dilute to semidilute regions for η^0 at the constant molecular weight[5].

Although a few studies have been reported on viscoelastic properties of polyelectrolyte solutions at finite concentrations where the coil-overlapping becomes significant[6-8], a definite conclusion has not been reached because the samples had relatively broad molecular weight distributions, and/or the ranges of polymer concentration were limited.

In a previous paper[9], we reported the preparation of polyelectrolyte samples with relatively narrow molecular weight distributions over a wide range of molecular weight. In the present work, therefore, we measured η^0 and J_e of the polyelectrolyte sample solutions at finite concentrations in the absence and presence of added-salt to study the effect of electrostatic interactions on viscoelastic properties in comparison with those of non-ionic polymer solutions[10,11].

EXPERIMENTAL SECTION

Polyelectrolytes used here were poly(N-methyl-2-vinylpyridinium chloride)s (PMVP-Cl) with narrow molecular weight distributions. They were prepared by quaternizing anionically polymerized poly(2-vinyl-pyridine)s with dimethylsulfate in dimethylformamide at room temperature and by dialyzing the aqueous sample solutions against aqueous NaCl solutions[9]. The weight-average molecular weights M_w of samples, MVPK-11, MVPK-12 and MVPK-13, determined by light scattering are 6.9_4 x 10^5, $11._0$ x 10^5 and $27._4$ x 10^5, respectively. The molecular weight distribution indexes M_w/M_n, determined by GPC are less than 1.1. The degrees of quaternization determined by neutralization titration are around 85%.

Viscoelastic data of relatively high concentrated solutions were measured in oscillating and steady-shear flow at room temperature (ca. 25℃) by using Fluid Spectrometers RFS-8500 and RFS II, and a Mechanical Spectrometer RMS-800 of Rheometrics, Inc., with the cone-and-plate and the co-axial cylinder. At relatively low concentrations most of η^0 data was measured with capillary viscometers of Maron-Krieger-Sisko type at 25℃, and J_e data at two low concentrations were obtained from flow birefringence measurements[12] for comparison. To examine the degradation of the samples after the measurements of viscoelastic properties, we measured $[\eta]$ to confirm that there was no degradation.

Solvents used were salt-free, 0.5, 0.1 and 0.01M NaCl solutions. All the polymer solutions in the presence of added-salt were prepared by dialyzing the polymer solutions against the aqueous solutions with the respective NaCl concentration to attain the Donnan equilibrium. The polymer concentrations (g/cm^3) were determined by measuring the UV absorption of the solutions at 267 nm[9].

RESULTS

As shear strorage and loss moduli, G' and G" in low frequency ranges were proportional to the second and first powers of frequency ω, respectively, η^0 and J_e were evaluated from the following equations in the terminal regions.

$$\eta^0 = \lim_{\omega \to 0} G''/\omega$$
$$J_e = \lim_{\omega \to 0} G'/[\omega^2(\eta^0)^2]$$

As reported previously[13], the form anisotropy in flow birefringence is negligible in semidilute regions for thermodynamic properties so that we evaluated J_o in 0.01M NaCl aqueous solutions at two polymer concentrations from the data of flow birefringence by using the following relationships[14].

$$J_o = \lim_{\gamma \to o} \Delta n \cdot \cos 2\chi / (2C' \eta^2 \dot{\gamma}^2)$$

where Δn is the birefringence, χ is the extinction angle, γ is the shear rate, η is the viscosity of solution, and C' (1.9×10^{-9} cm^2/dyn for PMVP-Cl in NaCl solutions) is the stress-optical coefficient.

As shown in Figure 1, at the same molecular weight, the lower the added-salt concentration, the larger η^0_{sp} is, but the lower its polymer concentration dependence is. In high polymer concentration regions, the polymer concentration dependence of η^0_{sp} becomes the same at the same added-salt concentrations, regardless of the

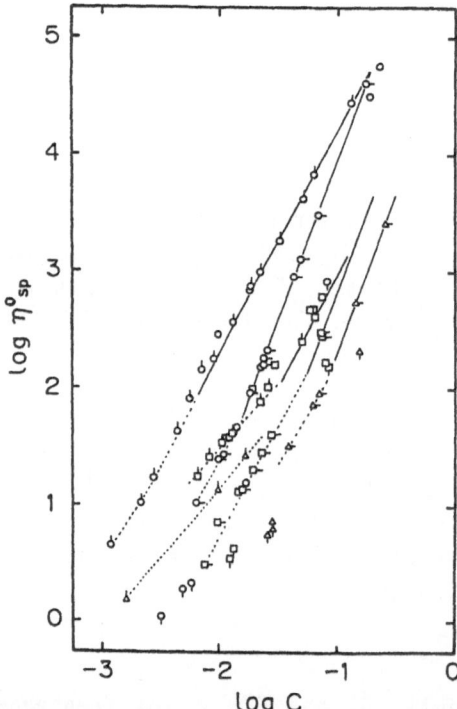

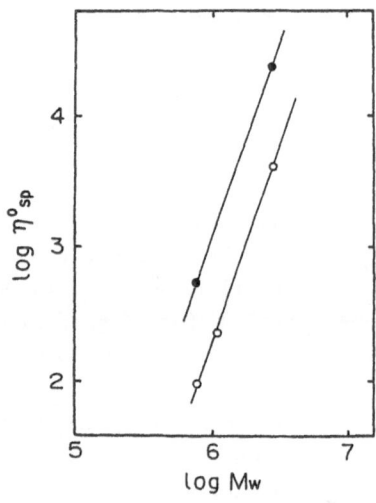

Figure 2. Molecular weight dependence of η^0_{sp} in 0.1M NaCl solutions. The closed and open circles denote the data at 0.15 and 0.06 g/cm^3, respectively. (Reproduced from ref. 10).

Figure 1. Polymer concentration dependences of η^0_{sp} in the absence and presence of added-salt. The triangles, squares and circles denote the data for MVPK-11, MVPK-12 and MVPK-13, respectively. The upward, rightward and downward pips indicate the data in 0.01, 0.1 and 0.5M NaCl solutions, respectively. The symbols without pip denote the data in the absence of added-salt. The dotted and solid lines were drawn to connect smoothly the data in dilute and semidilute regions, respectively. (Reproduced from ref. 10).

molecular weights, as denoted by the solid lines. Figure 2 shows double logarithmic plots of η^0_{sp} vs. M_w in 0.1M NaCl solutions at C=0.06 and 0.15g/cm³ where the molecular weight dependence of η^0_{sp} is independent of polymer concentration. This figure reveals that the molecular weight dependence of η^0_{sp} in 0.1M NaCl solutions is given by $\eta^0_{sp} \propto M^3$ denoted by the solid lines. Figure 3 shows double logarithmic plots of η^0_{sp}/M^3 vs. C in salt-free, 0.01, 0.1 and 0.5M NaCl solutions in the high polymer concentration regions. In each added-salt solution, the η^0_{sp}/M^3 data in the high polymer concentration region called the semidilute region are almost independent of molecular weight, and they are approximately given by

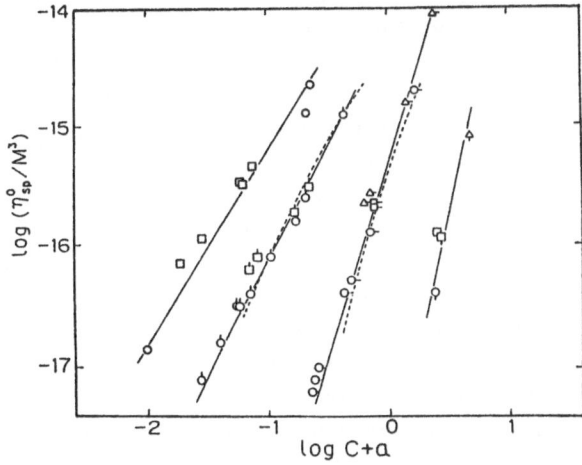

Figure 3. Double logarithmic plots of η^0_{sp}/M^3 vs. C in salt-free, 0.01, 0.1 and 0.5 NaCl solutions. a is the arbitrary number to shift the data in different added-salt concentration along the abscissa. Symbols are the same as in Figure 1. The solid and broken lines denote eqs 6 and 21, respectively. Both lines coincide with each other in salt-free and 0.5M NaCl solutions.

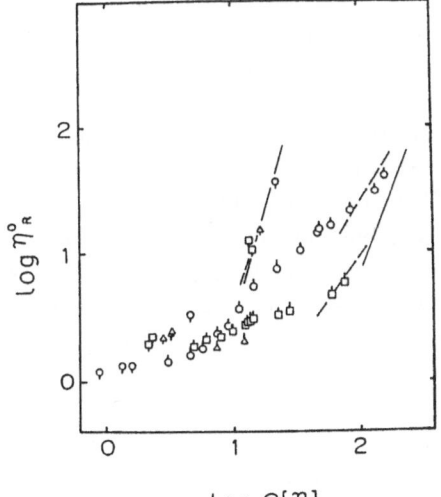

Figure 4. Double logarithmic plots of η^0_R vs. C[η] in 0.01 and 0.5M NaCl solutions. Symbols are the same as in Figure 1. The solid and broken lines denote the values calculated by eqs 7 and 21, respectively.

$$\eta^0_{sp} \propto M^3 C^{1.5} \qquad (C_s = 0M) \qquad (6a)$$

$$\eta^0_{sp} \propto M^3 C^{1.8} \qquad (C_s = 0.01M) \qquad (6b)$$

$$\eta^0_{sp} \propto M^3 C^{2.8} \qquad (C_s = 0.1M) \qquad (6c)$$

$$\eta^0_{sp} \propto M^3 C^{4.6} \qquad (C_s = 0.5M) \qquad (6d)$$

In the following discussion, therefore, we assume that η^0_{sp} is proportional to M^3 in the semidilute region measured here.

In Figures 4, the η^0_{sp} data in 0.01 and 0.5 M NaCl solutions are replotted in the double logarithmic form of η^0_R vs. $C[\eta]$. If we assume that $\eta^0_{sp} \propto M^3$ in semidilute regions, eq 3 can be rewritten by

$$\eta^0_R \propto (C/C^*)^{(4-3\nu)/(3\nu-1)} \qquad (7a)$$

$$\propto (C[\eta])^{(4-3\nu)/(3\nu-1)} \qquad (7b)$$

Introducing the ν values evaluated by assuming $\nu = (a+1)/3$ from the exponent a in the Mark-Houwink-Sakurada equations, $[\eta] = KM^a$ [9], into eq 7, we have the calculated results denoted by the solid lines in Figure 4. As shown in the figure η^0_R appears to be expressed as a universal function of $C[\eta]$ and its experimental dependence appears to agree with the calculated line in 0.5M NaCl solutions, whereas η^0_R is not expressed as a universal function and its dependences are lower than the calculated lines in 0.01M NaCl solutions.

Figure 5 shows double logarithmic plots of J_e vs. C in salt-free, 0.5, 0.1 and 0.01M NaCl solutions. As shown in the figure, the data from the mechanical and flow birefringence measurements are in good agreement with each other. At low polymer concentrations, the higher the added-salt concentration, the larger J_e is at the same molecular weight, and J_e decreases with increasing molecular weight at the same polymer and added-salt concentrations. At high polymer concentrations, on the other hand, J_e appears to depend on polymer concentration only, regardless of added-salt concentration and molecular weight, in the same manner as that of non-ionic polymer solutions. If we assume that the region where J_e depends on molecular weight is the dilute one in the same manner as that of non-ionic polymer solutions, the dependences of J_e on polymer concentration and molecular weight in the dilute region are very much different from those of non-ionic polymer solutions. Considering the J_e dependence on molecular weight and added-salt concentration in the dilute region in Figure 5, we plotted $J_e M^2/C_s$ vs. C in 0.01, 0.1 and 0.5M NaCl solutions in Figure 6. From the figure, we have the following empirical formula in the dilute region.

$$J_e \propto C^{-3} M^{-2} C_s \qquad (8)$$

The slopes of solid lines in the dilute region in Figures 5 and 6 correspond to the polymer concentration dependence of eq 8.

As shown by the broken lines in Figures 5 and 6, on the other hand, the deviation from eq 8 is observed in the entangled region,

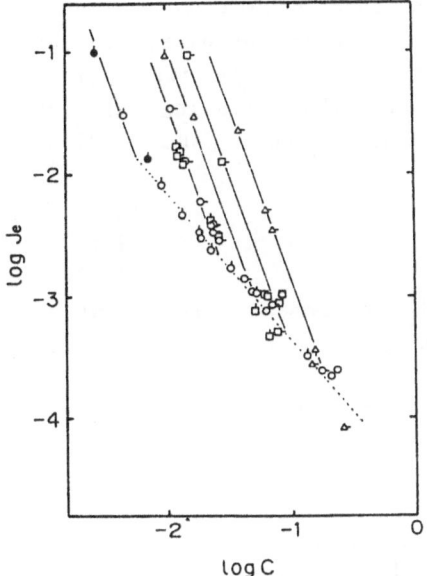

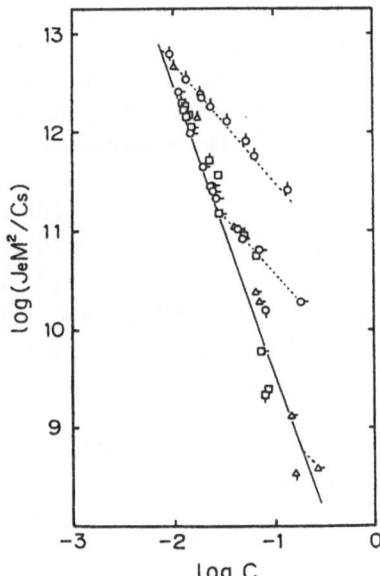

Figure 5. Polymer concentration dependences of J_e in the absence and presence of added-salt. Symbols are the same as in Figure 1. The closed circles denote the data from the flow birefringence measurements. The solid and broken lines are drawn to connect smoothly the data in dilute and entangled regions, respectively. (Reproduced from ref. 11).

Figure 6. Double logarithmic plots of $J_e M^2/C_s$ vs. C. Symboles are the same as in Figure 1. The solid and broken lines correspond to the data in dilute and entangled regions, respectively. (Reproduced from ref.11).

where the molecular weight dependence is not observed. In Figure 7, all the data in the entangled region are plotted in the double-logarithmic form of J_e vs. C. From this figure, we have the following empirical formula.

$$J_e \propto C^{-1.3} \tag{9}$$

Figure 5 indicates that the higher the molecular weight, or the lower the added-salt concentration, the lower the cross-over polymer concentration between the two regions is. Although the polymer concentration and molecular weight dependences of J_e are different from those of non-ionic polymer solutions, the order of cross-over concentrations is reasonable.

The weight-average relaxation time τ_w is given by

$$\tau_w = \eta^0 J_e \tag{10}$$

It is to be noted that the longest relaxation time is almost proprtional to τ_w in entangled non-ionic polymer systems[15].

Since the cross-over concentration from dilute to entangled regions for J_e is higher than that from dilute to semidilute regions for η^0, the semidilute region for η^0 is divided into two regions. Introducing eqs 6 and 8 into eq 10, we have the following empirical formulas for τ_w in the semidilute region for η^0 and the dilute region for J_e.

$$\tau_w \propto MC^{-1.2} \qquad (C_s = 0.01M) \qquad (11a)$$
$$\tau_w \propto MC^{-0.2} \qquad (C_s = 0.1M) \qquad (11b)$$
$$\tau_w \propto MC^{1.6} \qquad (C_s = 0.5M) \qquad (11c)$$

The experimental results are entirely different from that of non-ionic polymers in good solvents given by $\tau_w \propto M^{4.4}C^{3.3}$[16], and we have no theory to explain eq 11.

Introducing eqs 6 and 9 into eq 10, we have the following empirical formulas for τ_w in the semidilute region for η^0 and the entangled region for J_e.

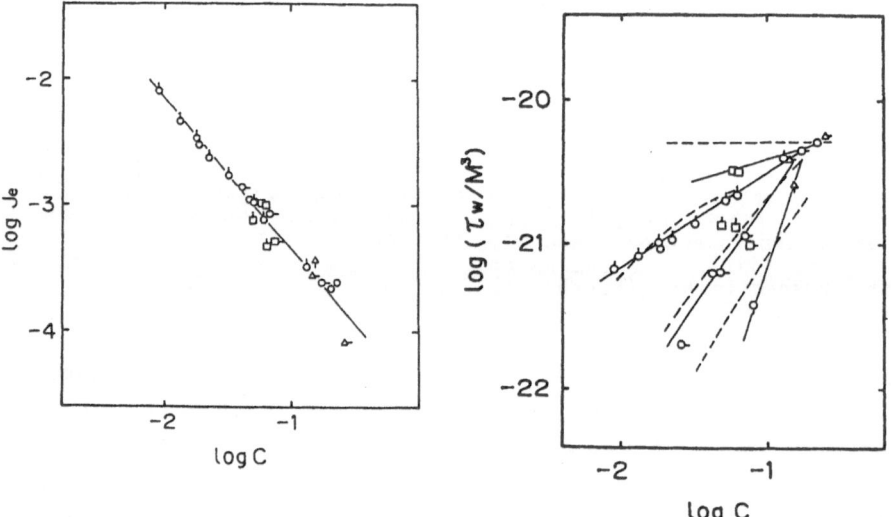

Figure 7. Double logarithmic plots of J_e vs. C in entangled regions. Symbols are the same as in Figure 1. The line denotes eq 9. (Reproduced from ref. 11).

Figure 8. Double logarithmic plots of τ_w/M^3 vs. C in the absence and presence of added-salt in the semidilute region for η^0 and the entangled region for J_e. Symbols are the same as in Figure 1. The solid and broken lines denote eqs 12 and 28, respectively, in salt-free, 0.01, 0.1 and 0.5M NaCl solutions from the top to the bottom. (Reproduced from ref. 11).

$$\tau_w \propto M^3 C^{0.2} \qquad (C_s = 0M) \qquad (12a)$$
$$\tau_w \propto M^3 C^{0.5} \qquad (C_s = 0.01M) \qquad (12b)$$
$$\tau_w \propto M^3 C^{1.5} \qquad (C_s = 0.1M) \qquad (12c)$$
$$\tau_w \propto M^3 C^{3.3} \qquad (C_s = 0.5M) \qquad (12d)$$

Figure 8 shows double logarithmic plots of τ_w/M^3 vs. C in the absence and presence of added-salt in this region, together with eq 12 denoted by the solid lines. The experimental results indicate that the polymer concentration dependence of τ_w increases with increasing added-salt concentration, and almost agrees with that for non-ionic polymers in good solvents at high added-salt concentrations[16].

DISCUSSION

Approximately, η^0 in entangled polymer systems is given by the product of the longest relaxation time τ and elastic modulus G[1,17].

$$\eta^0 \propto \tau G \qquad (13)$$

According to the reptation model, τ and G in semidilute regions can be given by[1,18]

$$\tau \propto (\eta_s/k_BT)(Na/n)^3 \qquad (14)$$
$$G \propto ck_BT/n \qquad (15)$$

where N is the number of segments per polymer chain, n and a are the number of segments and the dimension of chain between entanglement points, respectively, c is the number of segments per unit volume, which is proportional to C, k_B is the Boltzmann constant and T is the absolute temperature.

Using eqs 13, 14 and 15, we have

$$\eta^0 \propto \eta_s(N/n)^3 a^3 c/n \propto \eta_s(N/n)^3 \qquad (16)$$

Here, the following relationship was used.

$$n \propto ca^3 \qquad (17)$$

Moreover, if we assume that a is proportional to the thermodynamic correlation length ξ, which is expanded as a single polyion chain in dilute solutions by an expansion factor α, we have

$$a \propto \xi \sim bn^{1/2}\alpha \qquad (18)$$

where b is the segment length.

Applying the procedure of the theory of Flory on the expansion factor of a polyelectrolyte in dilute solutions, based on the Donnan equilibrium[19], to the polyion chain between entanglement points, we have

$$\alpha - 1/\alpha \propto n^{3/2}\alpha^2[(i^2c^2+4C_s^2)^{1/2}-2C_s] \qquad (19)$$

where i is the effective charge of charged segments, in other words, ic is the concentration of free counter-ions from polyelectrolytes. To make the discussion simpler, we assume that $\alpha \gg 1$, which usually holds for polyelectrolytes, so that we have the following approximate expression for α from eq 19.

$$\alpha^{-1} \propto n^{3/2}[(i^2c^2+4C_s^2)^{1/2}-2C_s]$$ (20)

Using eqs 16, 17, 18 and 20, we have the following equation for polymer concentration dependence of η^0 as a function of added-salt concentration.

$$\eta^0 \propto N^3[(i^2c^2+4C_s^2)^{1/2}-2C_s]^{9/4}/c^{3/4}$$ (21)

If the added-salt concentration is high, i.e., $C_s > ic$, eq 21 is rewritten as

$$\eta^0 \propto C_s^{-9/4}N^3C^{15/4}$$ (22)

Using the theory of Flory on the expansion factor of a polyelectrolyte in dilute solutions[19], we have the scaling law.

$$\eta^0_R \propto (C/C^*)^{11/4}$$ (23a)
$$\propto (C[\eta])^{11/4}$$ (23b)

Eq 23 corresponds to eq 7 with $\nu = 0.6$. If the polymer concentration is high, and/or the added-salt concentration is low, i.e., $C_s < ic$, on the other hand, eq 21 is expressed as

$$\eta^0 \propto N^3c^{1.5}$$ (24)

The comparison between observed and calculated polymer concentration dependences of $\eta^0_{s,p}$ in the absence and presence of added-salt is shown in Figure 3. Here, we assumed the effective charge $i=0.2$[20] and shifted the calculated lines along the vertical axis to fit the data in the respective added-salt concentration. Apparently, the theory well explains the experimental results given by eq 6. As shown in Figure 4, moreover, the theory explains the result that η^0_R obeys the scaling law at the high added-salt concentration (0.5M), but not at the low added-salt concentration (0.01M).

Using eqs 15, 17, 18 and 20, and assuming that $J_e \propto 1/G$, we have

$$J_e \propto 1/\{[[(i^2c^2+4C_s^2)^{1/2}-2C_s]^{3/4}c^{3/4}k_BT\}$$ (25)

If $C_s > ic$, we have

$$J_e \propto 1/C^{9/4}$$ (26)

This dependence corresponds to the prediction of scaling law for non-ionic polymers in good solvents. If $C_s < ic$, on the other hand, we have

$$J_e \propto 1/C^{3/2}$$ (27)

Since we can assume that $C_s < ic$ in the entangled region of the present samples, the polymer concentration dependence is predicted to be lower than that of non-ionic polymers in entangled regions given by eq 5. The experimental dependence (eq 9) agrees with this prediction, though it is slightly weaker than eq 27. The reason is not clear at present.

From eqs 21 and 25, we have the following equation for τ_w in the semidilute region for η^0 and the entangled region for J_e

$$\tau_w \propto \tau \propto N^3[(i^2c^2+4C_s^2)^{1/2}-2C_s]^{3/2}/c^{3/2}$$ (28)

If $C > ic$, we have

$$\tau_w \propto N^3 c^{3/2} \tag{29}$$

This result corresponds to τ_w for non-ionic polymer in good solvents according to the scaling law, though the experimental molecular weight and concentration dependences are slightly different[16]. If $C < ic$, on the other hand, we have

$$\tau_w \propto N^3 \tag{30}$$

Thus, the theory predicts that τ_w becomes independent of polymer concentration in salt-free solutions, and its polymer concentration dependence increases with increasing added-salt concentration. As shown in Figure 8, the experimental data at the various added-salt concentrations almost agree with the calculated results denoted by the broken lines. Here, we assumed the effective charge $i=0.2$ in eq 28 and shifted the calculated lines along the vertical axis to fit the data at the respective added-salt concentration.

In summary, we conclude that the linear viscoelastic properties of entangled-polyelectrolyte solutions in the terminal regions can be well explained by the reptation model taking into account the electrostatic interactions evaluated by the Donnan equilibrium [10,11].

ACKNOWLEDGMENT.   This work was supported by a Grant-in Aid for Scientific Research (No. 63430019) from the Ministry of Education, Science and Culture of Japan.

REFERENCES

1 de Gennes P -G (1979) Scaling concepts in polymer physics. Cornell Univ press, Ithaca, NY
2 Noda I (1988) In: Nagasawa M (ed) Molecular conformation and dynamics of macromolecules in condensed systems. Elsevier, Tokyo
3 de Gennes P -G (1976) Macromolecules 9:587
4 Takahashi Y, Isono Y, Noda I, Nagasawa M (1985) Macromolecules 18: 1002
5 Takahashi Y, Noda I, Nagasawa M (1985) Macromolecules 18:2220
6 Sakai M, Noda I, Nagasawa M (1972) J polym sci A-2 10:1047
7 Rinaudo M, Graebling D (1986) Polym bull 15:253
8 Morris E R, Rees D A, Welsh E J (1980) J mol biol 138:383
9 Yamaguchi M, Yamaguchi Y, Matsushita Y, Noda I (1990) Polym j (Tokyo) 22:1077
10 Yamaguchi M, Wakutsu, Takahashi Y, Noda I (1992) Macromolecules in press
11 Yamaguchi M, Wakutsu, Takahashi Y, Noda I (1992) Macromolecules in press

12 Suzuki F, Hori K, Kozuka N, Komoda H, Katsuro K, Takahashi Y, Noda I, Nagasawa M (1986) Polym j(Tokyo) 18:911

13 Takahashi Y, Yamaguchi M, Hori K, Suzuki F, Noda I (1989) Polym j (Tokyo) 21:815

14 Janeschnitz-Kriger H J (1969) Adv polym sci 6:170

15 Takahashi Y, Wakutsu M, Noda I (1990) Macromolecules 23:242

16 Takahashi Y, Umeda M, Noda I (1988) Macromolecules 21:2257

17 Doi M, Edwards S F (1986) The theory of polymer dynamics. Clarendon press, Oxford

18 de Gennes P.-G. (1976) Macromolecules 9:594

19 Flory P J (1953) Principles of polymer chemistry. Cornell Univ press, Ithaca, NY

20 Nagasawa M (1974) In: Selegny E (ed) Polyelectrolytes. D Reidel publishing company, Boston

Improved Polymers for Medical Implants-Polyurethanes

Gordon F. Meijs*, Pathiraja A. Gunatillake, Ezio Rizzardo,
Simon J. McCarthy, Ronald C. Chatelier,

CSIRO Division of Chemicals and Polymers,
Private Bag 10, Clayton, Victoria 3168 Australia,

Arthur Brandwood, Klaus Schindhelm,

Centre for Biomedical Engineering, The University of New South Wales,
P. O. Box 1, Kensington, NSW 2033, Australia.

ABSTRACT

Polyurethane elastomers, prepared from: (i) polyether macrodiols that contain a reduced number of ether linkages compared with PTMO [poly(tetramethylene oxide)], (ii) the diisocyanate MDI [4,4'-diphenylmethanediisocyanate], and (iii) the chain extender BDO [1,4-butanediol] offer enhanced stability towards oxidation and hydrolysis over their PTMO-based counterparts. Polyurethane-ureas prepared from the diisocyanate TMXDI (m-tetramethylxylene diisocyanate), however, show decreased stability. *In vivo* subcutaneous implant experiments (sheep; 90 days and 180 days), show that the new MDI-based ether-reduced polyurethanes do not undergo stress cracking while the PTMO-based materials do. The TMXDI materials performed poorly when implanted.

INTRODUCTION

Segmented polyurethane elastomers, prepared from diisocyanates, macrodiols and chain extenders, are frequently used in the construction of implantable medical devices, such as cardiac pacemakers, heart valves, catheters, and heart assist devices, because of their excellent mechanical properties and haemocompatibility. Polyether macrodiols, such as poly(tetramethylene oxide) (PTMO), are used to prepare polyurethanes for implant, since they offer an increased resistance to enzymatic hydrolysis compared with polyester-based polyurethanes.

Nevertheless, polyether-based polyurethanes have been found to degrade when implanted for long periods (1). The degradation is manifested in terms of surface or deep cracking, erosion, or deterioration of mechanical properties, such as flex life.

Y. Imanishi (Ed.)
Progress in Pacific Polymer Science 2
© Springer-Verlag Berlin Heidelberg 1992

Several major pathways have been postulated for *in vivo* degradation (1-7); they are:
(i) environmental stress cracking, (ii) metal ion induced oxidation, and (iii) calcium ion
induced embrittlement of the soft segment. Of these, arguably, environmental stress
cracking is the most important, but the least understood. Oxidative attack of the
polyether soft segment, which is believed to accumulate at the implant surface, has
been implicated.

We have prepared novel polyurethane elastomers with modified soft segment chemistry that has been designed to enhance stability. The new thermoplastic polyurethanes
are based on conventional diisocyanates and chain extenders, but contain macrodiols
with a reduced number of ether linkages when compared with PTMO. As part of the
development process, improved synthetic procedures involving condensation polymerization were devised (8,9) to prepare the macrodiols, poly(hexamethylene oxide) 1,
poly(octamethylene oxide) 2, and poly(decamethylene oxide) 3.

| 1 | 2 | 3 |

EXPERIMENTAL SECTION

Pellethane™ 2363-80A was obtained from Dow Chemicals, while Biomer™ was obtained from Ethicon. PTMO (M_n = 1000) was obtained from DuPont under the trade
name, Terethane™. Its hydroxyl number was determined to verify that it was
bis(hydroxyl) terminated. PTMO was dried *in vacuo* for 15 h at 105 °C before use.

The macrodiols, α,ω-dihydroxypoly(hexamethylene oxide) (PHMO),
α,ω-dihydroxypoly(octamethylene oxide) (POMO), and
α,ω-dihydroxypoly(decamethylene oxide) (PDMO) were prepared and dried as
described by us previously (8). MDI and TMXDI were distilled under vacuum and the
middle fractions were used for polymerizations. Dimethylformamide was dried over
activated 3 Å molecular sieves and distilled under nitrogen. 1,4-Butanediol and 1,6-
diaminohexane were dried over activated 3 Å molecular sieves and purified by
distillation.

General Method for the Two Stage Polymerization Procedure (10)

Step 1 (synthesis of the prepolymer). The macrodiol was placed in a flask under a
nitrogen atmosphere. The appropriate amount of freshly distilled MDI (150 % molar
excess), in sufficient anhydrous dimethylformamide to make a 50 % w/w solution and

stannous octoate (0.01% of total solids) were added to the flask and the mixture was heated between 60 and 90 °C to end-cap the macrodiol. The isocyanate content of the prepolymer was determined using ASTM method D1638-74.

Step 2 (chain extension of the prepolymer). The prepolymer was diluted to 25 % (w/v) with anhydrous dimethylformamide. The chain extender 1,4-butanediol (BDO) (amount based on the free isocyanate content of prepolymer) was then added over about 20 min at ambient temperature with stirring under a nitrogen atmosphere. The solution was then heated to 80-90 °C for 2-4 h to complete the chain extension reaction. Chain extension with 1,6-diaminohexane (HDA = hexamethylene diamine) was carried out by dropwise addition of a 10 % solution in anhydrous dimethylformamide to the prepolymer at 5 °C over 20 min and completing the reaction by stirring at ambient temperature for 1 h.

The polymer was then isolated by dropwise addition of the diluted (dimethylformamide) polymer solution to a large volume of deionized water with stirring. The precipitated polymer was filtered, thoroughly washed with fresh deionized water and dried in a vacuum oven (~ 0.1 torr) at 45 °C.

Sample Fabrication

After drying for 15 h *in vacuo*, polyurethanes were compression moulded at temperatures between 100 and 200 °C and at a nominal load of either 8 or 12 tons. Biomer™ was solvent cast in 3 layers directly from the 30% dimethylacetamide solution supplied and the sheets were dried at 40 °C in a flow of dry nitrogen for 7 days. The flat sheets had dimensions of 60 mm x 100 mm and were 1 mm thick. They were cut into several pieces or punched into dumbbells 3 cm in length. The straight (testing) area of the dumbbell was 13 mm x 4 mm. All samples showed no birefringence under cross polarizers, indicating that there was no detectable residual stress.

Degradation Experiments

Dumbbells of polyurethane were encased in a PTFE sheet that was bent once and tied loosely with PTFE tape. The PTFE sheet had holes punched in it to facilitate the circulation of the test solution; each PTFE sheet also had a binary code punched into it for sample identification. This arrangement weighted the sample so that it did not float, and prevented samples from adhering to each other; the enveloping was sufficiently loose that exposure of the material to the reagent was unhindered. The treatment time in all degradation experiments was 24 h.

The hydrolysis experiments (100 °C) were carried out by immersing the weighted dumbbells for 24 h separately in deionized water, 2 M hydrochloric acid, or 5 M sodium hydroxide, all at 100 °C. Treatment of the samples with water at 120 °C was carried out by placing the weighted dumbbells into a metal pressure vessel that was lined with glass. The pressure vessel was then filled to one third of its capacity with water, which was sufficient to immerse the test specimens. It was then placed in an oven, thermostatted to 120 °C. After 24 h, the pressure vessel was cooled rapidly to room temperature and its contents were examined.

Oxidative tests were carried out in a similar manner on dumbbells for 24 h by using either 25 % aqueous hydrogen peroxide (prepared by dilution of 30 % hydrogen peroxide from May and Baker) or by using sodium hypochlorite from freshly opened bottles (Ajax Chemicals), nominally containing 4 % available chlorine.

After these treatments, the dumbbells were rinsed thoroughly with deionized water and dried *in vacuo* at 40 °C overnight before being subjected to tensile testing.

In Vitro Evaluation

Mechanical testing was carried out in triplicate with an Instron model 4032 universal testing machine. A 1 kN load cell was used and the crosshead speed was 500 mm/min.

Gel permeation chromatography was carried out at 80 °C with 0.05 M lithium bromide in dimethylformamide as eluent on a Waters Associates chromatograph with 10^5, 10^3, and 50 Å μ-Styragel columns. The system was equipped with a refractive index detector and was calibrated with narrow distribution polystyrene standards. Results are expressed, therefore, as polystyrene-equivalent molecular weights.

In vivo Evaluation

180 day implant. Polyurethane rectangles (1.0 mm thick) were cut and sterilized with ethylene oxide and subsequently degassed for 7 days. They were then implanted subcutaneously (6 replicates; for 180 days) in the dorsal thoraco-lumbar region of 1-2 year old crossbred wether sheep. After retrieval, the samples were soaked in 0.1 M sodium hydroxide to remove remaining biological material. They were then washed in water, dried and examined by scanning electron microscopy at magnifications up to 2000. Control samples were stored (a) in sterile phosphate buffered saline (PBS) at pH 7.2 at 37 °C and (b) in darkness at room temperature, and treated similarly, except that they were not implanted.

90 day implant. Dumbbell specimens were cut from polyurethane sheets (0.5 mm thick) using a specially manufactured die. They were stretched over poly(methyl methacrylate) holders to extend the central portion of the dumbbell to an additional 150% length. A 3-0 polypropylene suture was tied firmly (once) around the central portion of the stretched dumbbell, so as to act as a local stress raiser. The holders were then implanted subcutaneously for 90 days, and then retrieved and examined as above.

Table 1. Mechanical Properties of Polyurethane elastomers[a]

Sample Code	Hard segment composition	Soft segment macrodiol (M_n)		Fail stress (MPa)	Fail strain (%)	Stress at 100% strain (MPa)	Hardness (Shore A)
4	Pellethane 2363-80A [b]			30	540	8	82
5	MDI/BDO[c]	PTMO	(980)	11	470	7	86
6	MDI/BDO	1	(650)	21	240	17	90
7	MDI/BDO	2	(1690)	13	320	8	93
8	MDI/BDO	3	(1270)	20	160	18	100
9	TMXDI/HDA	PTMO	(980)	17	460	6	72
10	TMXDI/HDA	1	(570)	23	720	19	91
11	TMXDI/HDA	2	(1170)	23	920	13	86
12	TMXDI/HDA	3	(1270)	20	310	15	97

[a] All polyurethanes reported herein were prepared by two step solution polymerization (10), with a diisocyanate:macrodiol:chain extender ratio of 2.5:1.0:1.5.
[b] This commercial material (Dow Chemical Co.) is composed of MDI/BDO and PTMO, but differs from the other polyurethanes in that it is prepared by bulk polymerization.
[c] The mechanical properties of this material were enhanced by bulk polymerization, and fail strains exceeding that of Pellethane 2363-80A could be obtained by adjustment of the conditions of synthesis.

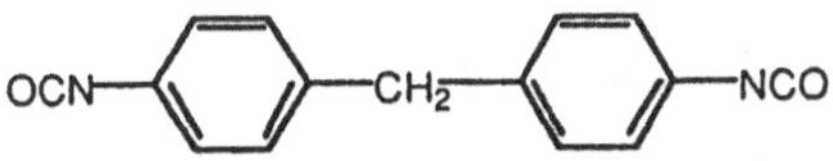

MDI

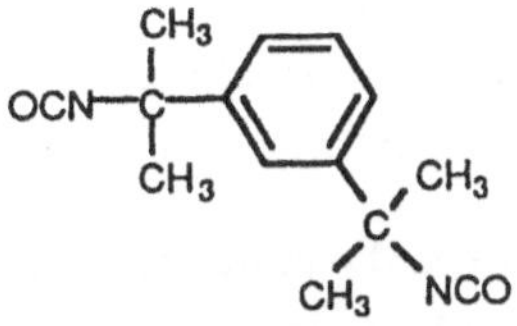

TMXDI

$$HO-(CH_2)_4-OH$$

BDO

$$H_2NCH_2(CH_2)_4 CH_2NH_2$$

HDA

Control specimens were stored both in PBS and in the dark, as above.

RESULTS AND DISCUSSION

The polyurethane elastomers 5-12 were prepared by two step solution polymerization using a procedure similar to that described by Lymann (10) and then screened for mechanical properties and processability. Table 1 shows the properties of several polyurethanes based on the novel macrodiols 1, 2 and 3 and hard segments composed of: (i) 4,4'-diphenylmethanediisocyanate (MDI), and 1,4-butanediol (BDO), or (ii) m-tetramethylxylene diisocyanate (TMXDI) and 1,6-hexanediamine (HDA). The properties of Pellethane™ 2363-80A—a PTMO-based polyurethane—and its laboratory synthesized equivalent 5 are also included for comparison. It is noteworthy that 5 had inferior mechanical properties to its commercial 'equivalent'. This is a consequence of the solution polymerization method and the stoichiometry used to prepare 5. The commercial material is prepared by bulk polymerization and contains allophanate linkages that contribute to improved mechanical properties. Its molecular weight is also higher. Recent experiments, not reported here, show that the mechanical properties of the MDI-based polyurethanes 5-8 can be improved by modifying the conditions of synthesis.

The processing temperatures for compression moulding were all in the range of 120-190 °C, comparing favorably with that of Pellethane 2363-80A, which moulds under a nominal load of 8 tons at 180-190 °C. Polyurethanes 6, 9, 10 and 11 were mouldable at temperatures below 150 °C, without the need for additives.

The new polyurethanes were subjected to tests to assay their stability to hydrolytic and oxidative conditions. This involved immersing 1 mm thick dumbbell-shaped specimens separately in water, 2 M HCl, 5 M NaOH, 25% hydrogen peroxide, and sodium hypochlorite solution (4% available Cl_2) for 24 h at 100 °C. Dumbbells of the polyurethanes were also subjected to water at 120 °C for 24 h. The decrease in fail stress relative to the untreated material was used to judge stability.

Figure 1 shows the cumulative decreases in fail stress of Pellethane™ 2363-80A (4) and polymers 5-12, when subjected to the hydrolysis and oxidation tests. The TMXDI-based polyurethane-ureas 9-12 performed poorly, while the MDI-based materials exhibited better stability. It is especially noteworthy that the MDI-based elastomers 6, 7, and 8 containing the macrodiols 1, 2, and 3 displayed a superior resistance to degradation (with the exception of 8 which became brittle in NaOCl). We speculate that the lowered hydrophilicity of 6, 7 and 8, containing the novel macrodiols, decreases both their rate of hydrolysis and oxidation compared with PTMO-based polyurethanes containing the same diisocyanate and chain extender. We also postulate that the reduced proportion of oxidizable methylene groups adjacent to oxygen in 6, 7 and 8 contributes to the enhanced peroxide stability. The situation with sodium hypochlorite is more difficult to

Figure 1. Cumulative Decrease in Fail Stress after Subjection to Oxidation and Hydrolysis

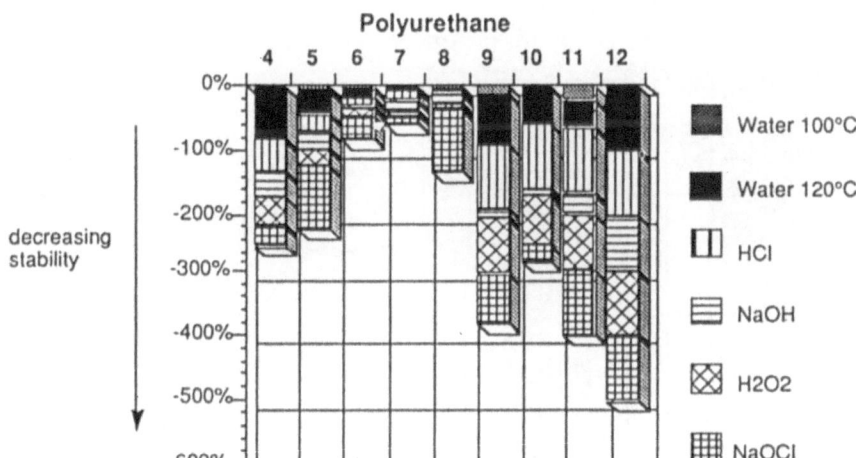

rationalise; there are indications, however, that this reagent reacts with the NH group in the hard segment (11), perhaps in addition to its capacity to oxidise the soft segment.

The good stability of 6, 7 and 8 was also reflected *in vivo*. Figure 2 shows the performance in two sets of subcutaneous implant experiments in sheep—in one set of experiments, the materials were implanted for 6 months in an unstressed configuration, while in the other, the materials were subjected to a 3 month implant with external strain (150%) applied by stretching over a special holder. It is well-known through the work of Stokes and his collaborators (12) that strain accelerates the rate of degradation. In each set of tests, the explanted samples were subjected to examination in a scanning electron microscope (at magnifications up to 2000) and ranked on a scale from 1-6 on the basis of the extent of cracking, pitting, or loss of material. Table 2 provides details of the ranking scale. When samples exhibited no signs of degradation, they were ranked 1, while ranking 6 implied severe surface cracking with loss of material.

The results in Figure 2 show the rankings as averaged over each of the two tests. It is noteworthy (data not shown) that both the strained and unstrained tests gave roughly the same amount of degradation, despite the shorter time of the former. Indeed, the correlation was such that the data did not differ by more than 1 ranking point between the two tests.

Figure 2 illustrates that the novel polyurethanes 6, 7, and 8 have improved biostability compared with Pellethane™ 2363-80A (4), as measured by subcutaneous ovine implant in stressed and unstressed configurations. Similar experiments (13) show that the

Figure 2. Average In Vivo Ranking from 90 and 180 day Implants

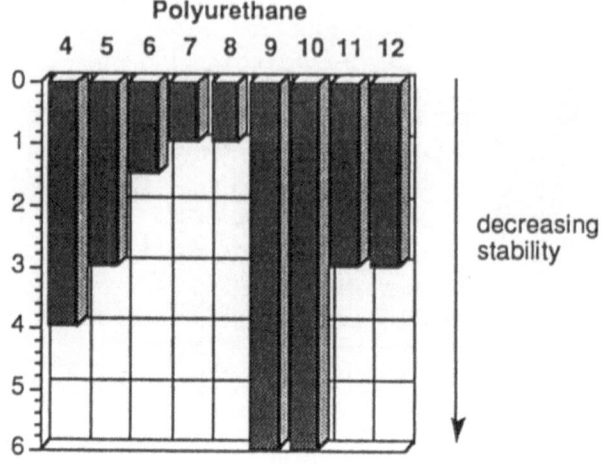

Table 2. Details of *In Vivo* Ranking Scheme

Ranking	Description
1	Specimen surface was smooth.
2	Slight cracking or pitting around the coding holes or ligature only. Remainder smooth.
3.	Fine cracks or pits over specimen surface.
4.	Coarse cracking or pitting around holes or ligature only. Fine cracking elsewhere.
5.	Coarse cracking or pitting over the entire surface.
6.	Extensive deep cracking or pitting & loss of material.

materials also degrade less than Biomer™ and Tecoflex™ EG-80A, which scored rankings of 5 and 4, respectively.

CONCLUSIONS

The above results show that the MDI-BDO polyurethanes, based on 1, 2, and 3 offer substantially enhanced resistance to hydrolysis and oxidation, while maintaining good thermal processability and satisfactory mechanical properties. It is particularly noteworthy that polyurethanes **6, 7**, and **8** display resistance to biologically induced environmental stress cracking, as evidenced by the results of strained (90 day) and unstrained (180 day) ovine implants (14).

ACKNOWLEDGEMENTS

Support of this research was provided under the Generic Technology component of the Industry Research and Development Act 1986 (Australia), in conjunction with Telectronics Pacing Systems, Terumo (Australia) Ltd., and Cyanamid Australia Pty. Ltd. We thank Donna McIntosh, Edith Cheong, Kate Noble, and Vicki Tatarinoff for technical assistance.

REFERENCES

1) Szycher M (1988) *J Biomat Appl* 3: 297

2) Stokes KB (1988) *J Biomat Appl* 3: 228

3) Coury AJ, Slaikeu PC, Cahalan PT, Stokes KB, Hobot CM (1988) *J Biomat Appl* 3: 130

4) Ratner BD (1989) In: (G. Allen and J. C. Bevington, (eds) *Comprehensive Polymer Science* , volume 7: Specialty Polymers and Polymer Processing, Pergamon, Oxford, pp 201-247

5) Ito Y, Imanishi Y (1989) *Crit Rev Biocompat* 5: 45

6) Phillips RE, Smith MC, Thoma RJ (1988) *J Biomat Appl* 3: 207

7) Thoma RJ, Tan FR, Phillips RE (1988) *J Biomat Appl* 3: 180

8) Gunatillake PA, Meijs GF, Chatelier RC, McIntosh DM, Rizzardo E *Polymer International* in press

9) Gunatillake PA, Meijs GF, Rizzardo E, *International Patent Application*, submitted.

10) Lyman, DJ (1960) *J Polym.Sci* 45: 49

11) Bamford CH and Middleton IP (1983) *Eur Polym J* 19: 1027

12) Stokes KB, Frazer AW, and Carter EA (1984) *Proceedings AANTEC 84* 1073

13) Meijs GF, Rizzardo E, Gunatillake PA, Brandwood A, Schindhelm, K (1991) *International Patent Application* PCT/AU91/00270

14) These materials are the subject of a CSIRO/UNSW patent application (13).

Effective Reconstitution of Cell Membrane Proteins into Artificial Cell Liposomes

J. Sunamoto, V. Rosilio, Y. Mori, K. Suzuki, M. Ishitobi, T. Sato, and K. Akiyoshi

*Department of Polymer Chemistry, Kyoto University,
Sakyo-ku, Yoshida Hommachi, Kyoto 606, Japan*

INTRODUCTION

Perfect extraction of intrinsic membrane proteins or enzymes from intact cells and reconstitution of them without any denaturation and deactivation are basic requirement in investigation of the function of membrane proteins and/or the utilization of them in membrane protein engineering. For this purpose, artificial cell, liposome, seems to be a desirable and promising tool in both basic investigation and application.

Very recently, we have developed a novel artificial boundary lipid, 1,2-dimyristoyl-amido-1,2-deoxyphosphatidylcholine (DDPC) [1] to make liposomes more stable and cell recognizable by reconstituting glycophorin into egg PC liposomes [2]. As expected, DDPC was better to keep glycophorin in egg PC liposomal membranes compared with a naturally occurring sphingomyelin. When such the specifically modified liposomes, which are containing an artificial boundary lipid, DDPC, are coincubated with intact cells, several membrane proteins and/or enzymes are able to be effectively transferred from the intact cells to the liposomes [3]. It has been found that the activity and the native orientation of the membrane proteins or enzymes are nicely retained in liposomes even after transferred from cells. In this paper, brief overview of our recent studies of direct transfer of membrane proteins from human erythrocyte, human platelet, B16 melanoma, and BALB RVD to various liposomes will be made.

PROTEIN TRANSFER FROM HUMAN ERYTHROCYTE

A 1.5 ml liposomal suspension (1.0×10^{-5} mol/ml) was mixed with a given volume of packed erythrocyte or erythrocyte ghost suspension (3.75 mg/1.5 ml) and kept to stand for 2 h at 37.0°C. Cell-liposome suspensions were incubated under agitation at 37.0°C on a shaking water bath. After the incubation, cells and liposomes were separated by centrifugation at $300 \times g$. Liposome suspensions in the supernatant (2.0

Y. Imanishi (Ed.)
Progress in Pacific Polymer Science 2
© Springer-Verlag Berlin Heidelberg 1992

ml) were further submitted to gel filtration through Sepharose 4B column ($\phi 1.8 \times 45$ cm) in order for them to be separated from soluble proteins. All the experiments were run in 310 mOsm phosphate buffered saline (pH 7.38) containing 140 mM NaCl, 5 mM KH_2PO_4, 2.5 mM Na_2HPO_4, 1 mM $MgSO_4$, 5 mM glucose, and 50 mM sucrose. Quantitative and qualitative studies of proteins transferred from the cells were carried out by fluorescamine method and SDS-PAGE. The transfer efficiency was governed by the relative fluidity of the participant membranes of both liposome and cell. When DMPC (40 mol%) / DDPC (60 mol%) liposome was employed, even Band 3 (95 kD), which is one of large membrane proteins and tightly interacts with cytoskeleton, could be transferred (Figure 1). Only in the case of DMPC(20)/DDPC(80) liposome, several kinds of lipids were additionally detected on TLC of recovered liposome: certainly, cholesterol and phosphatidylethanolamine and most probably, phosphatidylserine and sphingomyeline.

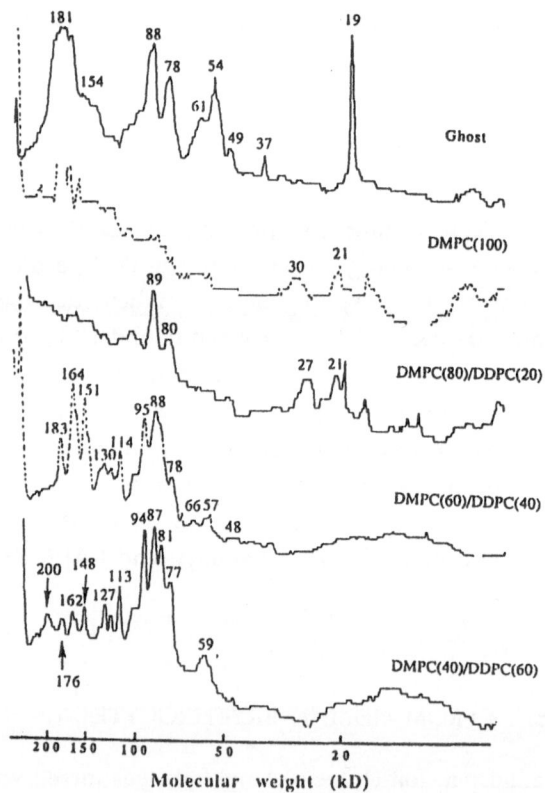

Figure 1. Two dimensional pattern of SDS-PAGE of proteins transferred from human erythrocyte ghost to various DMPC/DDPC liposomes.

As seen in Table 1, DMPC/DDPC liposome could extract more proteins than egg PC/DDPC liposome. Opposite to this, however, recovery of liposome determined as lipid concentration was less in the former liposome compared with the latter one. These results are not inconsistent with those of our preliminary study [3] and suggest that efficiency of transfer of membrane proteins is deeply correlated with rigidity and toughness of liposomal membrane.

Table 1. Amount of proteins transferred from human erythrocyte ghost[a] to two different liposomes[b] and recovery of liposome (as lipid concentration) after coincubation at 37.0°C for 2 hr.

Liposome	Protein transferred (%)	Lipid recovered (%)	Protein/Lipid (by wt)
DMPC/DDPC			
100 : 0	15.0	80.8	0.069
80 : 20	15.4	79.5	0.071
60 : 40	18.7	75.5	0.092
40 : 60	19.0	51.9	0.133
Egg PC/DDPC			
100 : 0	1.8	91.7	0.006
80 : 20	3.2	83.9	0.013
60 : 40	6.9	81.6	0.029
40 : 60	10.8	83.4	0.045

a) Initial lipid concentration of liposome was 1.0×10^{-5} mol/ml.
b) Total protein content of erythrocyte ghost was 2.50 mg/ml.

Using the same liposome, a sort of peripheral membrane proteins, *Acetylcholine Esterase* (AchE) was transferred and its enzyme activity was investigated. Results showed that more than 97 % of the activity was retained even after the transfer [3]. This means that this direct protein transfer can retain not only the original function but also the native orientation of the membrane proteins.

Another interesting finding is that various antigens of blood type determinant accompanied by other membrane proteins were also transferred from human erythrocyte to liposome. They were respectively assayed by the turbidity increase upon the aggregation of liposome with the specific antibody. All the modified liposomes, except simple DMPC liposome, extracted A-type antigen from erythrocyte (Table 2). The sensitivity of assay of the antigen on the liposome, which

Table 2. Blood type-determining antigens transferred from human erythrocyte to liposome. Antigens were assayed by each specific antibody (see text).

Liposome employed	Antigen[#]				
	A	Rh+	N	Lewis[b]	P1
DMPC(100)/DDPC(0)	–	±	±	±	+
DMPC(80)/DDPC(20)	+ +	±	±	±	+
DMPC(60)/DDPC(40)	+ +	±	±	±	+
DMPC(40)/DDPC(60)	+ +	±	±	±	+

A antigen was assayed by TSS Lot 55 antibody, Rh+ antigen was by Japan Red Cross monoclonal antibody, N antigen was by gamma monoclonal antibody, Lewis[b] antigen was by Bio-test monoclonal antibody, and P1 antigen was by Ortho antibody from goat, respectively.

was determined by dilution method, was found to be enhanced approximately 50-fold compared with the case when the assay was carried out using intact erythrocyte. This sensitivity enhancement was more or less observed also in cases of other antigens. Different from MN blood type antigen which is derived from protein portion of glycophorin or Rh blood type from Band 3 (?), ABO and Lewis[b] blood type antigens are derived from carbohydrate moiety of glycolipids and glycoproteins. P1 blood type antigen is, in addition, believed to be a sort of sphingoglycolipids. Taking account into these facts, it is reasonable to consider that not only membrane proteins but also glycolipids or other lipids are occasionally transferred due to the conditions employed, especially when a liposome which contains larger amount of DDPC was used.

TRANSFER OF HEPARIN BINDING RECEPTOR FROM HUMAN PLATELET

Using the same technique, transfer of heparin binding receptor proteins of human platelet was tried. The general procedure is shown in Scheme 2.

Different from erythrocyte, the cytoplasm membrane of human platelet is relatively less fluid and the membrane proteins are hardly transferred to liposome. In such the case, the cell membrane was made more fluid by adding a small amount of a-linolenic acid (ALA) (50.0 mM to 8.4×10^8 cells/ml) prior to exposure of cell to liposome (2.5×10^{-3} M). As seen in Figure 2, the membrane fluidity of platelet was significantly increased by adding unsaturated fatty acid. When such the cell of which membrane was fluidized in advance was employed, the amount of proteins transferred increased as expected (Table 3). When DMPC

(20)/DDPC (80) liposome was employed, following proteins were detected by SDS-PAGE: 350 (?), 194 (myosine), 89 (GP V), 67 (GP VI), 43 (GP IX), and 30 kD (?). Using the same liposome, when the cell was pre-treated by adding ALA, proteins of 268 kD (?) and 176 kD (GP Ib) were detected, and proteins of 145 kD and 67 kD were not detected (Table 4). When simple DMPC liposome was used, however, only GP Ib (177 kD) was detected.

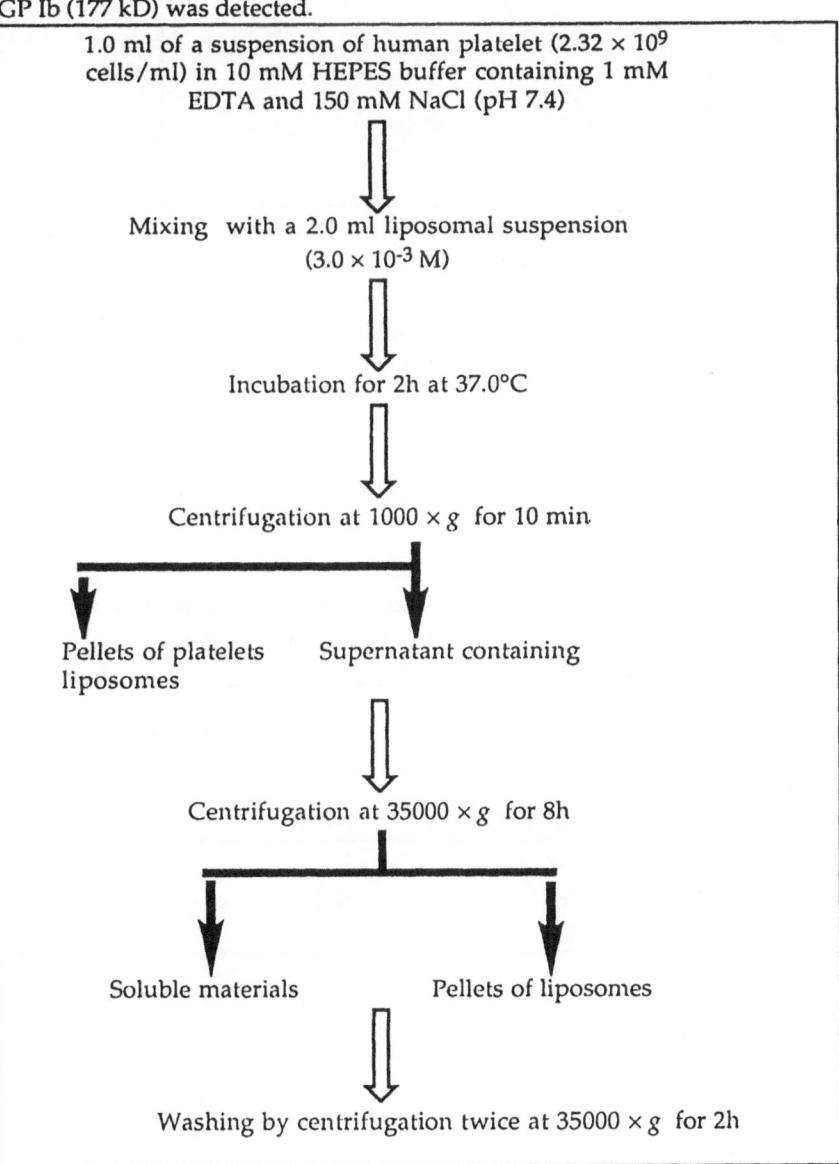

1.0 ml of a suspension of human platelet (2.32×10^9 cells/ml) in 10 mM HEPES buffer containing 1 mM EDTA and 150 mM NaCl (pH 7.4)

Mixing with a 2.0 ml liposomal suspension (3.0×10^{-3} M)

Incubation for 2h at 37.0°C

Centrifugation at $1000 \times g$ for 10 min

Pellets of platelets liposomes   Supernatant containing

Centrifugation at $35000 \times g$ for 8h

Soluble materials   Pellets of liposomes

Washing by centrifugation twice at $35000 \times g$ for 2h

Scheme 2. General procedure of the protein transfer from human platelet to liposome

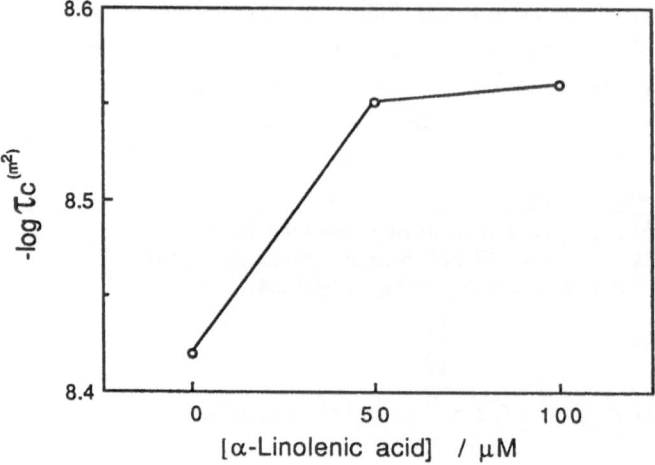

Figure 2. Membrane fluidity of human platelet as altered by adding a small amount of α-linolenic acid. The fluidity was determined by ESR spin probe method using 12-doxyl stearate.

Table 3. Amount of proteins after interaction at 37 °C for 1 hr between liposomes (2.5 × 10⁻³ M) and human platelets (8.4 × 10⁸ cells/ml).

Liposome	Amount of proteins transferred (mg/ml)	
	Without ALA	With ALA
DMPC(100)/DDPC(0)	1.4	8.3
DMPC(60)/DDPC(40)	9.9	32.2
DMPC(20)/DDPC(80)	37.1	62.7

Table 4 Proteins transferred from human platelet (2.52×10^9 cells/ml) to DMPC/DDPC liposome (2.5×10^{-3} M) upon the incubation at 37.0°C for 1 hr. Effect of addition of α-linolenic acid (ALA) to cell prior to the exposure to liposome.

Bands on SDS-PAGE (kD)		
Liposome I	Liposome II	Liposome III
353	358	331
268		268
232		
200	194	197
		176
157		
	145	
123		
94	89	88
	67	
44	43	42
30	30	
27		

Liposome I: DMPC(60)/DDPC(40) and no ALA was added to the cell.
Liposome II: DMPC(20)/DDPC(80) and no ALA was added.
Liposome III: DMPC(20)/DDPC(80) and ALA was added.

Viability of the cell after exposure to liposome also was investigated by measuring activity of extracellular LDH as the index. The higher release of LDH from the cell indicates the more decreased viability. As seen in Figure 3, cell viability drastically decreased when the liposome which contains more DDPC (liposome III) was employed. However, the pre-trearment of the cell with ALA did not affect much on cell viability.

In any event, proteins transferred can be controlled by altering membrane fluidity of either site, liposome or cell. This technique must be very convenient and useful for fractionation of membrane proteins and the more detailed investigation of them in cell and membrane biologies. Specific binding of these glycoproteins directly transferred to liposome is now under investigation by the use of fluorescent probe-labelled and fractionated heparins.

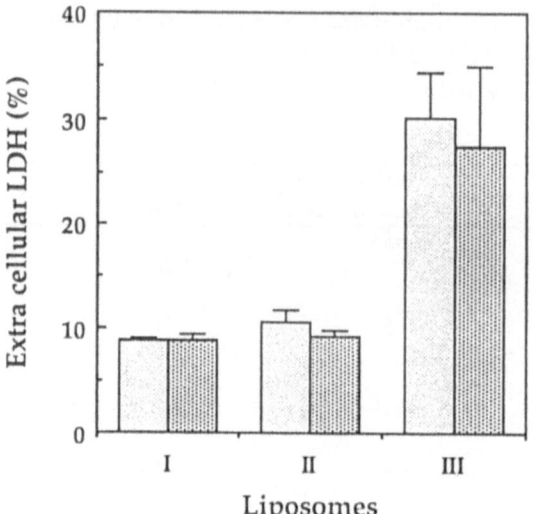

Figure 3. Viability of human platelet after exposure to liposome. Effects of the lipid composition of liposome and the pre-treatment of the cell with ALA before the exposure. I, DDMPC(100)/DDPC(0); II, DMPC(60)/DDPC (40); III, DMPC(20)/DDPC(80). The right-hand bar shows the case of pre-treatment of the cell with ALA, while the left-hand bar is the case without the ALA pre-treatment.

TRANSFER OF MEMBRANE PROTEINS FROM TUMOR CELLS

Similarly to the cases of erythrocyte and platelet, membrane proteins of tumor cells, mouse B16 melanoma [4] and BALB RVD [5], were directly transferred to the liposome.

B16 Melanoma

The most interesting finding was that, for B16 melanoma, both the kind and the amount of proteins transferred from the cell were different between the two culture systems of suspension and monolayer. Procedure of extraction of membrane proteins from B16 melanoma multiplied by monolayer and suspension cultures is given in Scheme 3 In the case of suspension culture, multiplication was carried out on a plastic dish (Iwaki SH-90-15) and cells were harvested simply by pipetting.

Other procedures were exactly the same with those adopted to the case of monolayer culture.

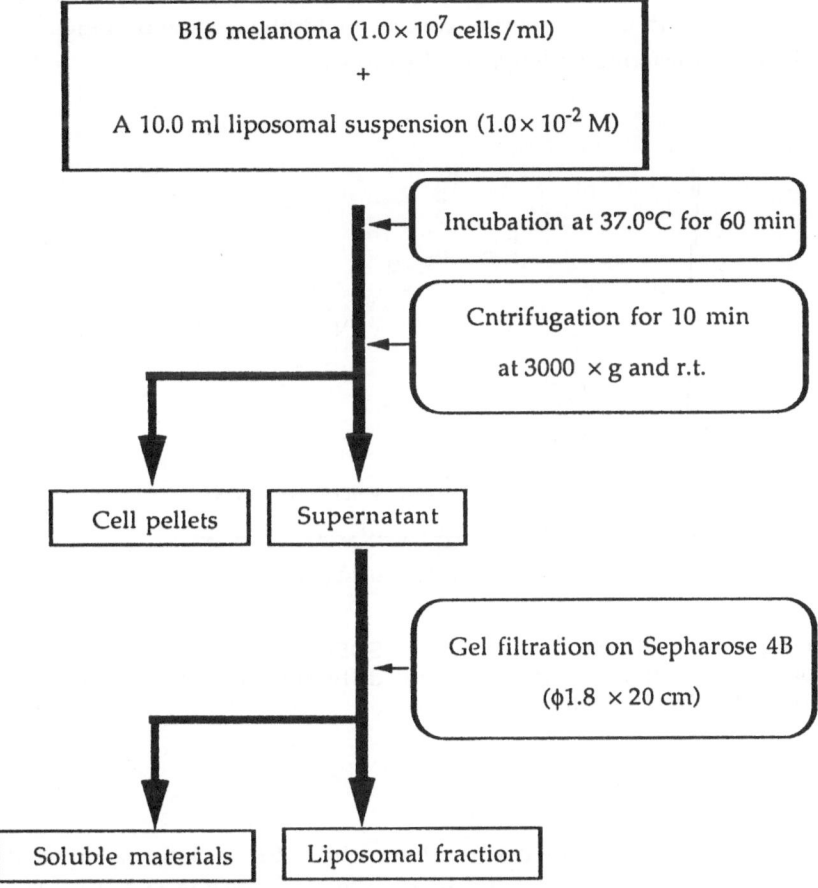

Scheme 3. Procedure of protein transfer from B16 melanoma to liposome in monolayer culture

The total amount of proteins transferred was relatively larger in suspension culture than that in monolayer culture (Table 5). Even after protein transfer, particle size of liposomes almost unchanged and efficiency of recovery of lipids was rather good. Viability of the cell after the exposure to liposome was largely affected by the method of culture. In the suspension culture, the cell viability was less than 20 % after the exposure. In the monolayer culture system, however, the viability of the cell was maintained more than 60 %. When simple DMPC liposome was used, the viability was almost 80 %.

For the monolayer culture, major proteins transferred were 95, 77, 68, 48 (MSH receptor ?), and 35 kD, while they were 112 (vitronectin receptor or adhesion protein ?), 101, and 51 kD for the suspension culture (Figure 4). When DMPC(20)/DDPC(80) liposome was used in the suspension culture, in addition, a tumor antigen ganglioside, GM3, was transferred accompanied by proteins.

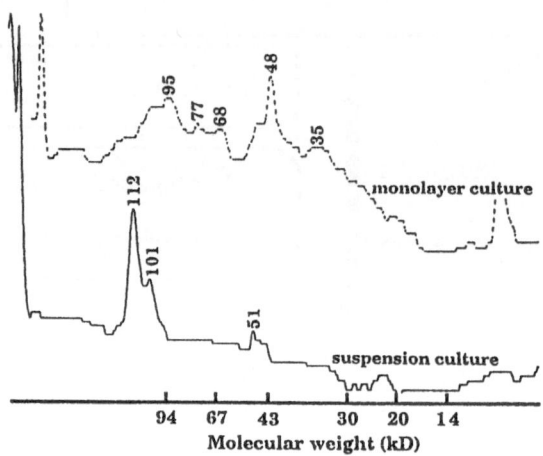

Figure 4. Two dimensional pattern of SDS-PAGE of proteins transferred from B16 melanoma in the two different culturing systems to DMPC(20)/DDPC(80) liposome.

Table 5 Amount of proteins transferred from B16 melanoma (1.0×10^7 cells/ml) to liposome (1.0×10^{-4} mol) under coincubation at 37.0°C for 60 min.

Liposome employed[a]	Suspension culture		Monolayer culture	
	Protein (mg)	Protein/Lipid (by wt)	Protein (mg)	Protein/Lipid (by wt)
I	480	0.008	30	0.001
II	1530	0.027	421	0.008
III	1890	0.078	1245	0.032

a) Liposome I, DMPC(100 mol%)/DDPC(0 mol%); Liposome II, DMPC(60 mol%)/DDPC(40 mol%), and Liposome III, DMPC(20 mol%)/DDPC(80 mol%).

BALB RVD Leukemia cell

For BALB RVD, tumor surface antigen presenting protein (TSAP) was certainly transferred from the tumor cell to liposome. Using this technique, therefore, we could make an effective liposomal vaccine against BALB RVD [6]. When male CB6F1 mice were challenged by RVD tumor cells after immunization by the liposomal vaccine so obtained, excellent rejection against tumor was observed. To our knowledge, this was the first success in the prophylaxis of cancer disease in animals [5].

Acknowledgment

This research was financially supported by Grant-in-Aid for Scientific Research from the Ministry of Education, Science, and Culture (Nos. 02250106, 03236106, and 02453099) and by the Mitsubishi Foundation. V. Rosilio is grateful for a grant from Japan Society for Promotion of Science for her staying in Kyoto University.

References:

1 Sunamoto J, Goto M, Iwamoto K, Kondo H, Sato T (1987) *Biochim Biophys Acta*, **1024** : 209
2 Sunamoto J, Goto M, Arakawa M, Sato T, Kondo H, Tsuru D (1987) *Nippon Kagaku Kaishi* 569
3 Sunamoto J, Goto M, Akiyoshi K (1990) *Chem Lett.* 1249
4 Sunamoto J, Shiku H (1989) IN: Sakurai Y, Rabischong P, Kawamura T (eds) Proceedings of the 3rd Japanese-French Biomedical Technologies Symposium, 82
5 Shibata R, Noguchi T, Sato T, Akiyoshi K, Sunamoto J, Shiku H, Nakayama E (1991) *Int J Cancer* **48** : 434
6 Sunamoto J, Noguchi T, Sato T, Akiyoshi K, Shibata R, Nakayama E, Shiku H (1992) *J Controlled Release* in press

Negative Cilia Concept for Enhanced Blood Compatibility: Grafting of PEO-Sulfonate onto Polyurethane

Y.H. Kim, D.K. Han, S.Y. Jeong, B.G. Min[*]

Polymer Chemistry Lab., Korea Institute of Science and Technology, P.O. Box 131, Cheongryang, Seoul 130-650, Korea, [*]Dept. of Biomedical Engineering, College of Medicine, Seoul National Univ., Seoul, Korea

Abstract: To improve the blood compatibility of commercial polyurethane (PU), PU surface was chemically grafted with a hydrophilic polyethyleneoxide (PEO) and further sulfonated to prepare PU-PEO-SO$_3$. The surface characteristics and blood compatibilities of surface modified PUs were investigated. PU-PEO-SO$_3$ surfaces showed very smoothness and complete wetting behaviors to be highly hydrophilic. The PEO-sulfonate grafted PUs were much more blood compatible than untreated PU and PU-PEO, but they exhibited a enhanced adsorption of fibrinogen compared with PU-PEO. The ex vivo occlusion times coincided well with the in vitro evaluation results: the less the adhesion and shape change of platelets and the larger the clotting times, the longer occlusion times. Such an enhanced blood compatibility of PU-PEO-SO$_3$ is due to the synergistic effect of the hydrophilic PEO and the negatively charged SO$_3$ groups on blood compatibility via a negative cilia concept.

INTRODUCTION

When blood is placed in contact with any foreign materials, it is known that protein adsorption is occurred first and followed by platelet adhesion and cascade reaction of coagulation factors and their subsequent activations, which lead to thrombus formation resulted into the formation of fibrin network structure.

In order to develop blood compatible polymeric surfaces many researches have been carried out on the basis of several hypotheses such as negative surface charge, surface or interfacial free energy, pharmacologically active surface, and surface motion.(1) It was reported that polymers grafted with hydrophilic polyethyleneoxide(PEO) showed less protein adsorption and platelet adhesion to improve the antithrombogenicity significantly.(2,3) On the other hands, many polymers containing negative charges, especially sulfonate groups, have also received much attention. Recently, Cooper et al.(4,5) have prepared sulfonated polyurethanes(PUs) and described that these anionic PUs exhibited a enhanced thromboresistance.

Accordingly, it is highly expected that the grafting of sulfonated PEO onto PU surface may enhance its blood compatibility significantly by means of a synergistic

Y. Imanishi (Ed.)
Progress in Pacific Polymer Science 2
© Springer-Verlag Berlin Heidelberg 1992

effect of the dynamic mobility of PEO chains and the electrical repulsion of negatively charged sulfonate groups.(6,7)

In this study, PU surface was chemically grafted with hydrophilic PEO and then sulfonated to investigate the relationship between surface characteristics and blood compatibility, and in particular, to study a possible synergistic effect of PEO and sulfonate groups.

MATERIALS AND METHODS

Preparation of Sulfonated Polyurethanes

Polyurethane (PU) bead, sheet (Pellethane 2363-80A, thickness 1 mm: Dow Chemical Co.), and tubing (Royalthene R-380 PNAT, ID 1.5 mm, OD 2.0 mm: Uniroyal Chemical Co.) were used in this experiment. Figure 1 illustrates the modification scheme of PU surfaces. The preparation methods of various sulfonated PUs have been described in detail elsewhere.(8)

Briefly, the surfaces of PU sheet and bead were treated with hexamethylene diisocyanate (HMDI) to introduce free isocyanate (-NCO) groups in toluene with stannous octoate at 40 °C for 1 h under nitrogen (PU-HMDI). PU-HMDI was further grafted with hydrophobic dodecanediol (DDO, MW=202) or hydrophilic PEO (MW=200, 1000, and 2000) to make PU surface hydrophobic or hydrophilic, respectively. The DDO, which used to compare with PEO 200, was grafted onto PU-HMDI in isopropyl ether with di-n-butyltin dilaurate for 30 h (8 h for PU tubing) at 45 °C (PU-DDO). The PEO was also grafted onto PU-HMDI in benzene with stannous octoate for 24 h (2 h for PU tubing) at 40 °C (PU-PEO). Sulfonations were accomplished onto modified PU surfaces by three different methods · using propane sultone. Propane sultone was coupled directly onto PU-NH$_2$ via PU-HMDI in acetonitrile for 4 h at room temperature (PU-SO$_3$). The hydroxyl end groups of grafted

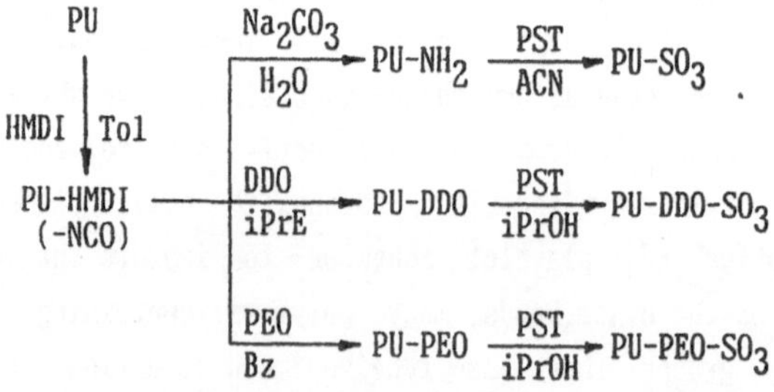

Figure 1. Modification scheme of PU surfaces:
HMDI = OCN(CH$_2$)$_6$NCO, PST = (CH$_2$)$_3$SO$_3$,
PEO = HO(CH$_2$CH$_2$O)$_n$H, DDO = HO(CH$_2$)$_{12}$OH.

DDO or PEO chains were also sulfonated by propane sultone in a mixture of isopropanol, Na_2CO_3, and dimethyl sulfoxide for 20 h (8 h for PU tubing) at 45°C to obtain $PU-DDO-SO_3$ or $PU-PEO-SO_3$ (PEO-sulfonate grafted PU), respectively.

Meanwhile, the above reaction conditions were applied for performing modification of inner surface of PU tubing, except circulating the reaction solution with a peristaltic pump. The modified PU tubings were then used for ex vivo rabbit test.

Surface Characterization

Attenuated total reflectance-Fourier transform infrared (ATR-FTIR) data were obtained from the surfaces of modified PU sheets using a Mattson Alpha Centauri FTIR spectrophotometer, coupled with an ATR accessory and 45° KRS-5 crystal.

The elemental compositions of modified PU surfaces were determined by using a Physical Electronic PHI 558 electron spectroscopy for chemical analysis (ESCA) spectrometer. The source was a 10 kV, 30 mA monochromatized X-ray beam from a magnesium anode. Surface charge build-up was corrected by considering the shift of C_{1s} peak at 285 eV.

The surface morphology of modified PU sheets was examined with a Hitachi S-510 scanning electron microscopy (SEM) at an accelerating voltage of 15 kV. Samples were mounted and then sputter-coated with gold using an ion coater.

The dynamic advancing and receding contact angles of modified PU sheets were evaluated with a Wilhelmy plate contact angle apparatus (WET-TEK F100, Biomaterials Int.) in double-distilled water at constant temperature (20 °C) and humidity (30% RH) The velocity of the translation stage was 20 mm/min.

In Vitro Protein Adsorption and Blood Compatibility

The adsorption method of fibrinogen is as follows: each of the different PU beads was first contacted with bovine plasma containing ^{14}C-labeled human fibrinogen as a function of adsorption time or plasma concentration. Sample was then rinsed with phosphate buffered saline (PBS, pH 7.4, 0.15M) twice and placed in 2% sodium dodecyl sulfate (SDS) solution for 2 days, and the surface concentration of fibrinogen was determined by counting the radioactivity.

Platelet adhesion test was carried out with human platelet-rich plasma (PRP). Sample sheet was immersed in PRP for 3 h at 37 °C and then withdrawn from PRP, rinsed with PBS. The adhered platelets were fixed with 2% glutaraldehyde solution in PBS buffer for 2 h at room temperature, dehydrated with several dilutions of ethanol and water, and then lyophilized. Sample was coated with an evaporated gold layer and the morphology of adhered platelets was observed with SEM.

Activated partial thromboplastin time (APTT) was determined using the Fibrometer method.(9) Sample sheet was incubated in 300 μl of control plasma (Control Plasma N, Beohring Co., Germany) for 1 h afterward the plasma was separated. The partial thromboplastin (0.1 ml, Neothrombin) was preheated for 2 min and the obtained test plasma (0.1 ml, 37 °C) as mentioned above was added, followed by the addition of 0.025M $CaCl_2$ (0.1 ml) exactly 30 s later. Then, the clotting time was measured using Fibrintimer (Beohring Co.).

The determination of prothrombin time (PT) was conducted via one-stage prothrombin time method.(10) After treating the sheet in the same manner as in APTT method, the mixture of 0.1 ml of thromboplastin (Sigma Chemical Co.) and 0.1 ml of 0.025M $CaCl_2$ solution were preheated to 37 °C. Test plasma (0.1 ml) was immediately added to the mixture and then the clotting time was measured using Fibrintimer.

Ex Vivo Blood Compatibility

An arterio-aterial (A-A) shunt test, a new ex vivo model was performed as the following: male rabbits (New Zealand White, 2-3 kg) were anesthetized with ketamine/urethane, and the right carotid arteries were carefully exposed surgically. Then, the modified PU tubings (2.0 mm OD x 1.5 mm ID, length 30 cm) equilibrated overnight with PBS were rinsed and carefully inserted into the clamped ligated carotid artery of the rabbit. The flow rate was controlled to 2.5 ml/min continuously using a suture tourniquet and measured with ultrasonic flow meter (ES-4100Z, ARS Electric). The occlusion time was defined as the time that the blood flow decreases to zero.

RESULTS AND DISCUSSION

Surface Characteristics

The surface reactions of all the modified PUs were confirmed by ATR-FTIR and ESCA. The characteristic peaks of -NCO, -OH, and $-SO_2$ on modified PUs were identified at 2250 cm^{-1} for PU-HMDI, 3300-3600 cm^{-1} for PU-PEO, and near 1030 cm^{-1} for PU-PEO-SO_3, respectively. As shown in Table 1, ESCA data allow us to support well IR results, as more O atomic % for PEO grafted PUs and higher S atomic % for all the sulfonated PUs were determined.

Figure 2 shows the surface morphology of modified PUs. Untreated PU surface was fairly smooth, whereas the surfaces of PU-DDO and PU-PEO were relatively smooth. When PEO200 and DDO, which have the same molecular weight, were grafted onto PU surface, PU-PEO200 displayed really smoother surface than PU-DDO owing to an inherent

Table 1. Surface characteristics[a] of modified PUs

Material	ESCA (atomic %)					Contact angle	
	C	O	N	S	Na	θ adv	θ rec
PU	85.5	10.3	4.3			104	59
PU, MeOH ext.	76.9	21.5	1.5			86	41
PU-DDO	80.6	14.5	4.8			66	46
PU-PEO200	69.0	24.8	6.2			30	20
PU-PEO2000	59.9	38.9	1.2			49	14
PU-SO$_3$	69.2	18.0	11.1	1.8	-	58	wetting
PU-DDO-SO$_3$	76.7	15.3	5.4	1.4	1.2	68	"
PU-PEO200-SO$_3$	65.6	24.3	3.9	3.1	3.1	39	"
PU-PEO2000-SO$_3$	58.3	37.9	1.2	1.5	1.2	51	"

a. all values are measured triplicate at least.

feature of PEO. All the sulfonated PUs exhibited the excellent smoothness, and in particular, PU-PEO-SO$_3$ surfaces were smoothest and most homogeneous.

As is seen in Table 1, PEO grafted PUs showed lower receding angles(more hydrophilized surface) than untreated PU, where higher MW of PEO indicated greater effect. PU-DDO, however, showed less change. In addition, all the sulfonated PU

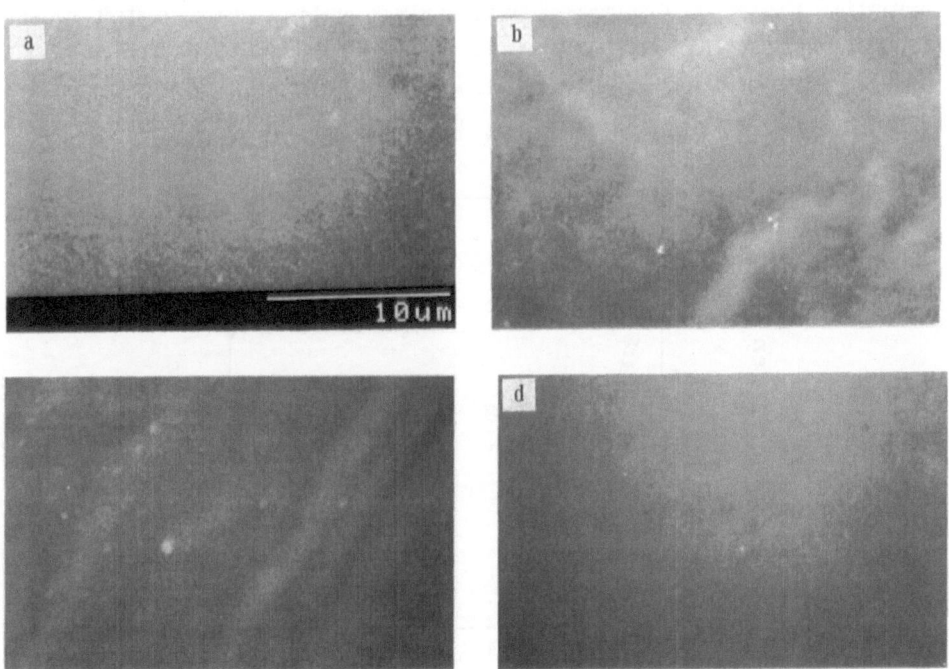

Figure 2. SEM micrographs of modified PU surfaces: (a) PU, (b) PU-PEO200, (c) PU-SO$_3$, (d) PU-PEO200-SO$_3$.

surfaces exhibited complete wetting behaviors regardless of their coupling agents on account of negatively charged SO_3 groups. Especially, PU-PEO-SO_3 which represents much higher hydrophilic surface is expected to show a more favorable response to blood.

Blood Compatibility

On the whole, initial adsorption of fibrinogen onto the surfaces was increased with increasing of adsorption time and plasma concentration, but after the plateau is reached, its adsorption amount was in inverse proportion to them. The adsorption amount of fibrinogen onto PU-PEO1000 was much less than that onto PU to confirm a characteristic effect of PEO grafted. However, PU-PEO1000-SO_3 exhibited a considerably increased fibrinogen adsorption to compare with PU-PEO1000, where the first adsorption was nearly same regardless of plasma concentration and adsorption time as shown in Figure 3. This may indicate a specific high affinity between sulfonate groups and fibrinogen. It is interesting to note that the sulfonated surfaces enhanced fibrinogen adsorption, which showed the same results as reported by Cooper et al.(4,5) Figure 4 shows typical adsorption kinetics for fibrinogen from 0.6% plasma on modified PU surfaces. All the surfaces exhibited the Vroman effect at about 0.6% plasma concentration, however the displacement by other plasma proteins such as contact phase clotting factors was relatively low.

The adhesion behaviors of platelets for modified PU surfaces have been previously described.(7) PEO grafted PUs and PU-SO_3 displayed less platelet adhesion than

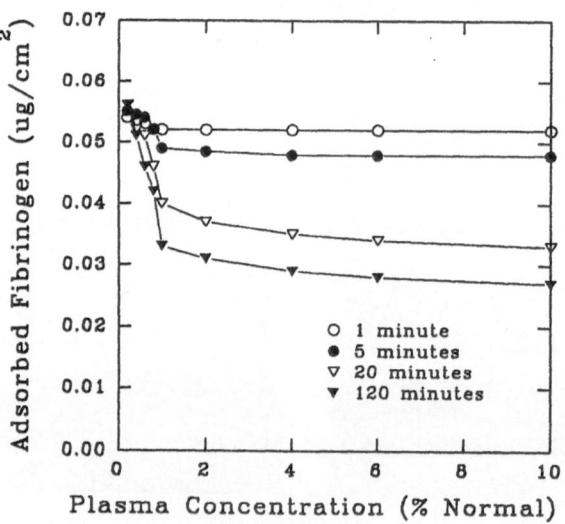

Figure 3. Adsorption isothermals for fibrinogen on PU-PEO1000-SO_3.

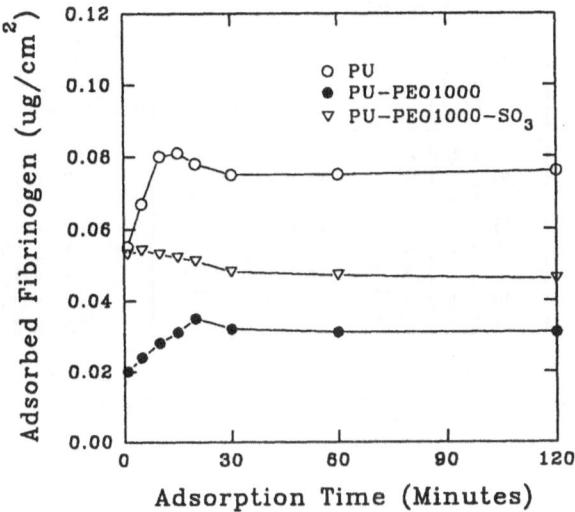

Figure 4. Adsorption kinetics for fibrinogen
from 0.6% plasma.

untreated PU, but in the case of PU-SO$_3$, the surface showed some shape change of the adhered platelets. The PEO-sulfonate grafted PUs exhibited a much lower degree of adhesion and shape change of platelet, demonstrating that PU-PEO-SO$_3$ may be the most blood compatible.

Table 2 lists the blood compatibility data of the various modified PUs. The APTT and PT, especially APTT, of sulfonated PUs were extended, while those of PU-PEO and PU-DDO did not show any significant change compared to PU control. This can be explained by an anticoagulant activity of SO$_3^-$ groups grafted, as reported in other sulfonated polymers.(11)

As shown in Table 2, the ex vivo A-A shunt occlusion time of PU was 50min, but that of PEO grafted PUs were extended to 120min(MW200) and 145min(MW2000), while hydrophobic PU-DDO showed a small increase up to 70min. It seems likely that the longer hydrated PEO chain, the greater chain motion, contributes to the enhanced thromboresistance by more suppressive protein adsorption and platelet adhesion. The occlusion time of PU-SO$_3$ was 90min, however that of PU-PEO-SO$_3$ and PU-DDO-SO$_3$ were 350min(PEO200), 370min(PEO2000), and 200min(DDO), respectively, indicating a synergistic effect of the hydrophilic PEO and negatively charged SO$_3$ groups.(6,7) These occlusion time results agree with in vitro platelet adhesion and blood clotting time data: the less adhesion and activation of platelet and the longer the APTT and PT, the more extended the ex vivo occlusion time.

Table 2. Blood compatibility data[a] of modified PUs

| Material | In vitro | | Ex vivo |
	APTT[b] (sec)	PT[c] (sec)	rabbit A-A shunt occlusion time (min)
PU	35.8 ± 0.2	13.3 ± 0.1	50 ± 5
PU-DDO	$36.2 \pm 1.0^*$	$13.5 \pm 0.2^*$	$70 \pm 10^{**}$
PU-PEO200	$33.1 \pm 0.5^{***}$	$13.9 \pm 0.2^{***}$	$120 \pm 15^{***}$
PU-PEO2000	$35.5 \pm 0.8^*$	$14.9 \pm 0.3^{***}$	$145 \pm 15^{***}$
PU-SO$_3$	$41.9 \pm 1.5^{***}$	$14.4 \pm 0.3^{**}$	$90 \pm 5^{***}$
PU-DDO-SO$_3$	$40.5 \pm 1.2^{***}$	$14.2 \pm 0.3^{***}$	$200 \pm 15^{***}$
PU-PEO200-SO$_3$	$49.7 \pm 2.5^{***}$	$15.2 \pm 0.6^{***}$	$350 \pm 30^{***}$
PU-PEO2000-SO$_3$	$41.8 \pm 1.4^{***}$	$14.5 \pm 0.4^{***}$	$370 \pm 30^{***}$

a. mean $\pm$ S.D.(n=3); significance level using an unpaired Student's t-test
   when comparing modified PUs to PU ($^*p > 0.1$, $^{**}p < 0.01$, $^{***}p < 0.005$).
b. APTT of pooled plasma was 36.0 sec.
c. PT of pooled plasma was 13.0 sec.

Negative Cilia Concept

The introduction of sulfonate groups at the end of the PEO chain grafted onto a PU surface, such as PU-PEO-SO$_3$, enhanced blood compatibility enormously. This can be ascribed to the synergistic effect of the hydrophilic PEO and negatively charged SO$_3$ groups as shown in the "negative cilia" model in Figure 5.

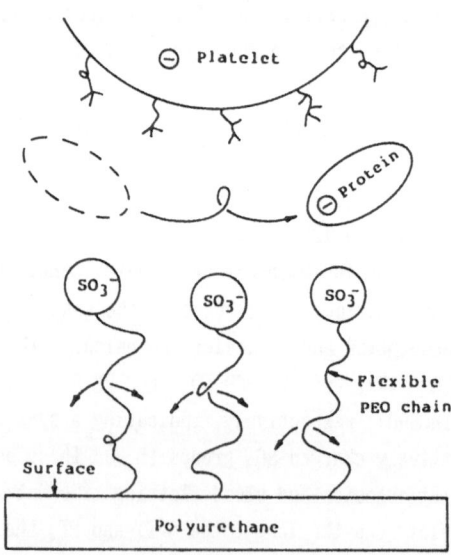

Figure 5. "Negative cilia" model on PU-PEO-SO$_3$ system.

In this model, the possible blood-material interactions are as follows; the hydrated flexible PEO chain motion suppresses protein adsorption and platelet adhesion, where the PEO chain motion will be increased by the electric repulsion between sulfonate end groups each other. In addition, the sulfonated end groups of PEO chain expel proteins and platelets further by electric repulsion, and moreover the sulfonate groups might contribute to better blood compatibility by inhibiting blood coagulation factors.

Besides, $PU-PEO-SO_3$ exhibited the smoothest surface, and the hydrated water structure at the interface may be changed due to the hydrophilic PEO and the negative SO_3 groups grafted, that might also responsible for improved blood compatibility. Hence PEO-sulfonate grafted PUs that displayed enhanced blood compatibility should be useful for blood contacting materials.

CONCLUSIONS

The commercial PU surface was chemically modified with PEO and then with propane sultone to produce negatively charged $PU-PEO-SO_3$. The PEO-sulfonate grafted PUs showed the excellent homogeneity and smoothness, and much higher hydrophilicity indicating complete wetting behaviors due to charged SO_3 groups. Meanwhile, the sulfonated PU-PEO displayed the enhanced adsorption of fibrinogen compared with PU-PEO. $PU-PEO-SO_3$ also exhibited a lower degree of adhesion and shape change of platelet, and considerably extended blood clotting times. The _ex vivo_ occlusion time of $PU-PEO-SO_3$ (350 min in PEO200 and 370 min in PEO2000) was even longer than that of PU (50 min), PU-PEO (120-145 min), and $PU-SO_3$ (90 min), demonstrating a synergistic effect of the hydrophilic PEO and negative SO_3 groups, as can be explained with a negative cilia concept on blood compatibility. The _ex vivo_ occlusion times corresponded well to the _in vitro_ platelet adhesion and blood clotting times results. Therefore, $PU-PEO-SO_3$ surfaces exhibited excellent blood compatibility to be useful for biomedical applications.

Acknowledgment: This work was supported by Korean Ministry of Science and Technology Grant N6390 and N7331.

References

1 Andrade JD, Nagaoka S, Cooper SL, Okano T, Kim SW (1987) Am Soc Artif Intern Organs J 10:75

2 Nagaoka S, Mori Y, Takiuchi H, Yokota K, Tanzawa H, Nishiumi S (1984) In: Shalaby SW, Hoffman AS, Ratner BD, Horbett TA (eds) Polymers As Biomaterials, Plenum, New York

3 Han DK, Jeong SY, Kim YH (1989) J Biomed Mater Res Appl Biomat 23(A2):211

4 Grasel TG, Cooper SL (1989) J Biomed Mater Res 23:311

5 Okkema AZ, Visser SA, Cooper SL (1991) J Biomed Mater Res 25:1371

6 Kim YH, Han DK, Jeong SY, Ahn KD (1990) Macromol Chem Macromol Symp 33:319

7 Han DK, Jeong SY, Kim YH, Min BG, Cho HI (1991) J Biomed Mater Res 25:561

8 Han DK, Jeong SY, Ahn KD, Kim YH, Min BG, Macromolecules, to appear

9 Mason RG, Shermer RW, Rodman NF (1972) Am J Pathol 69:271

10 Miale JB (ed) Laboratory Medicine Hematology, Mosby Co, St Louis, MO

11 Jozefonvicz J, Jozefowicz M (1990) J Biomater Sci Polym Edn 1:147

Orientation and Conductivity of Polyacetylene

Mamoru Soga

Central Research Laboratories, Matsushita Electric Industrial Co., Ltd., Yagumo-nakamachi, Moriguchi, Osaka 570, Japan

ABSTRACT

Relationship between orientation and conductivity of polyacetylene films has been studied. Polyacetylene films were synthesized by employing both trialkyl aluminiums and tetraalkyl titanates having long alkyl chains such as hexyl, octyl or decyl groups as catalysts. These polyacetylene films showed good stretchability. The polyacetylene films are oriented by stretching. Orientation degree of polyacetylene films was evaluated by means of polarized infrared spectra. Ca. twenty time of conductivity on the polyacetylene films can be obtained by stretching compared with as-grown films. Orientation is one of the most important factors to enhance the conductivity of pi-conjugated polymers like polyacetylene.

Y. Imanishi (Ed.)
Progress in Pacific Polymer Science 2
© Springer-Verlag Berlin Heidelberg 1992

INTRODUCTION

Conductivity of polyacetylene increases more than eight orders of magnitude when doped with iodine [1]. More interests have been paid to attention of a doped polyacetylene, since Naarmann and Theophilou synthesized highly conducting iodine doped polyacetylene [2].

On the other hand, orientation is one of the most important factors to enhance the conductivity of one dimensional pi-conjugated polymers like polyacetylene. Several attempts were made to obtain an oriented polyacetylene [3-5]. Akagi and Shirakawa applied magnetic force into a polymerization system consisting of nematic liquid crystals and obtained a highly oriented polyacetylene. Kahlert and Leising synthesized an oriented polyacetylene film by pyrolyzing prepolymers. Ozaki polymerized acetylene on the surface of benzene crystals and prepared an oriented polyacetylene.

Stretching is one of the easiest ways to orient fibrils of polyacetylene and to obtain oriented polyacetylene. Naarmann and Theophilou synthesized highly stretchable polyacetylene films by using a thermally aged Ziegler catalyst dissolved in silicone oil medium [2]. Polyacetylene with high stretchability can be obtained by employing trialkyl aluminium or tetraalkyl titanate having long alkyl chains as catalysts [6].

The author was interested in how much the conductivity of oriented polyacetylene films can be enhanced by stretching compared with that of as-grown films. Polyacetylene films were prepared by using both trialkyl aluminium and tetraalkyl titanate having long alkyl chains. Orientation and conductivity of stretched polyacetylene films have been reported here.

EXPERIMENTAL

Synthesis of Polyacetylene

Polyacetylene films were synthesized by a non-solvent polymerization method using a thermally aged Ziegler catalyst of trialkyl aluminium and tetraalkyl titanate [7]. Trialkyl aluminiums used were Trihexyl aluminium, Trioctyl aluminium and Tridecyl aluminium. Tetraalkyl titanates used were Tetrahexyl titanate,

Tetraoctyl titanate and Tetradecyl titanate. The Al/Ti molar ratio was two and aging time of the catalyst was 2 h at 120 °C. The polymerizations were carried out at -78 °C for 14-23 h. The initial acetylene pressure was ca. 600 Torr. Polyacetylene films of 7-25 microns in thickness were obtained. All operations of the preparation procedure are described in detail elsewhere [6].

The polyacetylene films thus obtained were, thoroughly washed several times to remove the residual catalyst, and dried in an argon gas. These polyacetylene films were cut into rectangular pieces of 15 mm x 30 mm, which were stretched by a multistage stretching method using a uniaxial stretching machine in an argon gas environment at room temperature .

Observation of Polyacetylene Structure
Fibril structures of the polyacetylene films were observed by a Hitachi S-800 SEM. A platinum and palladium alloy (80:20) layer with a thickness of 5 nm was sputter-deposited on the surface of specimens before the observation.

Orientation of polyacetylene polymer chains were observed by a X-ray camera and a fourier transform infrared spectrometer (FTIR) (Nicolet, Model 740) with a DTGS detector of a resolution of 8 cm^{-1}, the measuring ranges were 2000 to 400 cm^{-1} and interferograms accumulated were 32.

Measurement of Electric Conductivity
Electric conductivity was measured for the iodine-doped specimens by a four-probe method. The measurements were conducted at room temperature in a direction parallel to the stretching direction.

RESULTS AND DISCUSSION
Stretchability of Polyacetylene Films
Polyacetylene films with high stretchability were obtained by employing mixed catalysts of trialkyl aluminium and tetraalkyl titanate having long alkyl chains. Table 1 shows stretchability of the polyacetylene films. Maximum stretching ratio of the polyacetylene films were 8-12. The polyacetylene films, synthesized by Tridecyl aluminium/Tetraoctyl titanate and Tridecyl alu-

minium/Tetradecyl titanate, were too thin and weak, because the activity of these catalysts was very low.

Table 1 Stretching ratio of polyacetylene films prepared by catalysts of trialkyl aluminium and tetraalkyl titanate.

AlR_3	$Ti(OR')_4$	Stretching ratio $(1/1_o)$
hexyl	hexyl	12
hexyl	octyl	12
hexyl	decyl	12
octyl	hexyl	12
octyl	octyl	8
octyl	decyl	8
decyl	hexyl	10
decyl	octyl	_ a)
decyl	decyl	_ a)

a) self-supported polyacetylene films were not obtained.

Orientation of Stretched Polyacetylene

Figure 1 shows SEM photographs of the surface of as-grown and stretched polyacetylene films prepared by a mixture catalyst Trioctyl aluminium and Tetradecyl titanate. Fully stretched polyacetylene films were highly oriented films, as shown in Fig.1. Fibrils of about 50 nm in diameter were oriented along the stretching direction in the stretched polyacetylene film. On the other hand, fibrils arrangement was random in the as-grown film. The stretched polyacetylene films prepared by the other catalysts also showed oriented fibril structures.

Polyacetylene polymer chains, which constitute fibrils, were also oriented along the stretching direction in the stretched polyacetylene films. Figure 2 shows a photograph of X-ray diffraction patterns of the stretched polyacetylene film prepared by a mixture catalyst of Trioctyl aluminium and Tetrahexyl titanate. Diffraction patterns were observed along the equator. Clear reflection spots along the equator were located at 2 θ = 23.7°, corresponding to the (200) plane [8]. This suggests that polyacetylene polymer chains are highly oriented.

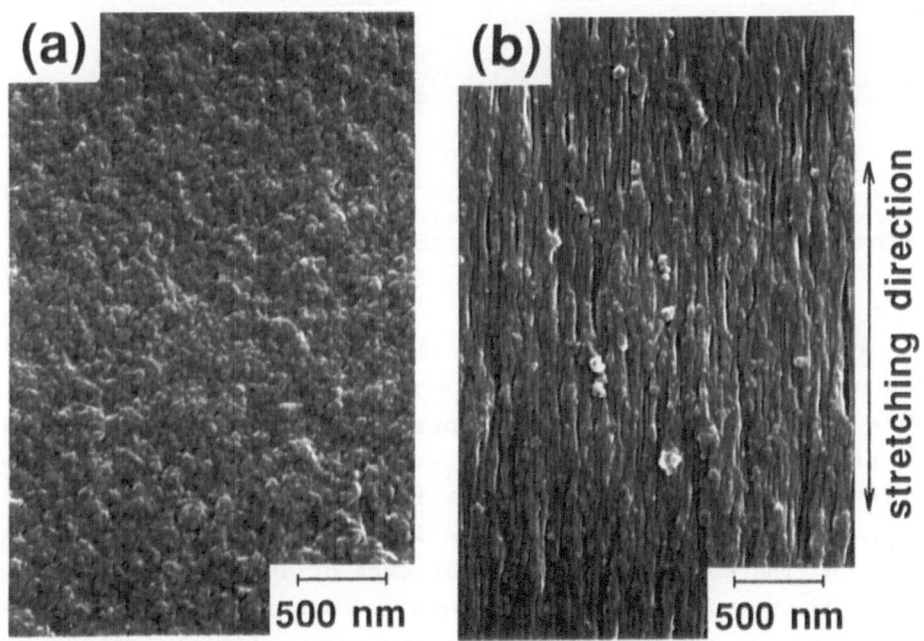

Fig. 1 SEM photographs of the surface of polyacetylene films prepared by a mixture catalyst of Trioctyl aluminium and Tetradecyl titanate: (a) As-grown film; (b) Stretched film ($1/1_o$=8).

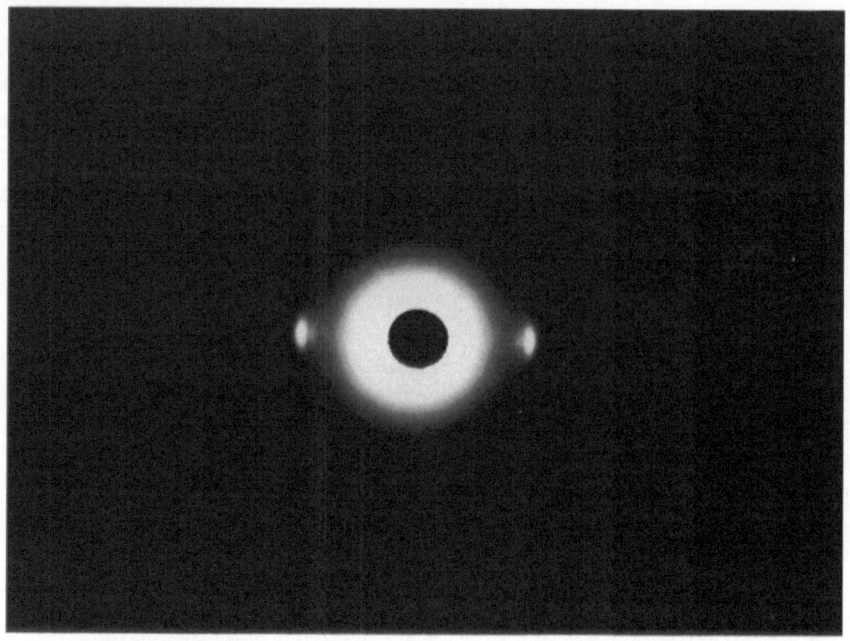

Fig. 2 X-ray diffraction patterns of photograph of the stretched
polyacetylene film prepared by catalyst of Trioctyl aluminium and
Tetrahexyl titanate: Stretching direction vertical.

Orientation of polymer chains for stretched polyacetylene
films was also confirmed by polarized infrared spectroscopy.
Figure 3 shows polarized infrared spectra of stretched polyacetylene films prepared by a mixture catalyst of Trihexyl aluminium
and Tetradecyl titanate. Polarized light was irradiated perpendicular to the stretching direction in Figure 3a and parallel in
Figure 3b. The polarized infrared spectra of stretched polyacetylene films showed dichroism. The peak, which indicates a characteristic of cis-type C-H out of plane deformation vibration at
740 cm^{-1}, was very strong as shown in Figure 3a and very weak in
Figure 3b.

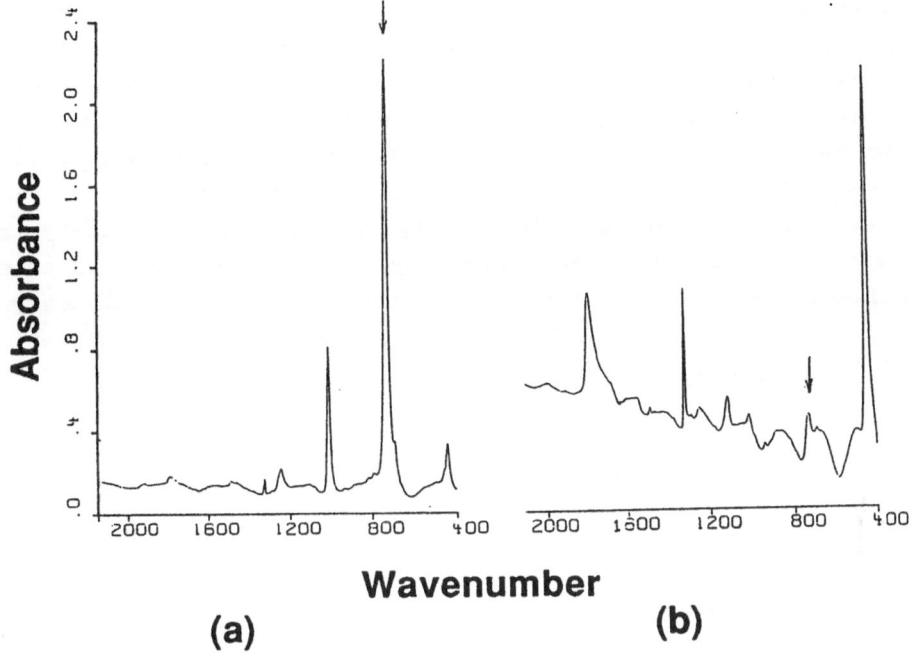

Wavenumber

(a)   **(b)**

Fig. 3   Polarized infrared spectra of stretched polyacetylene films prepared by a mixture catalyst of Trihexyl aluminium and Tetradecyl titanate: (a)perpendicular; (b) parallel.

Orientation degree of polyacetylene polymer chains in fibrils could be estimated from dichroic ratio. Figure 4 shows relationship between dichroic ratio and stretching ratio of the polyacetylene films prepared by a mixture catalyst of Trihexyl aluminium and Tetradecyl titanate. Dichroic ratio was calculated from the absorbances at the peaks of 740 cm^{-1}. The larger stretching ratio became, the larger dichroic ratio became. Dichroic ratio was ca. 15 when the polyacetylene film was fully stretched ($1/1_o=12$). This value was at the highest value of all the oriented polyacetylene films prepared by mixtures catalysts of trialkyl aluminiums and tetraalkyl titanates used in this experiment. The value is larger than that on oriented polyacetylene films polymerized in liquid crystal mediums under magnetic field (dichroic ratio: 4-5) [3].

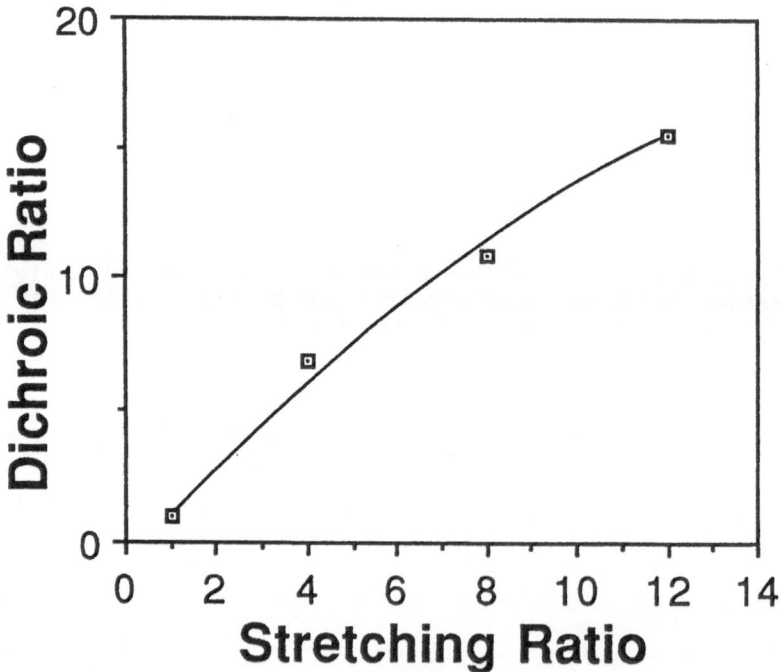

Fig. 4   Relationship between dichroic ratio and stretching ratio of polyacetylene films prepared by a mixture catalyst of Trihexyl aluminium and Tetradecyl titanate.

The axis of polyacetylene polymer chains in the fibrils may not be not parallel to the direction of fibrils, even in fully stretched polyacetylene films. If it is so, the peak at 740 cm^{-1} should disappear in the parallel polarized FTIR spectrum on fully stretched polyacetylene films (Fig. 3b). But the weak peak at 740 cm^{-1} appeared. Akagi estimated the average orientation angle between the axis of polyacetylene polymer chains and the direction of fibrils by the following equation [9],

$$\theta = \cos^{-1}[(2R-1)/(2R+1)]^{1/2} \qquad (1)$$

where R is dichroic ratio. When R = 15 is substituted in equation (1), θ = 15° is obtained. This indicates that the axis of polyacetylene polymer chains tilts to the direction of fibrils or stretching direction at 15°.

Conductivity of Stretch-Oriented Polyacetylene

Stretch-oriented polyacetylene films show one order higher electrical conductivity than as-grown films. Table 2 shows the conductivity of polyacetylene films synthesized by mixtures catalysts of trialkyl aluminium and tetraalkyl titanate having long alkyl chains. As-grown polyacetylene films show the conductivity as high as 10^2-10^3 S/cm and highly oriented polyacetylene films as high as 10^3-10^4 S/cm.

Electrical conductivity of polyacetylene films depends on orientation of polymer chains. Figure 5 shows the relationship between relative electrical conductivity and dichroic ratio of polyacetylene films prepared by a mixture catalyst of Trihexyl aluminium and Tetradecyl titanate. Conductivity of fully stretched or oriented polyacetylene films showed twenty times higher than that of as-grown films. Conductivity of polyacetylene films, prepared by other catalysts systems of trialkyl aluminium and tetraalkyl titanate, showed similar relationship, although increment of the conductivity was lower.

Table 2 Conductivity of as-grown and stretched polyacetylene films prepared by catalysts of trialkyl aluminium and tetraalkyl titanate.

		Conductivity (S/cm)		
AlR_3	$Ti(OR')_4$	As-grown	Stretched	ratio
hexyl	hexyl	950	11,000	12
hexyl	octyl	620	8,600	14
hexyl	decyl	600	12,000	20
octyl	hexyl	650	9,200	14
octyl	octyl	1,300	15,000	12
octyl	decyl	730	6,500	9
decyl	hexyl	850	10,000	12

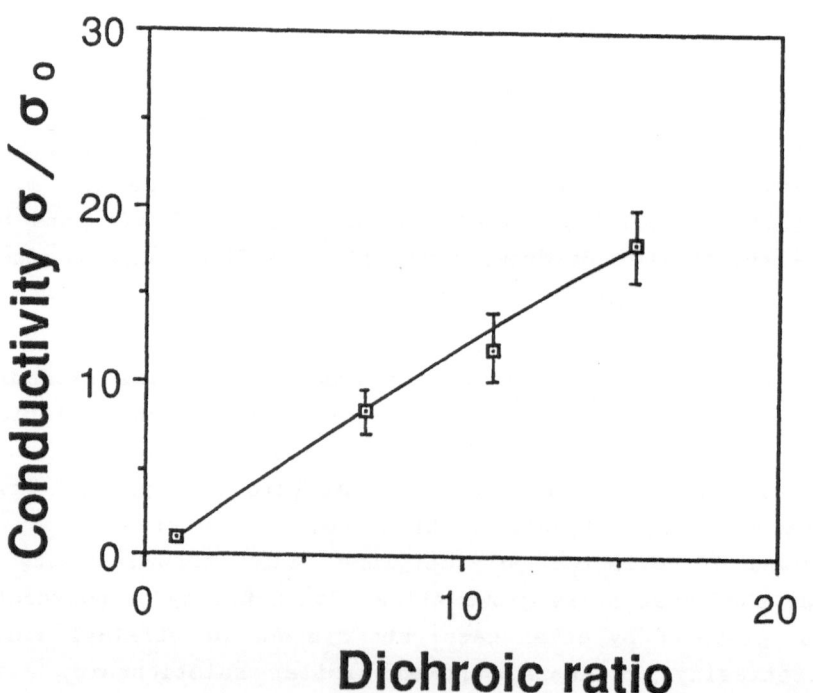

Fig. 5 Relationship between relative electrical conductivity and dichroic ratio of polyacetylene films prepared by a mixture catalyst of Trihexyl aluminium and Tetradecyl titanate.

Tsukamoto presented conductivity of stretched polyacetylene films 10 times higher than that of as-grown polyacetylene films, that is, 10^4 S/cm to 10^5 S/cm [10]. Naarmann-Thiophilou reported that the conductivity of stretched polyacetylene films prepared by $AlEt_3/Ti(OBu)_4$ in silicone oil increased 9-20 times higher than that of as-grown polyacetylene films [2]. The similar results were obtained as shown in Table 2. Conductivity of polyacetylene increased 10 to 20 times higher compared with as-grown films with increasing the orientation of polymer chains.

CONCLUSION
Polyacetylene films with high stretchability have been synthesized by employing mixtures catalysts of trialkyl aluminiums and tetraalkyl titanates having long alkyl chains such as hexyl, octyl or decyl groups. Fully stretched polyacetylene films have oriented fibrils, in that, polymer chains are very highly oriented along the stretching direction. The conductivity of the polyacetylene films can be increased up to twenty times as large as that of as-grown films by orientation.

ACKNOWLEDGEMENTS
It is my pleasure to acknowledge Dr. K. Ogawa for his extensive comments after carefully reading the first script of this paper.

REFERENCES
[1] C. K. Chiang, C. R. Fincher Jr., Y. W. Park, A. J. Heeger, H. Shirakawa, E. J. Louis, S.C. Gau and A. G. MacDiarmid, Phys. Rev. Lett. 39, 1098 (1977).
[2] H. Naarmann and N. Theophilou, Synth. Met., 22, 1 (1987).
[3] K. Akagi, S. Katayama, H. Shirakawa, K. Araya, A. Mukoh and T. Narahara, Synth. Met., 17, 241 (1987).
[4] H. Kahlert and G.Leising, Mol. Cryst. Liq. Cryst., 117, 1 (1985).
[5] M. Ozaki, Y. Ikeda and T. Arakawa, J. Polym. Lett. Ed., 21, 989 (1983).
[6] M. Soga, S. Hotta and N. Sonoda, Synth. Met., 30, 251 (1989).
[7] K. Akagi, M. Suezaki, H. Shirakawa, H. Kyotani, M. Shimomura and Y.Tanabe, Synth. Met., 28, D1(1989).

[8] J. C. W. Chien, F. E. Karasz and K. Shimamura, Macromolecules, 15, 1012 (1982).

[9] K. Akagi, Kagaku, 44, 408(1989).

[10] J. Tsukamoto, A. Takahashi and K. Kawasaki, Jap. J. Appl. Phys. 29, 125(1990).

Functional Water Soluble Polymers

D. N. Schulz, J. Bock, J. Maurer, E. Berluche, J. Kaladas, and
I. Duvdevani

Exxon Research and Engineering Company
Route 22 East, Clinton Township Annandale, NJ   08801

INTRODUCTION

Functional polymers are macromolecules containing functional
groups (e.g. -OH, -COOH, hydrophobes) which have
polarity/reactivity differences from backbone chains.
Alternatively, functional polymers may be viewed as materials
that have a function or a use. Functional polymers often show
unusual or improved properties by virtue of enhancements in
phase separation, reactivity or associations.

This paper examines two classes of associating functional
groups (i.e. hydrophobes and zwitterions) on water soluble
polyacrylamide (PAM) homo- and copolymer backbones.

RESULTS AND DISCUSSION

Previously, we [1-4] and others [5-10] have shown the
benefits of using hydrophobically associating polyacrylamides to
control the viscosity of aqueous and brine solutions. For
example, acrylamide copolymers containing small amounts of long
chain C_8-C_{20} N-alkyl acrylamides exhibit enhanced
viscosification efficiency, shear thickening rheology, as well
as shear and salt stability. However, these polymers like other
PAM's [11] suffer the debit of relatively poor thermal
hydrolytic stability. Also, these materials require special
micellar polymerization techniques, requiring large amounts of
external surfactant to solubilize the alkyl acrylamides in the
aqueous polymerization mixture [1-3].

Copolymers of acrylamide and surfactant macromonomers [e.g.
alkyl polyetheroxyacrylates (R-PEO-AC)] do not require
external surfactant for their preparation [4]. However, they
too exhibit poor hydrolytic stability because of the hydrolysis
of the amide and ester linkages upon heating in water or brine.

Stahl and co-workers [12] noted that copolymers of NVP and
acrylamide have improved hydrolytic stability compared to homo
polyacrylamide. However, the NVP copolymers have reduced
viscosification efficiency compared with acrylamide homopolymers
because NVP containing polymers intrinsically go to lower
molecular weights than polyacrylamide.

Consequently, as part of this study, we elected to combine
the viscosifying effect of hydrophobic associations with the
thermal stability of NVP and the high molecular weight
capability of acrylamide. Sepcifically, we synthesized
terpolymers of acrylamide, N-alkyl acrylamide, and N-vinyl
pyrrolidone (NVP-RAM) and tetrapolymers of acrylamide,
N-alkylacrylamide, sodium acrylate, and N-vinyl pyrrolidone

Y. Imanishi (Ed.)
Progress in Pacific Polymer Science 2
© Springer-Verlag Berlin Heidelberg 1992

(NVP-HRAM). We examined both synthetic variables and the aqueous (brine) viscometry of the products. The NVP-RAM polymers were synthesized by terpolymerization of acrylamide (AM), N-octylacrylamide (RAM), and N-vinylpyrrolidone (NVP) in water with AIBN initiator and SDS surfactant. Since NVP is a moderately good solvent for N-octyl acrylamide, only low levels of SDS are required. The effects of polymerization variables on product solution properties was studied.

It was determined that an NVP-RAM polymer of 30% NVP/ 69% AM/ 1%C$_8$ AM gave the best balance of desired properties. This material was then subjected to base catalyzed hydrolysis and and RAM of similar compositions.

The product of the base catalyzed hydrolysis of NVP-RAM is NVP-HRAM. It has superior viscosification ability than its nonassociationg counterpart NVP-HRAM, and its viscofication ability compares favorably with other high molecular weight aqueous (brine) viscosifiers.

Previously, we[1] had shown that hydrophobically associating polymers can be isoviscous in salt solutions of increasing brine level. We [13,14] and others [15,16] has also shown that aqueous poly zwitterions are antipolyelectrolytes; i.e. their solubility and viscosity increases with increasing salt content.

In the second part of this study, we combine hydrophobes and zwitterions together on the same water soluble polymer backbone and examine the solution properties of such materials. Terpolymers were prepared from acrylamide, N-octylacrylamide and a zwitterion monomer [i.e. (2-hydroxyethyl) dimethyl (3-sulfopropyl) ammonium methacrylate (SPE)] by aqueous terpolymerization, using the micellar method. Alternatively, terpolymers of acrylamide, surfactant macromonomers and zwitterion (SPE) monomers were prepared. These terpolymers show an even more pronounced anitpolyelectrolyte behavior than either the hydrophobically associating acrylamide or the zwitterion acrylamide polymer themselves [17] (Figure 1). Salamone [18] has observed similar behavior for other polymers containing both hydrophobic and zwitterionic groups.

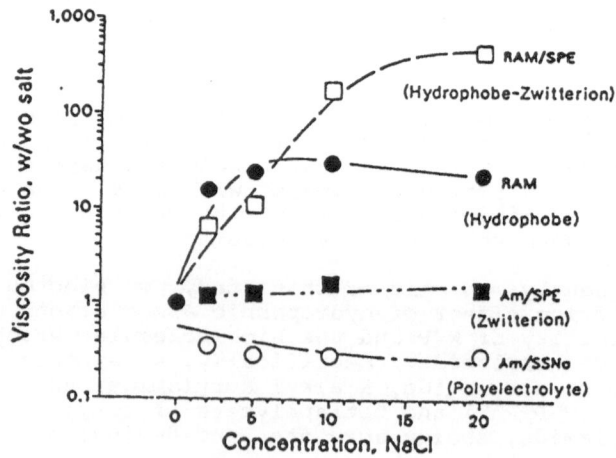

EXPERIMENTAL

Terpolymers of acrylamide, N-alkyl acrylamide, and zwitterionic monomers [e.g. (hydroxyethyl) dimethyl (3-sulfopropyl) ammonium methacrylate (SPE)] were prepared according via micellar polymerization, according to our earlier reports. [17] Terpolymers of acrylamide, surfactant macromonomer [e.g. nonyl phenyl polyetheroxyacrylates] and zwitterion monomers [e.g. (2-hydroxyethyl) dimethyl (3-sulfopropyl) ammonium methacrylate (SPE)] were synthesized via solution polymerization by our previously published methods. [17].

REFERENCES

1.   Bock J, Valint Jr P, Pace SJ, Siano DB, Schulz DN, Turner S R (1988) in Water Soluble Polymers For Petroleum Recovery, Stahl GA and Schulz D N Eds. Plenium, New York, ppp 147-160
2.   Valint Jr P L, Bock J, Schulz D N (1989) Polymers in Aqueous Media: Performance Through Association, Advnances in Chemistry Series No. 223, Glass JE, Eds., American Chemical Society, Washington, D. C., pp 399-410
3.   Bock J, Siano DB, Valint PL Jr, Pace SJ (1989) Polymers in Aqueous Media: Performance Through Association, Advances in Chemistry Series No. 223., Glass JE, Ed., American Chemical Society, Washington, D. C., pp 412-424
4.   Schulz DN, Kaladas JJ, Maurer J, Bock J, Pace S J, Schulz WW (1987), Polymer 28:2110
5.   Emmons WD, Stevens TE, (1983) U.S. Pat. 4,395,524
6.   Evani S, (1984) U. S. Pat. 4,432,801
7.   Costein VG, King T, (1985) U. S. Pat. 4,541,935
8.   Costein VG, (1986) Am. Chem. Soc. Div of Polym. Chem. Polym. Prepr. 27, 1:243
9.   McCormick CL, Johnson CB, in Water Soluble Polymers For Petroleum Recovery, (1988) Stahl GA, Schulz DN, Eds. Plenium, New York, pp 161-180
10. McCormick C L, Johnson C B (1989) in Polymers in Aqueous Media: Performance Through Association, Advances in Chemistry Series No. 223, Glass JE, Ed., American Chemical Society, Washington, D. C., pp. 437-456
11. Maurer JJ, Harvery DG, Klemann LP, (1988) in Water Soluble Polymers for Petroleum Recovery, Stahl GA, Schulz DN, Eds., Plemium, New York, pp. 261-270
12. Stahl GA, Moradi-Araghi A, Doe PH, (1988) Water Soluble Polymers for Petroleum Recovery, Stahl GA, Schulz DN, Eds., Plenium, New York, pp. 121-130

◄Fig 1. Viscosity ratio, with/without added salt as a function of added salt. O is a copolymer of acrylamide and sodium styrene sulfonate (5%), a polyelectrolyte control. ■ is a copolymer of acrylamide and SPE, a zwitterionic polymer. ● is a copolymer of acrylamide and C_8-alkyl acrylamide, a hydrophobe containing polymer. □ is a terpolymer of acylamide, C_8-alkyl acylamide and SPE, a polymer containing both hydrophobic and zwitterionic groups.

13. Schulz DN, Peiffer DG, Agarwal PK, Larabee PK, Kaladas J, Soni L, (1986) *Polymer* 27:1734

14. Schulz, DN Kitano K, Danik JA, Kaladas JJ (1989) in *Polymers in Aqueous Media: Performance Through Association*, Advances in Chemistry Series No. 223, Glass JE, Ed., American Chemical Society, Washington, D. C., pp. 105-114

15. Salamone JC, Voksen JC, Olsen P, Israel SC, (1978) *Polymer* 19:1157

16. Monroy Soto VM, Galin JC, (1984) *Polymer* 25:121, 154

17. Schulz DN, Bock J, Maurer JJ, Duvdevani I, Berluche E, Kaladas J (1988), U.S. Patents 4,742,135, 4,788,247

18. Salamone JC, Thompson AM, Su CH, Waterson AC (1989) *PMSE Prepr.* 61:518

Thermotropic Liquid Crystalline Polyarylates

Toshihide Inoue

Research Association for Polymer Basic Technology
(Toray Industries,Inc.Plastics Research Laboratories
9-1,Oe-cho,Minato-ku,Nagoya,455 JAPAN)

INTRODUCTION

Recently thermotropic liquid crystalline polyarylates(TLCPs)
have been attracting much attention because of their very high
modulus,excellent processability,thermal resistance,dimentional
stability,flame resistance and chemical resistance,and the
possibility of being a substitute for Al.However,TLCPs have
also some serious problems such as mechanical anisotropy,low
weldline strength and fibrization(delamination),and the modulus
is still lower than that of Al.We have been working on the
preparation of novel TLCPs with higher weldline strength and
resistance to fibrization and found that TLCPs derived from
4,4'-diphenyldicarboxylic acid(BB) would be promising materials
as a substitute for Al.
The goal of this paper is to report the development of a novel
class of TLCPs derived from BB through the investigation on the
relationship between the chemical structure and the mechanical
properties of TLCPs.

POLYMERS

Polyarylates were prepared by melt polycondensation involving
removal of acetic acid from substituted hydroquinone diacetates
and aromatic dicarboxylic acids such as BB.

FIBER PREPARATION

Polyarylates were melt spun with an koka type flow tester by
using a capillary of 0.3 mm diameter.Modulus of as-spun fiber
was measured on REOVIBRON at a frequency of 110 Hz,a heating
rate of 2°C/min and with an interchuk distance of 40 mm.

INJECTION MOLDING

The test pieces of polyarylates were obtained by injection
molding using Sumitomo NESTAL injection molding machine(0.5
ounce).ASTM D790 was used for measuring flexural modulus of
injection molded test pieces(thickness, 1/32'').

RESULT AND DISCUSSION

THE EFFECT OF THE CHEMICAL STRUCTURE ON THE STABILITY OF LIQUID CRYSTALLINE STATE

At first,we studied the effect of the chemical structure
of polymer chain on the stability of liquid crystalline
state.The liquid crystalline state of polyether-esters derived
from substituted hydroquinone and substituted 1,2-bis(phenoxy)
ethane-4,4'-dicarboxylic acid could be controled by the

Y. Imanishi (Ed.)
Progress in Pacific Polymer Science 2
© Springer-Verlag Berlin Heidelberg 1992

substituents.Polyether-esters derived from chlorohydroquinone
(Cl-HQ) or methylhydroquinone(Me-HQ) and 1,2-bis(2-chloro-
phenoxy)ethane-4,4'-dicarboxylic acid(Cl-PEC)(Cl-HQ/Cl-PEC or
Me-HQ/Cl-PEC) showed liquid crystallinity and polyether-esters
derived from t-butylhydroquinone(tBu-HQ) or phenylhydroquinone
(Ph-HQ)and 1,2-bis(phenoxy)ethane-4,4'-dicarboxylic acid(PEC)
(tBu-HQ/PEC or Ph-HQ/PEC) showed decreased stability of the
liquid crystalline state because of the steric effect of the
bulky substituents on HQ units.Polyether-esters derived from
tBu-HQ or Ph-HQ and Cl-PEC(tBu-HQ/Cl- PEC or Ph-HQ/Cl-PEC)
didn't show liquid crystallinity.Polyether-esters derived from
Cl-HQ and 1,2-bis(2,6-dichlorophenoxy)ethane-4,4'-dicarboxylic
acid or 1,2-bis(2-methoxyphenoxy)ethane-4,4'-dicarboxylic acid
also did'nt show liquid crystallinity because of the steric
effect of the substituents on PEC units.The effect of the
substituents on the stability of liquid crystalline state
is summarized in Table 1 and 2 (1,2).From these data,we can
conclude that the stability of liquid crystalline state can be
controlled by the lateral substituents on HQ or PEC units.

Table 1   Thermal properties and modulus of polyether-esters.

| Polymer | | | | Thermal properties | | | Modulus | |
X	Y	Tg E''max (^{O}C)	(DSC) (^{O}C)	Anisotropic melt temp. (^{O}C)	Tm (DSC) (^{O}C)	ΔH (DSC) (cal/g)	d (mm)	M (GPa)
Cl	H	86	87	$\geq$282	268	0.6	0.09	48
Me	H	-	104	$\geq$265	291	1.3	0.06	42
Ph	H	134	124	235-265	- ,(270)	- ,(5.7)	0.07	26
tBu	H	-	146	240-316	233,(305)	0.2,(3.3)	0.06	36
Cl	Cl	127	120	$\geq$247	236,265	2.3	0.06	68
Me	Cl	138	129	$\geq$303	310	5.1	0.11	72
Ph	Cl	124	115	-	(191)	(3.8)	0.08	4
tBu	Cl	-	141	-	(247)	(7.0)	0.06	5
Cl	OMe	-	110	-	(237)	(5.3)	-	-
Ph	OMe	-	118	-	(-)	(-)	0.09	4
tBu	OMe	-	121	-	(170)	(3.1)	-	-

Tm and ΔH in ( ) denote the melting temp. and enthalpy in the
isotropic state,respectively.

$$\left[-O \text{—}\langle O \rangle\text{—} OC\text{—}\langle O \rangle\text{—} O(CH_2)_2 O\langle O \rangle\text{—} C- \right]_n$$

Table 2   Thermal properties of polyether-esters.

| Polymer | | | Tg | | Thermal properties | | |
X	Y	Z	E''max (^{o}C)	(DSC) (^{o}C)	Anisotropic melt temp. (^{o}C)	Tm (DSC) (^{o}C)	ΔH (DSC) (cal/g)
Cl	Cl	H	127	120	$\geq$247	236,265	2.3
Cl	Cl	Cl	-	116	-	(226)	(0.7)
Ph	Cl	H	124	115	-	(191)	(3.8)
Ph	Cl	Cl	-	118	-	(-)	(-)

Tm and ΔH in ( ) denote the melting temp. and enthalpy in the isotropic state, respectively.

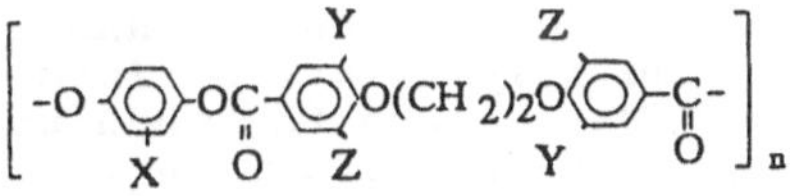

$$\left[-O-\underset{X}{\underset{\|}{\overset{Y}{\bigcirc}}}-OC-\underset{Z}{\overset{Z}{\bigcirc}}O(CH_2)_2O\underset{Y}{\bigcirc}-\underset{O}{\overset{Z}{\underset{\|}{C}}}- \right]_n$$

We expected that polyarylates derived from 4,4'-diphenyl-dicarboxylic acid(BB) would show increased stability of the liquid crystalline state compared with PEC due to the longer mesogen unit, but at the same time, we were afraid that melting temperatures(Tms) of these polyarylates would be very high because Tm of poly(ethylene 4,4'-diphenyldicarboxylate) was much higher than that of poly(ethylene terephthalate) and poly(methylhydroquinone terephtalate)(Me-HQ/TA) showed no liquid liquid crystallinity even at the temprature above 450°C. However, poly(methylhydroquinone 4,4'-diphenyldicarboxylate) (Me-HQ/BB) melted at 372°C and showed liquid crystallinity(3). We clarified the reason of the lower Tm of Me-HQ/BB than that of Me-HQ/TA by investigating the stability of liquid crystalline state of the model compounds derived from BB and TA as shown in Fig. 1. The reason that Me-HQ/BB showed lower Tm is probably the lower enthalpy of Me-HQ/BB due to the twisted chemical structure of biphenylene unit as reported by K.Tashiro (4).

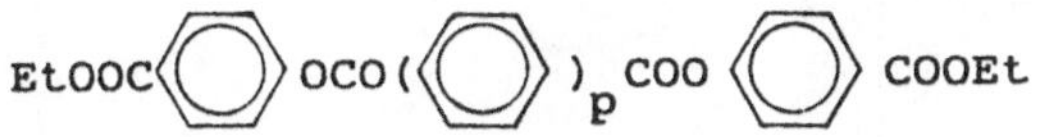

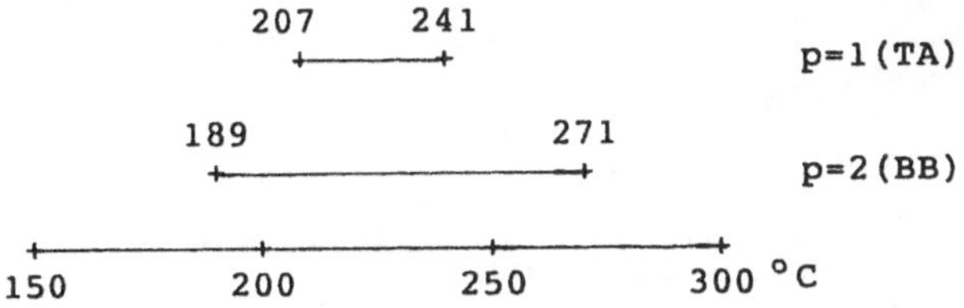

Fig. 1 Stability of liquid crystalline model compounds derived from BB and TA.

THE EFFECT OF THE CHEMICAL SRUCTURE ON THE MODULUS OF AS-SPUN FIBERS

The moduli of as-spun fibers of Me-HQ/Cl-PEC or Cl-HQ/Cl-PEC were 72 and 68 GPa,respectively which were higher than those of Me-HQ/PEC or Cl-HQ/PEC.The reason of higher modulus appeared to be the increased rigidity of polymer chain caused by the restricted rotation of the ether-linkage of Cl-PEC due to the steric hindrance as shown in Fig. 2(1,5).Ph-HQ/PEC or tBu-HQ/PEC showed the moduli of 26 and 36 GPa, respectively which were lower than those of Me-HQ/PEC or Cl-HQ/PEC because of their decreased stability of liquid crystalline state by the presence of the bulky substituent as shown in Fig. 3.Ph-HQ/Cl-PEC and tBu-HQ/Cl-PEC showed low moduli of 4 and 5 GPa, respectively because they did'nt show liquid crystallinity as shown in Table 1(1,2)

In order to obtain high modulus as-spun fibers,the stability of liquid crystalline state and the rigidity of polymer chain are influential factors.Therefore,we expected that the polyarylates derived from BB would show higher modulus than those from PEC or Cl-PEC.As-spun fiber of the polyarylate derived from Cl-HQ and BB modified with TA(Cl-HQ/BB/TA) showed the modulus of 95 GPa(m/n= 7/3) which was higher modulus compared with that of Me-HQ/Cl-PEC. However,as-spun fiber of Cl-HQ/BB/TA showed the modulus of only 11 GPa(m/n=8/2) in spite of the rigid chemical structure,and this is referred to the decreased elongational flow orientation as shown in Fig. 4(3).

The elongational flow orientation could be obsereved as the orientation of fibrils on the cross section of the tensile fractured fibers as shown in Fig. 2,3 and 5(1,3).

 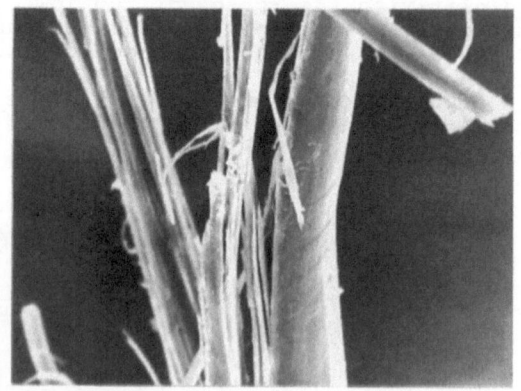

(I);X=H   (II);X=Cl

Fig. 2 SEM of tensile fractured as-spun fibers of Me-HQ/PEC (42 GPa)(I) and Me-HQ/Cl-PEC(72 GPa)(II).

$$\left[-O-\bigcirc\!\!-OC-\bigcirc\!\!-O(CH_2)_2O-\bigcirc\!\!-C-\right]_n$$

(I);X=Ph   (II);X=Me

Fig. 3 SEM of tensile fractured as-spun fibers of Ph-HQ/PEC (26 GPa)(I) and Me-HQ/PEC(42 GPa)(II).

$$\left[-O \underset{X}{\bigcirc} -O\underset{O}{\overset{\|}{C}} -\bigcirc -O(CH_2)_2O\bigcirc -\underset{O}{\overset{\|}{C}} - \right]_n$$

 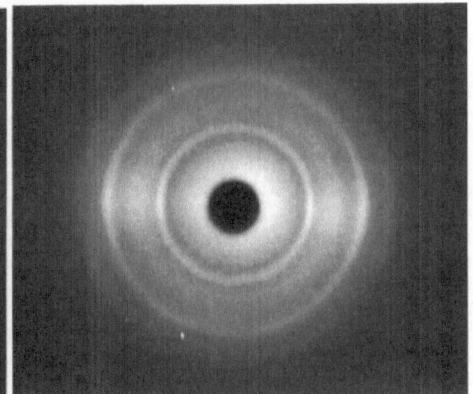

(I)(95 GPa)   (II)(11 GPa)

Fig. 4 X-ray diffraction patterns of as-spun fibers of Cl-HQ/ BB/TA(m/n=7/3)(I) and Cl-HQ/BB/TA(m/n=8/2)(II).

$$\left[O\underset{Cl}{\bigcirc} -O\underset{O}{\overset{\|}{C}} -\bigcirc-\bigcirc -\underset{O}{\overset{\|}{C}} \right]_m \Big/ \left[O\underset{Cl}{\bigcirc} -O\underset{O}{\overset{\|}{C}} -\bigcirc -\underset{O}{\overset{\|}{C}} \right]_n$$

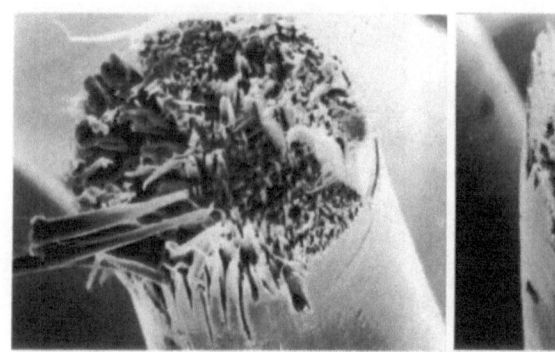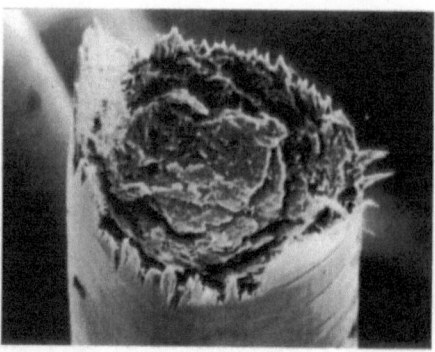

(I)(95 GPa)   (II)(11 GPa)

Fig. 5   SEM of tensile fractured as-spun fibers of Cl-HQ/BB/
TA(m/n=7/3)(I) and Cl-HQ/BB/TA(m/n=8/2).

Although as-spun fiber of the polyarylate derived from Me-HQ and
BB modified with NDA(Me-HQ/BB/NDA) showed high modulus of 112 GPa
(m/n=9.25/0.75),as-spun fiber of Me-HQ/BB/NDA(m/n=9.5/0.5) showed
low modulus of 19 GPa because of the decreased elongational flow
orientation.In order to obtain high modulus as-spun fibers,not
only the stability of liquid crystalline state and the rigidity
of polymer chain but also the elongational flow orientation is
important factor,and the elongational flow orientation is the
most important factor among them.

THE EFFECT OF THE CHEMICAL STRUCTURE ON THE FLEXURAL MODULUS OF
INJECTION MOLDED PARTS

Although as-spun fiber of Me-HQ/Cl-PEC showed high modulus of 72
GPa,the modulus of the injection molded test pieces of Me-HQ/Cl-
PEC was only 26 GPa.Ph-HQ/Cl-PEC and tBu-HQ/Cl-PEC showed no
liquid crystallinity,but the moduli of the injection molded test
were 7.5 and 15.1 GPa,respectively,which were 2-5 times higher
than that of isotropic polymers.tBu-HQ/Cl-PEC showed many
fibrils on the cross section of the flexural fractured test
pieces as shown in Fig. 6(2).We assumed that tBu-HQ/Cl-PEC was
a quasi liquid crystalline polymer which could orient at the
high shear rate.Although W.B.Black reported a lyotropic quasi
liquid crystalline polymer which could orient at the high shear
rate,tBu-HQ/Cl-PEC is the first example of the thermotropic
quasi liquid crystalline polymer(6).

Fig. 6   SEM of tensile fractured injection molded test pieces
of tBu-HQ/Cl-PEC(15.1 GPa).

As previously mentioned,the modulus of as-spun fiber of Cl-HQ/BB/
TA(m/n=7/3) was 95 GPa and those of polyarylates derived from Me-
HQ and BB modified with Cl-PEC(Me-HQ/BB/Cl-PEC) were 72 GPa(m/n=
8.5/1.5) and 65 GPa(m/n=9/1).The flexural moduli of injection
molded test pieces of Cl-HQ/BB/TA(m/n=7/3),Me-HQ/BB/Cl-PEC(m/n=
8.5/1.5) and Me-HQ/BB/Cl-PEC(9/1) showed 28 GPa,32 GPa and 40 GPa,
respectively as shown in Table 3(7,8).
Thus,the relationship between the modulus of as-spun fibers and
the modulus of injection molded test pieces is not clear,but the
flexural modulus is depending upon the BB mol% of polyarylates
and the BB mol% might be used as the parameter of the rigidity
of polymer chains.We observed many plate-like fibrils and a few
needle like fibrils on the cross section of the flexural
fractured test pieces of Me-HQ/BB/Cl-PEC(m/n=9/1).Then we
prepared TLCPs with more rigid chemical structure,Me-HQ/BB/Cl-
PEC(m/n=9.5/0.5),which showed a similar flexural modulus of 39
GPa as Me-HQ/BB/Cl-PEC(m/n=9/1),but X-ray diffraction pattens
and SEM of these 2 polymers were quite different.The injection
molded test pieces of Me-HQ/BB/Cl-PEC(m/n=9.5/0.5) showed lower
degree of orientation and fewer fibrils than those of m/n=9/1 as
shown in Fig.7 and8(8).

Table 3   Modulus of as-spun fibers and flexural modulus of injection molded test pieces of TLCPs derived from BB

Polymer	Modulus			
	as-spun fiber		injection molded test piece	
	diameter (mm)	modulus (GPa)	thickness (mm)	modulus (GPa)
Me-HQ/BB/Cl-PEC(m/n=0/10)	0.11	72	0.8	26
Cl-HQ/BB/TA(m/n=7/3)	0.12	95	0.8	29
Me-HQ/BB/Cl-PEC(m/n=8.5/1.5)	0.20	72	0.8	32
Me-HQ/BB/Cl-PEC(m/n=9/1)	0.10	65	0.8	40
Me-HQ/BB/Cl-PEC(m/n=9.5/0.5)	0.14	91	0.8	39

Me-HQ/BB/Cl-PEC

Cl-HQ/BB/TA

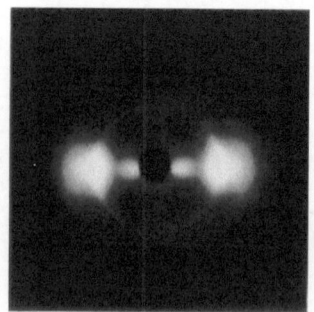

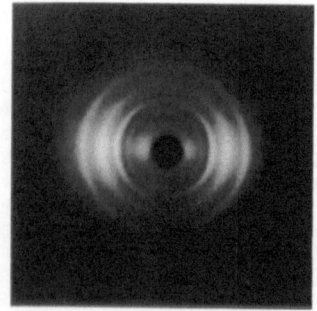

(I)(40 GPa)   (II)(39 GPa)

Fig. 7   X-ray diffraction patterns of injection molded test pieces of Me-HQ/BB/Cl-PEC(m/n=9/1)(I) and Me-HQ/BB/Cl-PEC(m/n=9.5/0.5)(II).

 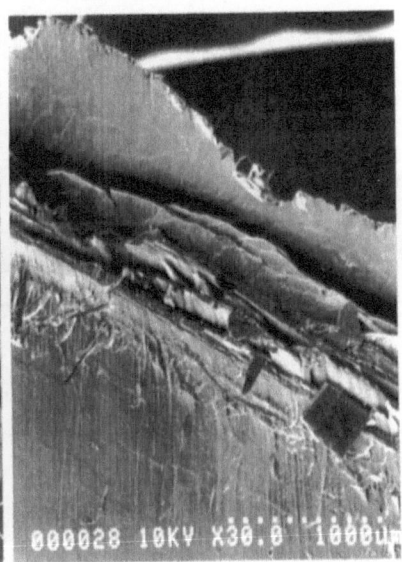

(I)(40 GPa)   (II)(39 GPa)

Fig. 8   SEM of flexural fractured   injection molded test
pieces of Me-HQ/BB/Cl-PEC(m/n=9/1)(I) and Me-HQ/BB/Cl-PEC(m/n=
9.5/0.5)(II).

Me-HQ/BB/Cl-PEC

As previously mentioned,as-spun fiber of Me-HQ/BB/NDA (m/n=9.5/
0.5) showed lower modulus of 19 GPa compared with that of Me-HQ/
BB/NDA(m/n=9.25/0.75)(112 GPa) because of the decreased
elongational flow orientation.However,the flexural modulus
depended upon the BB mol% and we expected that the injection
molded test pieces of Me-HQ/BB/NDA(m/n=9.5/0.5) would show higher
flexural modulus than that of Me-HQ/BB/NDA(m/n=9.25/0.75).Me-HQ/
BB/NDA(m/n=9.5/0.5) showed higher flexural modulus of 47 GPa and
lower degree of orientation and fewer fibrils compared with
Me-HQ/BB/NDA(m/n=9.25/0.75)(34 GPa) as shown in Fig. 9.
Thus,in order to obtain high modulus injection molded parts,
the rigidity of   polymer chain is the most important factor
and the high modulus injection molded parts show low degree
orientation and few fibrils.

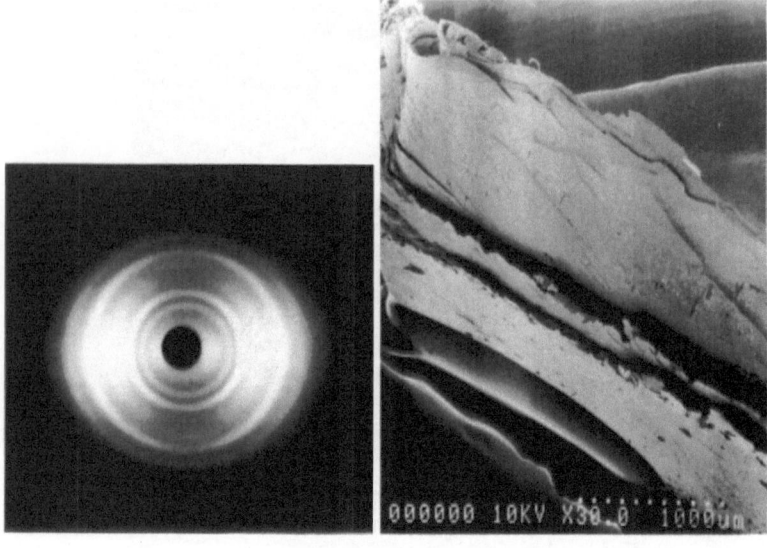

(I)   (II)

Fig. 9 X-ray diffraction pattern of injection molded test
piece(I) and SEM of flexural fractured injection molded
test piece(II) of Me-HQ/BB/NDA(m/n=9.5/0.5).

CHARACTERISTICS OF ME-HQ/BB/NDA INJECTION MOLDED PARTS

DENSITY

The density of the injection molded test pieces of polyarylates
derived from BB was low.The density of Me-HQ/BB/NDA(m/n=9.5/0.5)
was 1.3 .We assumed that the reason of the low density was the
twisted structure of biphenylene unit(4).

MECHANICAL ANISOTROPY

We expected that the injection molded parts of Me-HQ/BB/NDA(m/n=
9.5/0.5) would show not only higher flexural modulus along the
flow direction,but also higher flexural modulus along the
transversed direction than those of the TLCPs on the market.
The flexural modulus of injection molded test pieces of Me-HQ/BB/
NDA(m/n=9.5/0.5) along the transversed direction(1/8''thickness)
was higher than that of ‹VECTRA›(Hoechst Celanese) along the
flow direction(1/8''thickness).
Thus,the mechanical anisotropy which was one of the most
disadvantageos properties of TLCPs was overcome.

THERMAL ANISOTROPY

Although the coefficient of thermal expansion(CTE) of injection molded parts of TLCPs along the flow direction is below $1 \times 10^{-5} c^{-1}$ the CTE along the transversed direction is high.
The CTE along the transversed direction of the injection molded parts of Me-HQ/BB/NDA(m/n=9.5/0.5) was 50 % lower than that of ‹VECTRA› because of the higher flexural modulus along the transversed direction.

WELDLINE STRENGTH

The low weldline strength is an another disadvantageous problem of injection molded parts of TLCPs.The injection molded test pieces of Me-HQ/BB/NDA(m/n=9.5/0.5) showed twice higher weldline strength than those of the TLCPs on the market.

RESISTANCE TO FIBRIZATION

The poor resistance to fibrization(delamination) is also one of the disadvantageous problems of injection molded parts .As the injection molded parts of Me-HQ/BB/NDA(m/n=9.5/0.5) showed few fibrils,the resistance to fibrization(peeling off) of outer skin layer was superior to the TLCPs on the market.

THERMAL RESISTANCE

The injection molded test pieces of Me-HQ/BB/NDA(m/n=9.5/0.5) showed high thermal resistance and the DTUL(1.82MPa) was $272^{\circ}C$.

IZOD IMPACT STRENGTH

Izod impact strength is very important characteristic of the engineering plastics.The izod impact strength(notched,J/m) of injection molded test pieces was 92 J/m which was lower than ‹VECTRA› ,but higher than those of poly(butylene terephthalate) and glass fiber reinforced ‹VECTRA›.

CONCLUSIONS

In order to obtain high modulus injection molded parts ,the rigidity of polymer chain is the most important factor.
The injection molded parts of Me-HQ/BB/NDA(m/n=9.5/0.5) showed high flexural modulus,low mechanical anisotropy ,high weldline strength and excellent resistance to delamination and we consider that this material is a very promising material as a substituent for aluminium.

ACKNOWLEDGMENTS

This work was performed under the management of Research Association for Basic Polymer Technology as a part of R&D of Basic Technology for future Industries sponsored by NEDO.
I would like to acknowledge the numerous contributions of M.Okamoto,T.Yamanaka and N.Goto who prepared the thermotropic liquid crystalline polyarylates.

REFERENCES

(1)   Inoue T,Okamoto M,Hirai T (1986) Kobunshi Ronbunshu 43;261
(2)   Inoue T,Yamanaka T,Okamoto M (1988) Kobunshi Ronbunshu 45;
       661
(3)   Inoue T,Okamoto M,Hirai T (1986) Kobunshi Ronbunshu 43;253
(4)   Tashiro K, Hou Jian-an,Kobayashi M,Inoue T (1990) J Am
       Chem Soc.112;8273
(5)   Inoue T,Komatsu H,Yanagi M (1984) Kobunshi Ronbunshu 41;685
(6)   Black W.B; Miller R.L(ed) FLOW-INDUCED CRYSTALLIZATION IN
       POLYMER SYSTEMS GORDON AND BREACH SCIENCE PUBLISHERS
(7)   Inoue T,Okamoto M (1987) Kobunshi Ronbunshu 44;151
(8)   Inoue T,Yamanaka T (1988) Kobunshi Ronbunshu 45;783
(9)   Inoue T,Tanaka T (1989) Proc. 1st Japan International SAMPE
       Symposium ,557
(10)   Inoue T (1990) Inter Chem 90 Conference on Plastics
       Technology; New Challenges Official Proceedings 5

Recent Development Status of LCP Applications

Tsuneyoshi Okada

Polyplastics Co., Ltd., Vectra Marketing Department
3-2-5 Kasumigaseki, Chiyoda-ku, Tokyo 100

1. MARKET STATUS OF LCP

In 1985, thermoplastic LCP was introduced by Mitsubishi Kasei (registered name: Novaccurate), Unitika (Rodrun) and Polyplastics (Vectra) in Japan. Hoechst Celanese also officially introduced Vectra in the US at the same time, thus 1985 can be recognized as the first year of LCP worldwide.

At the beginning of LCP introduction, LCP was considered as a super engineering plastics looking for a niche in the market due to the high price, but recently LCP is changing its category from a mere super engineering plastic to a second generation of engineering plastics because of its rather unique characteristics which only LCP reveals.

The growth rate of LCP in these past five years was over 50% and sales volume for 1991 is expected to reach 500 tons in Japan and 1,800 tons worldwide.

This favourable LCP growth has induced many new entries into the LCP business especially in Japan. (Table 1)

Table 1   LCP Manufacturers in Japan

Manufacturer	Registered Name	Plant Capacity (T/Y)
Polyplastics	Vectra	*
Sumitomo	Ekonol	800
Nippon Petrochemical	Xydar	*
Mitsubishi Kasei	Novaccurate	150
Ueno Fine Chemical	Ueno LCP	300
Idemitsu Petrochemical	Idemitsu LCP	50
Mitsubishi Gas Chemical	–	50
Toray	–	50
Kawasaki Steel	–	50
Toso	Toso LCP	50

* Import from US

Y. Imanishi (Ed.)
Progress in Pacific Polymer Science 2
© Springer-Verlag Berlin Heidelberg 1992

2. TYPES OF LCP

LCPs should be distinguished from conventional polymers because of its different polymer chain structure, that is, extended polymer chain.

Many unique LCP characteristics come from this structure. LCP is also classified into thermotropic and lyotropic LCP, just as conventional polymers, namely thermoplastics and thermoset. Thermotropic LCP corresponds to thermoplastics because it reveals its liquid crystal nature in the molten stage.

Figure 1   Polymer Tree

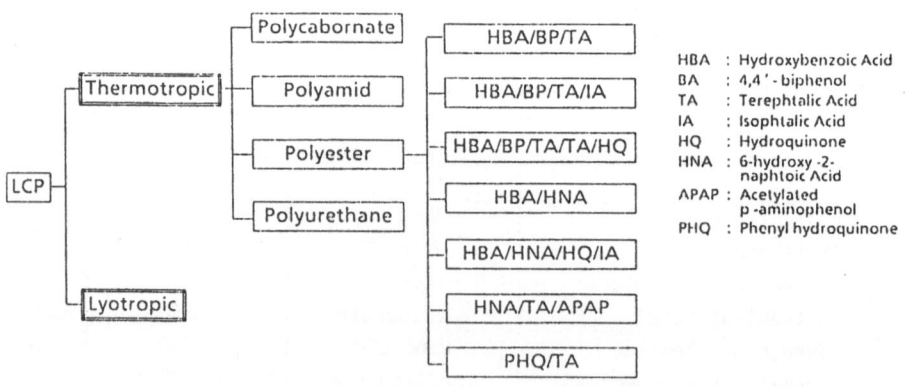

There are several kinds of thermoplastic LCP, for example, polyester type, polyesteramide type and polyestercarbonate type. (Table 2) This paper will mainly describe the polyester type LCP because this is the only commercial type available at this moment.

Table 2   **LCP Variation**

HBA : Hydroxybenzoic Acid
BA : 4,4′ - biphenol
TA : Terephtalic Acid
IA : Isophtalic Acid
HQ : Hydroquinone
HNA : 6-hydroxy -2-naphtoic Acid
APAP : Acetylated p -aminophenol
PHQ : Phenyl hydroquinone

Thermotropic:
- Polycabornate
- Polyamid
- Polyester
- Polyurethane

Variations:
- HBA/BP/TA
- HBA/BP/TA/IA
- HBA/BP/TA/TA/HQ
- HBA/HNA
- HBA/HNA/HQ/IA
- HNA/TA/APAP
- PHQ/TA

Lyotropic

When LCPs first entered the market in 1985, it was possible to clearly distinguish into three categories: [Type I] extreme high temperature resistance, [Type II] high temperature resistance and [Type III] moderate temperature resistance. (Table 3) Recently, there have been improvements made on Types I and III. By slightly rearranging the configuration of Type I, it has reduced heat resistance temperature making molding easier. Also, changing polymerization conditions of Type III has improved heat resistance to a level of Type II. Mechanical properties, heat resistance and moldability differ significantly among the various types of LCP.

Table 3 — Classification of LCPs

Type	Configuration	HDT*	Registered Name
I		275~350°C	Ekonol / Xydar
II		240~250°C	Vectra
III		180~210°C	Novaccurate / Rodrun

* GF Reinforced Grade

3. CHARACTERISTICS OF LCP

LCPs show several unique characteristics which conventional polymers do not possess. This nature comes from LCPs extended polymer chain structure.

3.1 High Strength/High Modulus

LCPs maintain a highly-oriented molecular structure in both the molten and solid states. In general, such polymers are known as rodlike polymers, and exhibit extremely strong mechanical properties in the direction of molecular orientation. For example, unfilled grades of Type II LCPs exhibit tensile strength of 200 MPa and flexural modulus of 9 GPa (both values are of 3 mm thickness), which are values that cannot be reached even with glass filled conventional polymers.

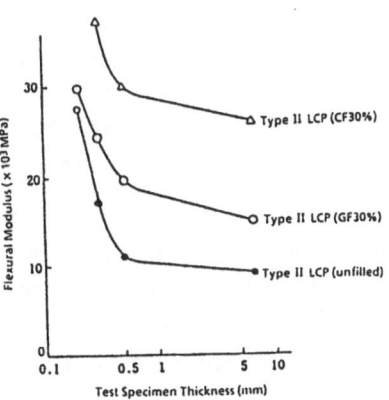

Figure 2   Flexural Modulus vs Thickness

Flexural Modulus ($\times 10^3$ MPa)

Type II LCP (CF30%)
Type II LCP (GF30%)
Type II LCP (unfilled)

Test Specimen Thickness (mm)

On top of that, another unique feature is that flexural modulus shows higher values as thickness becomes thinner. (Fig. 2) This is due to the existance of a distinct skin/core structure, that is, a randomly oriented core and a skin layer which is uniformly oriented. Mechanical strength is high in the oriented skin portion, but not so in the randomly oriented core. As the thickness of molded products become thinner, the ratio of the skin/core increases, and accordingly, mechanical properties are increased. This skin/core structure has been identified to exist in extreme thinness of 0.2 mm thickness. This highly ordered characteristic is also LCPs weakness. That is, this strong orientation, put differently, is high degree of anisotropy. The ratio of mechanical properties between flow direction (MD) and transverse direction (TD) (MD/TD) of unfilled grades is circa 3 to 4, which is a much larger value than conventional plastics. Another interesting feature is that by adding fiberglass, the ratio is reduced to 2 or 3 and falls within the range of conventional plastics. For such reasons, fiberglass reinforced grades are mainly used in the area of injection molding.

3.2 Low Coefficient of Linear Thermal Expansion (CTE)

It is widely known that most materials expand when heated. CTE of conventional plastics is rather large and ranges between $5 - 8 \times 10^{-5}$ cm/cm/°C; molded product of 100 mm expands and shrinks 0.5 - 0.8 mm when there is a temperature change of 100°C. An unusual characteristic of LCP is that it shows negative CTE in the direction of orientation. This

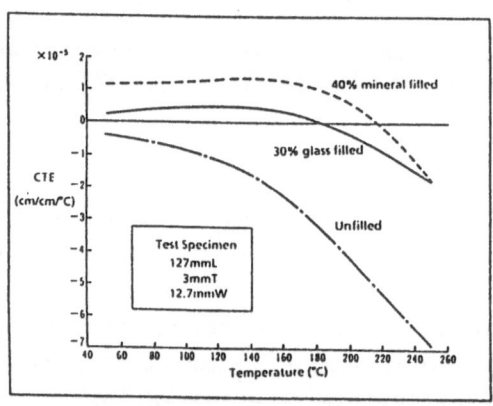

Figure 3   CTE of LCP (Flow direction)

unique characteristic is seen in rodlike polymers, but by adding fillers such as fiberglass, the molecular orientation is disrupted and CTE gradually becomes larger. In other words, CTE can be tailored by controlling the amount of filler. For such reasons, LCP is used for optical fiber coupler cases. The amount of fiberglass is controlled to create CTE equivalent to that of optical fiber.

This characteristic, though is only seen in the flow direction and CTE in the transverse direction is equivalent to conventional plastics. CTE for molded parts cannot be easily calculated because the flow pattern is complex. CTE for such parts are commonly expressed by the average CTE of flow and transverse. CTE of 30% fiberglass reinforced LCP is 2.0×10^{-5} cm/cm/°C and is almost equal to that of aluminum; for 30% carbon fiber reinforced LCP is 0.8×10^{-5} cm/cm/°C and is similar to the CTE value of steel. LCP is becoming a popular materials for parts requiring high dimensional stability such as FDD carriages, DAT chassis, CD pickups, etc.

3.3 Wide Range Temperature Usage

As stated previously, heat resistance varies among the different types of LCP. The molecular structure of Types I and II are formed by a benzene ring that heat resistance is the greatest among existing thermoplastics. (Fig. 4)

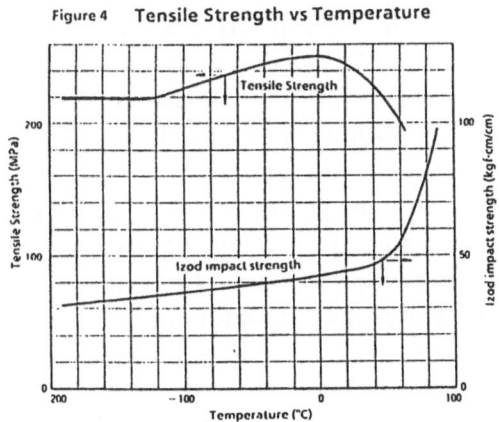

Figure 4   Tensile Strength vs Temperature

The recent miniaturization trend of equipment is bringing about a need for high density mounted printed wiring boards (PWB). For the purpose of designing high density mounted PWBs, electronic components such as switches and connectors are changing over to SMT (Surface Mount Technology). This method directly exposes the plastic housing to the high temperature of infrared soldering. PC (polycarbonate) and PBT (polybuthyleneterephthalate) which have been the main materials used for housing of electronic components do not satisfy the minimum heat resistance temperature of 260°C which is required for SMT. Types I and II can be exposed to soldering for over 10 seconds without melting or blistering which makes LCP an ideal material for this technology.

Heat resistance can be divided into short term such as soldering temperature resistance and long term such as maximum continuous usage temperature, or so called UL (Underwriters Laboratory) thermal index. Types I and II have been granted a 240°C index by UL. This is equivalent to PEEK (polyetheretherketone), and is superior than PES (polyethersulfone), PEI (polyetherimide) and PPS (polyphenylenesulfide).

Not only do LCP exhibit superior performance under high temperatures, but are of the few plastic materials which perform well under cryogenic temperatures. For most plastics, there is a certain temperature between -50°C and -100°C where the physical properties reach a ductile-brittle transition point. Even under cryogenic temperatures as low as that of liquid nitrogen, the mechanical properties of LCP remain unaffected.

3.4 Damping

LCP's skin/core structure or "sandwich" structure which has been explained previously plays an active role in bringing out damping properties. In general, materials such as metal which are highly rigid, have poor damping or energy absorbing properties and soft materials such as rubber have good damping properties. Owing to this skin/core structure, LCP satisfies both stiffness and damping properties. (Fig. 5)

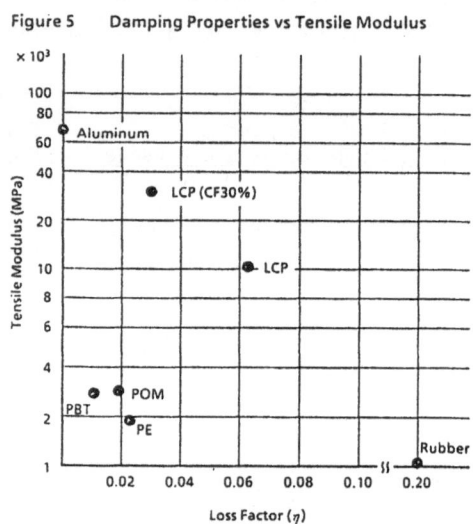

Figure 5   Damping Properties vs Tensile Modulus

Utilizing these excellent damping properties, LCPs are replacing POM (polyacetal) gears in measures to reduce noise. As Figure 6 shows, not only is forced vibration absorbed in low frequency ranges, but resonance sounds diminish in high frequency ranges. Precise (low mold shrinkage, low CTE), high transmission torque (high flexural modulus) and low noise gears are possible with LCP.

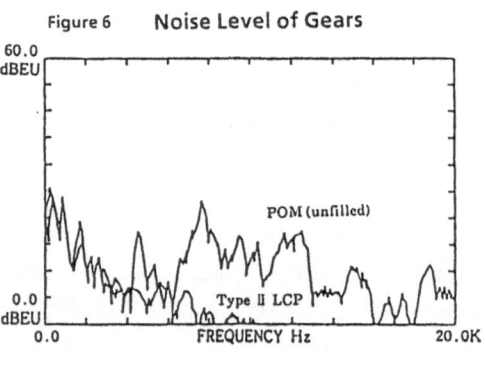

Figure 6   Noise Level of Gears

LCPs damping properties are contributing to noise reduction and high performance of speaker cones, CD actuators, dot-printer end plates and motor end brackets.

3.5 High Flow

In general, molecules move around actively in the molten phase and the molecular chains form a random coil which is the most stable energy state. The molecular chains entangle during flow causing high viscous flow. For improving flowability, the molecular chains are shortened to reduce entangling in most cases.

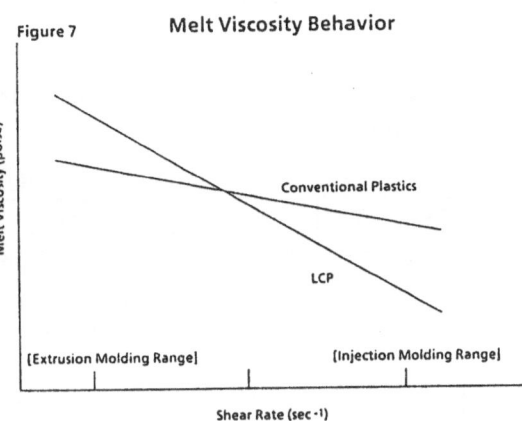

Figure 7   Melt Viscosity Behavior

LCP, on the other hand, maintain their rigid, rodlike structure even in the molten phase that entangling does not occur. LCPs have high flowability, and in practice, wall thickness of 2 mm will flow at least 1000 mm and thinness of 0.5 mm will flow 100 mm or more. Flash rarely is a problem even with this high flow characteristic. The flow length is 2 - 5 times longer than PPS and PA (polyamide), but the occurance of flash is less frequent. This can be explained by solidification speed and shear. Compared with other plastics, the solidification speed of LCP is more than 10 times faster that LCP solidifies before flash occurs. Under high shear rate, melt viscosity is low compared with conventional polymers, but in low shear rate, viscosity significantly increases.

Utilizing such characteristics, coil bobbins of 0.1 mm thickness and 10 mm length have been commercialized.

4.   APPLICATIONS OF LCP

4.1 Injection Molding

LCP finds its major market in the field of injection molding. In this field, electronic devices such as connectors, SMT switches and relays are favorable applications using LCPs sufficient soldering temperature resistance and excellent flowability at very thin wall thickness without flash. The next important field for LCP is business machine and precision industry, for example, hard disc drives, printers, cameras and watches.

These applications require exceptional dimensional stability, low CTE and sometimes damping properties when a replacement from metal to plastics is considered.

Table 4 summarizes which of LCPs characteristics is utilized for individual LCP applications.

Table 4 LCP Application Field

Properties / Applications	Thermal		Physical			Chemical	Molding	
	Soldering Resistance	Continued Usage	Stiffness	Damping Properties	Low CTE	Chemical Resistance	High Flowability	Low Mold Shrinkage
SMT Electronic Components	◎		△				○	
Audio/Visual			◎	○				△
Automotive		○	◎		△			
OA Equipment (printers, FDDs, etc.)			○	△	◎			
Precision Equipment (cameras, watches, etc.)			◎				○	∧
Medical/Chemical (sterilizing equipment)		○				◎	△	

4.2 Fibers

Aramid fiber (lyotropic LCP) is known as a super high modulus fiber. Recently, a fiber of thermotropic LCP has been introduced. This fiber is made by polyester and exhibits high modulus as that of aramid. (Table 5)

Table 5 Basic Properties of LCP Fiber

			Polyester Fiber	Polyaramide Fiber	
			VECTRAN HT	PPTA (REG.)	PPTA (HM)
Density		(g/cm³)	1.41	1.44	1.45
Decomposition Temp./Melting Point		(°C)	>400	>400	>400
Moisture Absorption Ratio		(%)	0	4.9	4.3
Limited Oxygen Index		(%)	37	42	39
Mono-Filament	Tensile Strength	(g/d)	29	25	25
		(kg/mm²)	368	324	326
	Breaking Elongation	(%)	3.9	3.6	3.0
	Initial Modulus	(g/d)	670	620	960
		(kg/mm²)	8500	8040	12530
Multi-Filaments Yarn	Deniers/Filaments		1500/300	1500/1000	1420/1000
	Tensile Strength	(g/d)	26	21	21
		(kg/mm²)	330	272	274
	Breaking Elongation	(%)	3.9	3.9	2.7
	Initial Modulus	(g/d)	600	558	877
		(%)	7610	7230	11440
	Knot Strength	(g/d)	6.8	7.1	6.5
	Tensile Strength at wet atmosphere	(g/d)	26	19	20
	Tensile Strength Ratio of Wet/Dry	(%)	98	91	95

4.3 Films

Excellent properties such as heat
resistance, strength and low CTE,
there are great expectations for
usage of LCP films for FPCs
(flexible printed circuits) and
FDs (floppy disks). Due to the
high degree of anisotropy, it is
difficult to produce films which
show sufficient bi-axial
strength. Attempts to lower the
degree of anisotropy by changing
the processing method and
molecular configuration, are
being taken into consideration.
Another excellent property of LCP
films is low gas permeability.

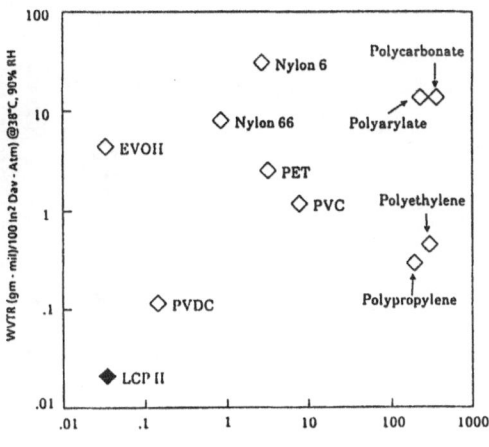

Figure 8   PERMEABILITY OF POLYMER FILMS

Oxygen Permeability (cc-mil)/100 in², - Dav - Atm) @23°C, 75% RH

5.   SPECIAL TOPICS

5.1 MID

Recently, MID (Molded Interconnection Device) technology was introduced by
utilizing LCPs characteristics to its fullest. MID is a concept which can
eliminate metal lead frame and metal contact pin on electronic devices by
plastic metalization. This new technology contributes to miniaturization
and reliability increase for various electronic devices.

There are two different processes in MID: one-shot process and two-shot
process. (Figs. 9 and 10)

For MID material, other than platability, soldering temperature resistance,
high flow under low pressure which is required for intricate designing and
CTE equivalent of metals which prevents delamination of metal portion from
the polymer layer during heat cycling are all necessary conditions.

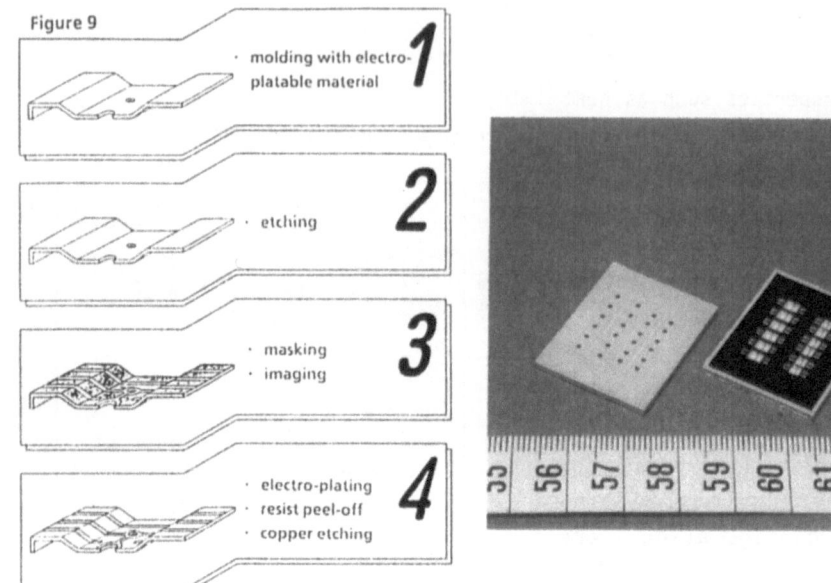

Figure 9

1 · molding with electro-platable material

2 · etching

3 · masking
· imaging

4 · electro-plating
· resist peel-off
· copper etching

MID by 1 Shot Molding Process

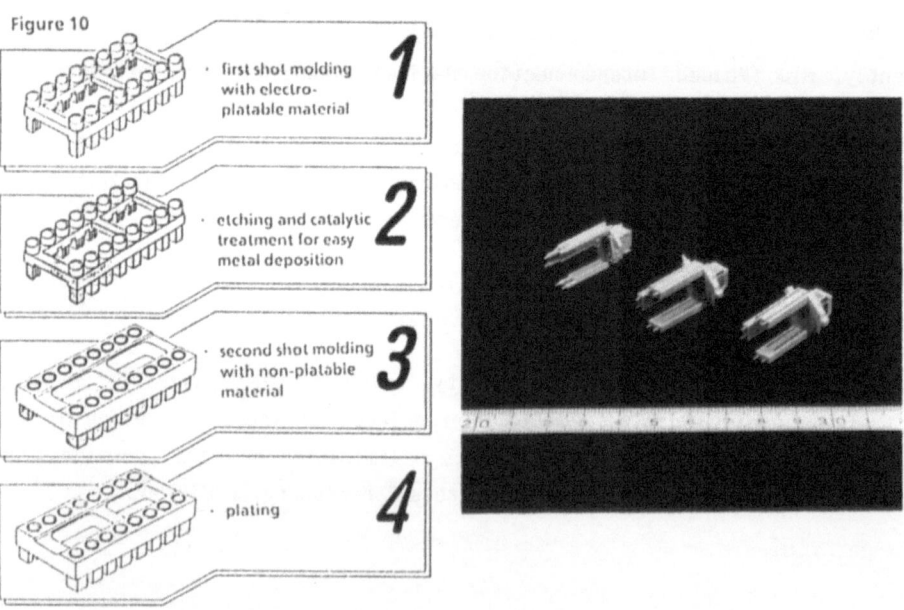

Figure 10

1 · first shot molding with electro-platable material

2 · etching and catalytic treatment for easy metal deposition

3 · second shot molding with non-platable material

4 · plating

MID by 2 Shot Molding Process

5.2 Polymer Blend/Alloy

Polymer blend/alloys which are mixed with LCPs can be divided into three categories according to the percentage used.

Table 6   **Category of LCP Blend / Alloy**

LCP Content (%)	Objective	Example
0 ~ 15	Processability Improvement	PET/LCP (solidification speed) PP/LCP (solidification speed) PES/LCP (flow aid)
15 ~ 50	Property Improvment	PES/LCP (mechanical property) PVC/LCP (low CTE)
50 ~ 100	Low Cost	LCP/PBT LCP/PET

The first category contains 0 - 15% LCP. This is to improve the characteristics of matrix resins. For example, crystallization speed is increased by adding a small quantity of LCP with PET and PP. This also improves flowability of amorphous polymers as PES which normally have poor flowability. (Fig. 11)[1]

Figure 11   Viscosity of LCP/PEI Blend

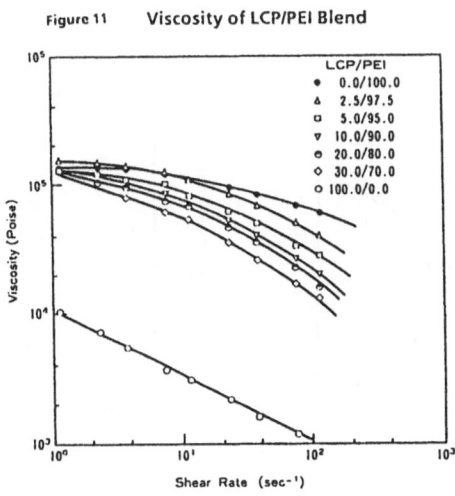

Category 2 contains 15 - 50% LCP.   Adding LCP to the matrix resin, it is expected to improve mechanical properties and lower CTE.   LCPs are incompatible with most polymers; the molecular chain of LCPs are stretched in the flow direction and perform as a reinforcement such as fiberglass. (Fig. 12)[2]

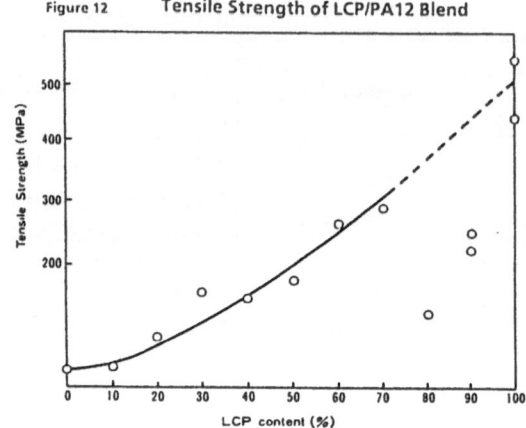

Figure 12   Tensile Strength of LCP/PA12 Blend

The third category contains 50 - 100% LCP.   The purpose of adding conventional engineering plastics into LCP is aimed at lowering costs without significantly changing the nature of LCP.   In this case, appropriate phase control should be considered. (Figs. 13 and 14)

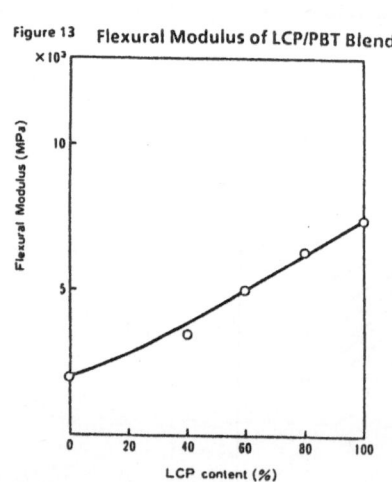

Figure 13   Flexural Modulus of LCP/PBT Blend

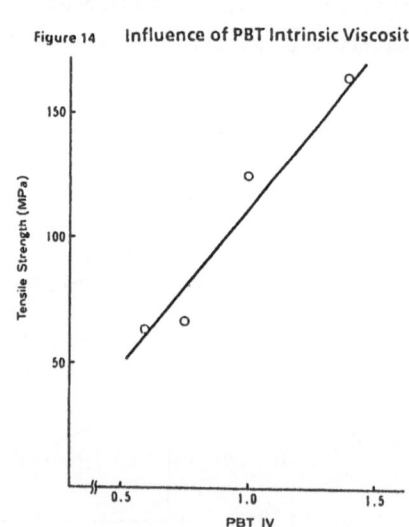

Figure 14   Influence of PBT Intrinsic Viscosity

References:
1 Kiss G, Polymer Engineering and Science, 27(6), 410 (1987)
2 Chung T, Plastics Engineering, 43 (10), 39 (1987)

Synthesis and Properties of Sequentially Ordered Copolyesters and Regioregularly Ring Substituted Aromatic Homopolyesters

Jung-Il Jin, Chung-Seock Kang, Il-Hoon Lee

Chemistry Department, Korea University
1-Anam Dong, Seoul 136-701, Korea

Abstract : Copolyesters having well-defined sequential order and aromatic polyesters having regioregularly positioned substituents were successfully synthesized via multistep routes. Their properties such as thermal transitions, thermotropic properties, crystallization tendency etc. were compared with those of random sequence copolyesters and randomly substituted counterparts. The polymers of regular microchemical structures were found to have much higher melting transition temperatures and greater degree of crystallinity than those having irregular structures. It also was learned that positional isomerism of the substituents leads to different crystal structure.

INTRODUCTION

In order to establish the structure-properties relationship, we first have to develop synthetic methods of the polymers having well-defined microchemical structures. Recently we[1-3] and others[4,5] observed that the properties such as thermal transition temperatures, liquid crystallinity, crystallization tendency etc. depend very strongly on the sequential order of comonomer units in aromatic copolyesters. Very often the melting point of an ordered sequence aromatic copolyester is higher by as much as 100°C than that of the corresponding random copolyester. For example, the mp of the following copolyester is 290°C whereas the mp of the random counterpart is only 191°C. Moreover, the former's degree of crystallinity was significantly higher than the latter's[1].

Mosophase-forming ability of a copolyester also strongly depends on its comonomer sequence. For instance, the ordered sequence copolyester whose structure shown below is not thermotropic, while the random copolyester having the same overall composition is liquid crystalline[1,6]. Moreover, the former is semicrystalline, while the latter is amorphous.

Y. Imanishi (Ed.)
Progress in Pacific Polymer Science 2
© Springer-Verlag Berlin Heidelberg 1992

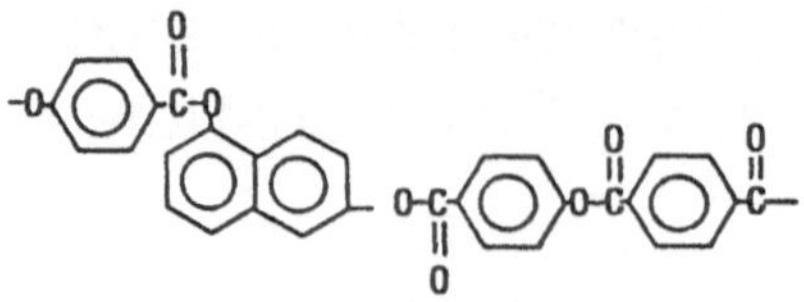

286

Tm ; ordered, mp 229°C-nonliquid crystalline
random, amorphous-nematic

Such a difference can be ascribed to the presence of dimeric or longer, repeated
p-oxybenzoyl units along the chain in the random copolyester, whereas in the ordered
sequence copolyester every p-oxybenzoyl unit exists in a monomeric unit connected to
the bent 1,6-naphthalenediol moiety on one side resulting in a too short linear
segment to render the polymer the ability to form a mesophase in melt.

More recently, we[7] observed that thermal sequence randomization of an ordered
sequence copolyester occurred rather rapidly at an elevated temperature through ester
exchange reactions.   The process could be followed quantitatively using C-13 NMR
analysis.

In this article, in connection with our previous endeavor, we would like to report
synthesis and properties of new copolyesters having ordered comonomer sequences
consisting of mesogenic units and polymethylene spacers, i.e., thermotropic aromatic-
aliphatic copolyesters having sequentially ordered structures :

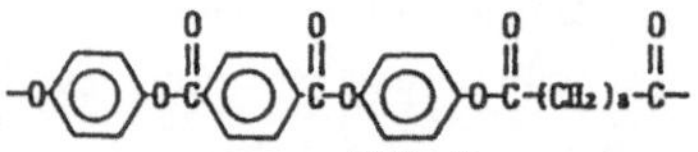

OS-Ar-8

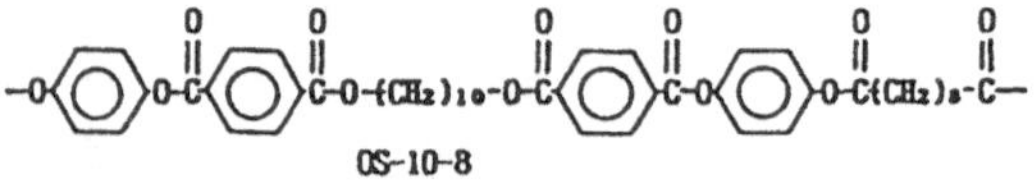

OS-10-8

In addition, synthetic approaches and general properties of regioregularly ring
substituted aromatic polyesters are described in this report :

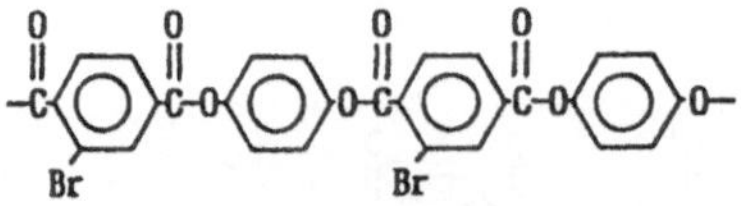

RR-BH

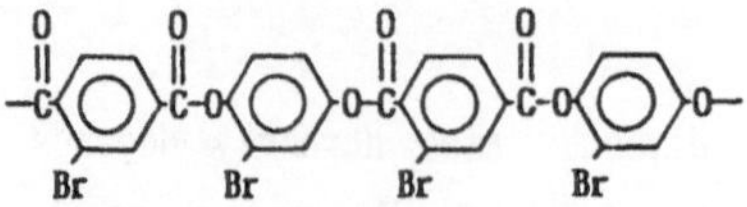

RR-BB

Comparison of the properties of these polymers with those of random sequence or
randomly substituted ones will be made.

EXPERIMENTAL

Synthesis of Polymers

Synthetic methods of the required monomers, OS-Ar-8, OS-10-8[8-11] and RR-BH[12,13], and their random sequence polyesters were the same as reported earlier by us and others. Therefore, only the synthetic procedures for RR-BB polymers will be briefly described in this article.

4-(Benzyloxy)-2-bromophenol was obtained by bromination of 4-(benzyloxy)phenol with Br_2 in CH_2Cl_2 at room temperature. Repeated recrystallization of the crude prodct from n-hexane was necessary to obtain pure product. Yield was 40%, mp 72℃. 4-(Benzyloxy)-2-bromophenol thus prepared was reacted with 3-bromo-4-(carbobenzoxy) benzoyl chloride in THF solution in the presence of pyridine. The reaction was run at room temperature for 12h followed by further reaction at 60℃ for 1h. The ethanol recrystallized product, 4-(benzyloxy)-2-bromophenyl 3-bromo-4-(carbobenzoxy) benzoate (yield, 80.7%) had a mp of 105-106℃. 4-(Benzyloxy)-2-bromophenyl 3-bromo-4-(carbobenzoxy)benzoate(3.0g; $5.0×10^{-3}$ mole) was dissolved in 40mℓ of N,N-dimethyl acetamide containing 6g of 10% Pd-C. 1,4-Cyclohexadiene(4.7mℓ) was added to the above solution under a dry N_2 atmosphere in an ice-bath. Reaction was allowed to proceed at room temperature for 4h[14]. The filtrate was poured into excess water and the precipitate was recrystallized from a mixture of dioxane/toluene(v/v=16/84). The yield of purified 4-(2-bromo-4-hydroxyphenyl)2-bromoterephthalate was 61.5%(1.28g), mp. 250℃. 4-(2-bromo-4-hydroxyphenyl)2-bromoterephthalate(1.50g; $3.6×10^{-3}$ mole) was dissolved in 20mℓ of pyridine containing 2mℓ of $SOCl_2$. The solution was stirred at 80 ℃ for 18h under N_2 atmosphere. And then the reaction mixture was poured into excess 0.1M HCl. The precipitate, after being thoroughly washed with ethanol, was subjected to ethanol extraction for 3 days using a Soxhlet extractor. The polymer yield(1.28g) was 89%.

Characterization

Elemental analyses were conducted by the Analytical Department of the Korea Research Institute of Chemical Technology. Solution viscosities of the polymers were measured at 25℃ or 30℃ using a Cannon-Ubbelohde type viscometer on 0.1g/dL or 0.2g/dL solution in a solvent. Differential scanning calorimeteric(DSC) analyses were performed under a N_2 atmosphere on a DuPont 910 DSC at a heating rate of 10℃/min. Wide angle X-ray diffractograms were recorded on a JEOL JDX-80 instrument using Ni-filtered Cu-Kα radiation. The optical textures of the melts and thermal behavior of the polymers were examined on a hot-stage attached to a polarizing microscope(Leitz, Ortholux).

RESULTS AND DISCUSSION

Synthesis of Polyesters

The two ordered sequence copolyesters, OS-Ar-8 and OS-10-8, were synthesized by direct polycondensation of a dicarboxylic acid and a diol at 80℃ under a N₂ atmosphere in the presence of the condensing agent pair of SOCl₂/pyridine[15].

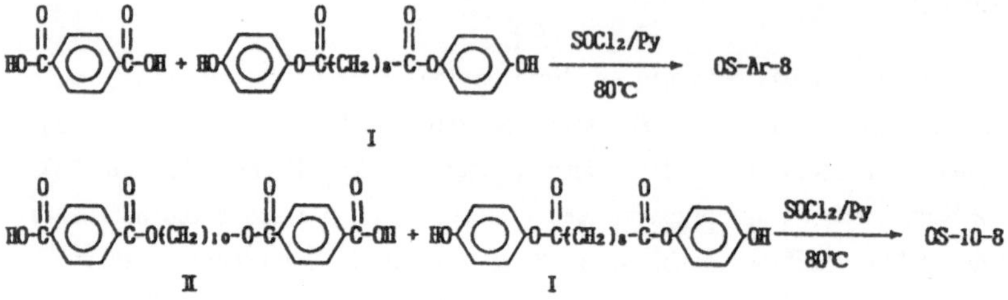

I

II   I

The compounds I and II[8,9] had to be prepared beforehand prior to the final polymerization.

The OS-Ar-8 polymer is an alternating copolyester consisting of oxy-1,4-phenyleneoxy-terephthaloyl(PT) and oxy-1,4-phenyleneoxydecanoyl(PD) units. In contrast, the OS-10-8 polymer is a terpolyester having a sequential order of oxy-1,4-phenyleneoxy-terephthaloyl(PT), oxy-1,10-decamethyleneoxyterephthaloyl (PT) and oxy-1,4-phenylene-oxydecanoyl (PD) units.

In order to prepare the corresponding random sequence copolyesters(RS-Ar-8 and RS-10-8), we employed transesterification polymerization reactions in melt using the same dicarboxylic acids and acetylated bisphenols.

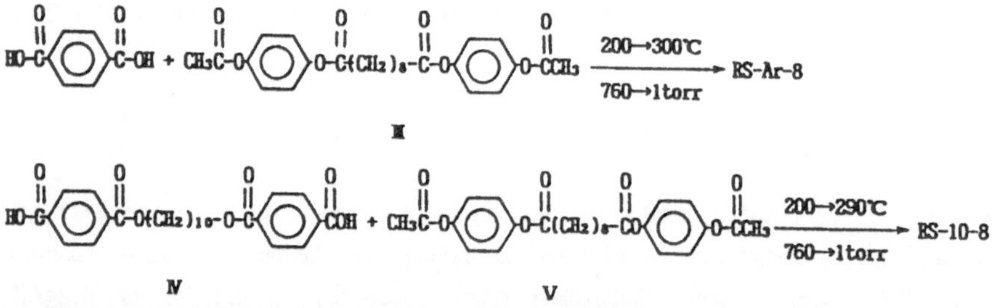

III

IV   V

Although the monomers III-V contain preset monomer sequences, they are destroyed and randomized during polymerization at elevated temperatures, leading to the formation of random copolyesters. Such randomizations were proved to occur in the preparation of other liquid crystalline copolyesters[4,5]. Moreover, utilization of high molecular weight monomers minimizes the loss of reactants through volatilization during polymerization.

The regioregularly substituted polyester RR-BB was prepared via the following multistep routes as described in Experimental:

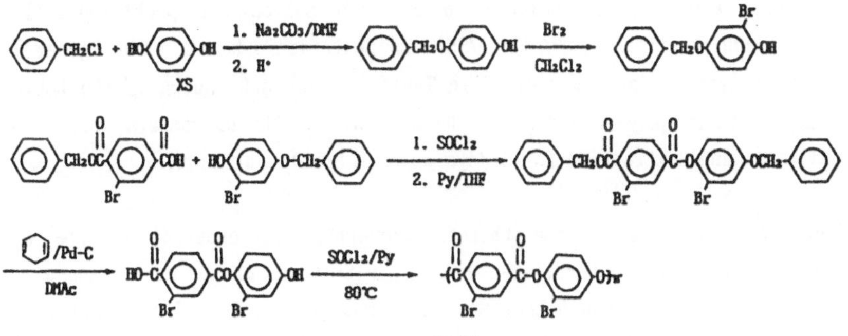

RR-BB

In regioregularly substituted RR-BH the bromine substituents are located only on the terephthaloyl unit, whereas in RR-BB both the two structural units contain the substituents. If one polymerizes bromoterephthalic acid with hydroquinone or bromohydroquinone, bromine substituents will be located in an irregular mode along the chain:

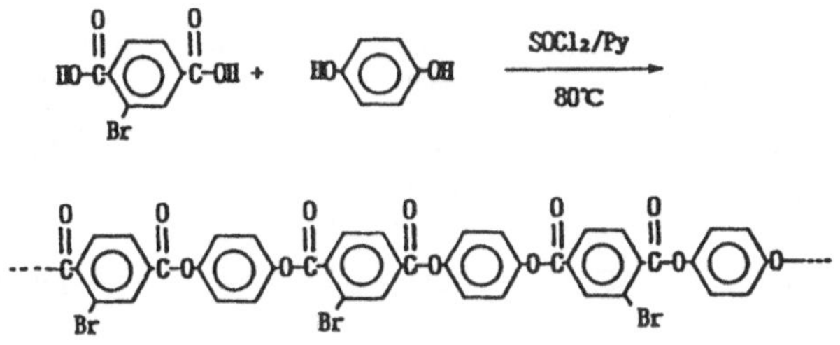

IR-BH

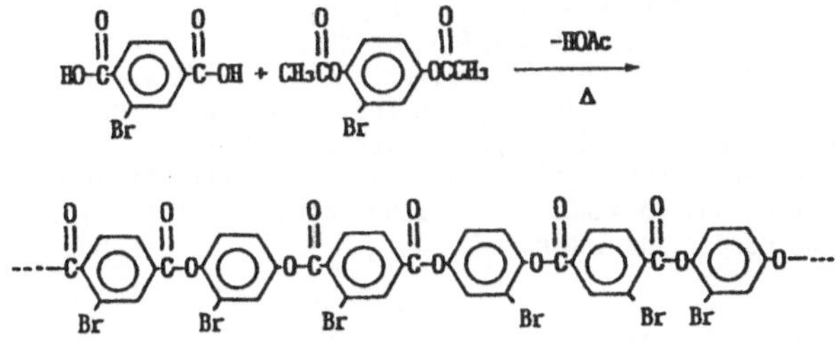

IR-BB

Therefore, these polyesters, in the strict sense, should be taken to be random sequence copolyesters arising from the possible positional isomerism of the substituents. For a comparison purpose, the random positional polymers were synthesized in solution or in melt by reaction bromoterephthalic acid and hydroquinone or diacetate of bromohydroquinone.

General Properties of Polymers

All of the present polyesters exhibit very poor solubility in common organic solvents. We found that they are slightly soluble in a mixed solvent of p-chlorophenol/phenol/1,1,2,2-tetrachloroethane(w/w/w=40/25/35) or pentafluorophenol. The general properties of the polyesters are summarized in Table 1. The data shown in the table are for the samples thoroughly washed with ethanol. All of the polyesters prepared were subjected to ethanol extraction for 3 days using a Soxhlet extractor.

The most contrasting feature in the thermal properties of ordered sequence or regioregularly substituted polyesters, when compared with those of the cor- responding random counterparts, is their much higher melting temperatures, T_m. For example, T_m of OS-10-8 is higher by 85°C than T_m of RS-10-8; 251° vs. 166°C.

Table 1. General Properties of Polyesters

Polymer	η_{inh}	T_g,°C[d]	T_m,°C[e]	T_i,°C[e]	D.C.,%[f]
OS-Ar-8	1.07[a]	77	280	>330	~5(36; 230°C/3h)
RS-Ar-8	0.82[a]	63	206	330	38(41; 170°C/6h)
OS-10-8	0.85[a]	63	251	279	~5(44; 200°C/3h)
RS-10-8	1.50[a]	62	166	195	33(35; 90°C/3h)
RR-BH	0.72[b]	93	365	dec.	38(46; 250°C/1h)
IR-BH	0.43[b]	91	286	dec.	26(37; 250°C/1h)
RR-BB	2.60[c]	110	284	333	13(14; 200°C/2h)
IR-BB	2.15[c]	88	211	348	12(13; 150°C/2h)

[a]Measured at 30°C on the solution of 0.1g/dL in a mixed solvent of p-chlorophenol/phenol/1,1,2,2-tetrachloroethane=40/25/35(w/w/w).
[b]Measured at 30°C on the solution of 0.1g/dL in pentafluorophenol.
[c]Measured at 25°C on the solution of 0.1g/dL for RR-BB or 0.2g/dL for IR-BB in a mixed solvent of pentafluorophenol/p-chlorophenol/chloroform=3/3/4(v/v/v).
[d]Temperature where a first change in slope on the DSC curve was observed.
[e]Temperature where peak minimum endothermic point was located on DSC thermogram.
[f]Approximate degree of crystallinity determined by the areas of crystalline and amorphous diffractions on WAXD. The values in the parentheses are those after annealing under the conditions indicated.

On the contrary, glass transition temperatures, T_g, of the ordered sequence copolyesters are comparable or slightly higher than those of random ones. And the dependence of T_g on the regioregularity of the substituents also is not as much clear as T_m. The significantly lower T_g of IR-BB, when compared with that of RR-BB, may be ascribed to the increased free volume resulting from the irregular positions of the rather bulky bromine substituents. When the substituent is located only on one of the rings of the repeating unit, this effect appears to be negligible and, thus, the T_g values of RR-BH and IR-BH are about the same.

Figure 1 compares the wide angle X-ray diffractograms of OS-10-8 and RR-BH respectively with those of RS-10-8 and IR-BH. It is evident that the crystalline structures of OS-10-8 and RR-BH are much different from those of RS-10-8 and IR-BH. In other words, crystalline structure of a polymer is strongly dependent on the comonomer sequence as well as positional regularity of substituent. It is also clear that the degree of crystallinity of OS and RR series is greater than that of RS and IR series. Structural regularity certainly favors easier packing of polymer chains. However, the presence of bulky substituents on both benzene rings of the repeating unit, irrespective whether they are regularly positioned or not, hinders three dimensional chain packing, which leads to rather low degree of crystallinity as observed for RR-BB and IR-BB.

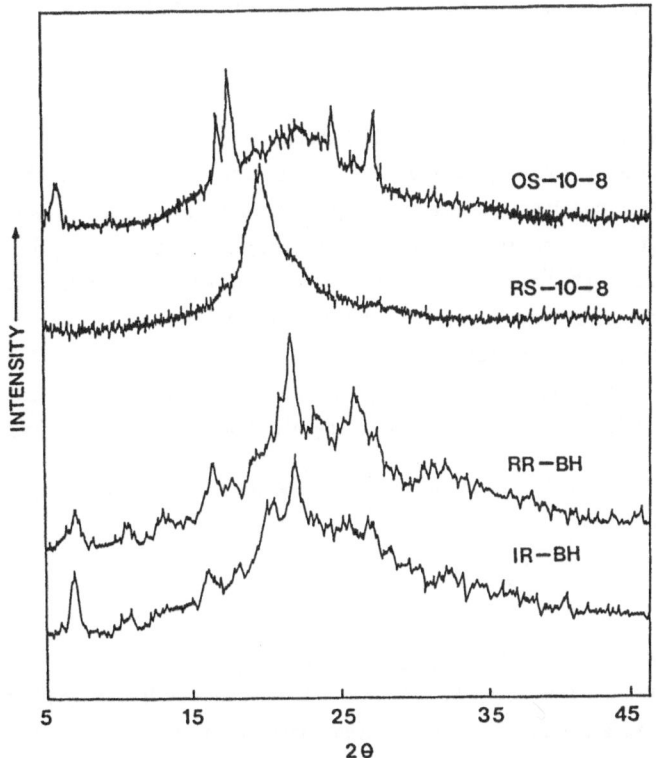

Figure 1. Wide angle X-ray diffractograms of polymers; OS-10-8 and RS-10-8 were annealed as indicated in Table 1, whereas RR-BH and IR-BH were not.

Liquid Crystalline Properties

All of the present polyesters are thermotropic and formed nematic liquid crystalline phase upon melting. The isotropization temperatures, T_i, of the ordered sequence copolyesters, OS-Ar-8 and OS-10-8, are significantly higher than those of the corresponding random copolyesters. The regular presence of uninterrupted triad(in OS-Ar-8) or dyad aromatic ester type mesogens(in OS-10-8) along the polymer chain definitely improves the thermal stability of the mesophase. On the other hand, in the case of random sequence copolyesters a part of aromatic moieties is expected to exist linked to aliphatic structures, which would result in the reduction of the concentration of mesogenic unit or long enough rigid segment in the chains. This, in turn, is to destabilize the mesophase resulting in lowering T_i values.

The dependence of T_i values on regioregularity of substituents is not so much clear, partly due to lack of enough data. Very often T_i values of wholly aromatic polyesters are so high that the polymers tend to undergo thermal decomposition before reaching isotropization temperature, which makes the comparison more difficult.

CONCLUSION

The following conclusions are drawn from the present investigation:
1. Multi-step synthetic routes have been devised for the preparation of thermotropic, sequentially ordered copolyesters and regioregularly ring substituted polyesters.
2. The crystalline melting temperature of the polyesters having regular chemical structures are much higher than those of irregular sequence or irregular placement of substituents. Moreover, the crystal structures of the formers appear to be much different from those of the latters.
3. The isotropization temperature of polyesters strongly depends on the comonomer sequence, that governs the distribution of mesogenic moieties along the chain.

Acknowledgement : This work was supported by the Korea Research Foundation, which is gratefully acknowledged.

References

1 Jin JI, Chang JH(1989) Macromolecules 22:4402
2 Jin JI, Chang JH, Jo BW, Sung KY, Kang CS Makromol Chem Macromol Symp(1990) 33:97
3 Jin JI(1990) in Weiss R, Ober CK(ed) Recent Advances in Liquid Crystalline Polymers ACS Symposium Series No 435 Chapter 3
4 Krigbaum WR, Kotek R, Ishikawa T, Hakemi H(1984) Europ Polym J 20:225
5 Martin PG, Stupp SI Macromolecules(1988) 21:1222
6 Jin JI, Chang JH, Shim HK Macromolecules(1989) 22:93
7 Jin JI, Chang JH, Hatada K, Ute K, Hotta M Polymer in press
8 Hässlin HW, Dröscher M, Wegner G Makromol Chem(1980) 181:131
9 Jin JI, Lee SC Polymer(Korea)(1985) 9:454

10 Jin JI Proceedings International Workshop on Polymeric Materials for Advanced Technology(1991) March 11-12 Seoul Korea The Polymer Society of Korea-IBM Korea Inc pp495-500

11 Jin JI, Kang CS, Chang JH J Polym Sic Polym Chem Ed Submitted

12 Lee IH MS Thesis(1991) Korea University Seoul Korea

13 Jin JI, Kang CS, Lee IH(1992) Polymer Preprints(ACS) Vol 33 in press

14 Felix M, Heimer EP, Lambros TJ, Tzougraki C, Meienhofer J J Org Chem(1978) 43:4194

15 Higashi F, Mashimo T, Takahashi I J Polym Sci Polym Chem Ed(1986) 24:97

Morphological Studies of High Performance Network Polymers

R.P. Burford and J.J. Jones

School of Chemical Engineering and Industrial Chemistry
University of NSW, Kensington NSW, Australia 2033

ABSTRACT

When styrene-butadiene-styrene (SBS) block copolymers are chemically cross-linked and subsequently converted to semi or full interpenetrating networks (IPNs) by the *in situ* polymerization of styrene, transparent tough products can be made. The process is complicated by changes in block copolymer morphology, grafting reactions and the form of interpenetration that the polystyrene takes within the thermoplastic elastomer continuous phase. Here, changes in morphology during IPN production are monitored by transmission electron microscopy, and associated thermal behaviour shown by dynamic mechanical analysis. In tough products, changes in domain structure in the SBS occur, together with an increased miscibility between the SBS and PS.

INTRODUCTION

Block Copolymers

Well characterized copolymers have been commercially available since the mid 1960's when Shell manufactured, by anionic polymerization, SBS and SIS copolymers. Both block and graft copolymers will chain segregate due both to low entropy contributions and to lack of interactions which would favour miscibility (1). The combination of melt processability, high strength and recoverable high deformability made these and related polymers attractive alternatives to elastomers which require chemical cross-linking. A key to the behaviour of the thermoplastic elastomers is the thermally reversible association of similar chain segments into domains, which for chains below the Tg can act as cross-links.

Y. Imanishi (Ed.)
Progress in Pacific Polymer Science 2
© Springer-Verlag Berlin Heidelberg 1992

The article by Brown *et al* (1) provides a good overview of chain segregation as shown by electron microscopy and scattering techniques.

Models of varying complexity have been developed by Meier to predict the dimensions of domains (2,3) and more recently the nature of the interface (4). Depending upon the relative amounts or lengths of each block, spheres, cylinders or lamellae can form, the respective dimensions being predictable and related to the square root of the molecular weight.

The predominant factor controlling domain size is claimed (4) to be the need to uniformly fill space by chain segments. Thermodynamic theories are also well established to explain the change from spherical to cylindrical to lamellar morphology as second monomer content increases (5). Equilibrium domain structure depends primarily upon component weight fractions and also upon segment molecular weight.

Morphologies of several commercial thermoplastic elastomers based on styrene and butadiene or isoprene have been comprehensively reviewed (6) and the role of copolymer composition and molecular geometry including graded and radial blocks illustrated. The morphology of SBS prepared by sectioning from the bulk polymer differs from that observed when thin films are cast from solvents, and solvent type also affects structure. Highly ordered domains were found in extruded block SIS (where the styrene blocks were of relatively low molecular weight) only after extended annealing (7) indicating that processing history also significantly alters microstructure. Other studies have included SAXS and SANS (ref 1, pp 164-167). The mutual diffusion of polystyrene and PS/polyisoprene block copolymer at the interface (8) has also been studied by electron microscopy.

Polymer Blends

Whilst SBS and related block copolymers have been widely used as interfacial agents (9), they are also used in large amounts as a major constituent of a binary blend. Thus "transparent HIPS", in which SBS lamellae modify polystyrene, has been described (10) and the morphology and other properties of various styrene-butadiene block copolymers in admixture with polystyrene are also known (11). In these cases although some modest improvements in toughness can be achieved, the overall combination of properties is less impressive than for ABS, for example.

The formation of SBS/PS blends may be complicated due to the dynamics of phase inversion as styrene polymerization proceeds and a range of morphologies including coils, shells and lamellae can arise. Furthermore, grafting of polystyrene onto the SBS occurs, to give a range of molecular structures. Even in the absence of cross-linker, a number of additional features may occur locally including gelation (as described for PS in a poor solvent (12)) and spinodal decomposition. The diffusion of linear polystyrene of varying molecular weights in highly entangled PS has also revealed added complexity (13).

IPNs Based on Block Copolymers

The first polystyrene/block copolymer IPNs were described by Sperling (14-16) and it was shown that the general phase-separated form of the block copolymer is preserved although domains will swell depending on interfacial tension, cross-link density and polymer weight fraction.

Semi and full IPNs have also been prepared by the *in situ* polymerization of styrene in chemically cross-linked SBR and SBS (17-20). Using the method of Plati and Williams (21) Gc values of up to 10 kJm^{-2} and breaking strains exceeding 50% can be obtained particularly when highly cross-linked SBS is used as polymer I. The addition of divinyl benzene cross-linker to the styrene causes an enhancement in stiffness, transparency and solvent resistance, with significant retention of toughness and ductility.

In this paper the factors which lead to SBS/PS IPNs having the above properties are discussed.

EXPERIMENTAL METHODS

Materials Preparation

Solprene 1205 and 416 block copolymers, products of Phillips Australia Chemicals, were used throughout. The former is stated by the manufacturer to be a styrene-butadiene linear diblock containing 25% PS and to have a molecular weight of 83,000. The 416 is a radial SBS triblock with 30% PS and a molecular weight of 140,000. To prepare IPNs the block copolymers were cross-linked thermally with up to 1% dicumyl peroxide (145°C, 1 hr) in a compression press with moulds conforming to ASTM D3182. The pads were then swollen in styrene containing 1% benzoyl peroxide with and without 5% divinyl benzene. Thermal curing in metal frames led to transparent or translucent samples with 70% additional polystyrene weight increase. More complete details are given elsewhere (17, 19, 22).

Electron Microscopy

Sections from 50 to 100 nm thick were stained with 5% aqueous OsO_4 for 30 minutes prior to viewing in a Hitachi 7000 TEM operating at either 75 or 100 kV. Stiff IPNs were sectioned using a Reichert-Jung Ultracut E at ambient temperatures using either glass or diamond knives. Staining prior to sectioning increased hardness and improved section quality. For the constituent block copolymers and more ductile IPNs, sections were prepared at -100°C using an Ultracut FC 4E cryo-ultramicrotome. Solvent cast samples from 0.01wt% toluene were mounted on Formvar supported grids.

Dynamic Mechanical Analysis

Test pieces 25 x 12 x 3 mm were subjected to dynamic mechanical analysis in a Du Pont 983 instrument, operating in resonant frequency mode. Samples were heated from -150 to 150°C at 20°C min^{-1}. The oscillation displacement amplitude used was 0.05 mm. Data was processed using TA 2100 software.

RESULTS AND DISCUSSION

Mechanical Behaviour

Fracture toughness and tensile properties are summarized in Table 1 below. The ductile IPNs showed substantial drawing and yielding in the impacted test-pieces compared with either smooth or sometimes locally crazed fracture surfaces for the more brittle polymers.

Polymer	DICUP (%)	DVB (%)	G_c (kJ/m²)	σ_y (MPa)	ε_b (%)
1205	0.2	0	1.4	21	15
	0.2	5	0.7	22	25
	1.0	0	11.4	14	70
	1.0	5	4.1	21	140
416	0.2	0	2.9	11	20
	0.2	5	1.4	18	30
	1.0	0	9.5	10	45
	1.0	5	5.6	17	80

Table I. Mechanical and toughness properties of IPNs.

For both block copolymers, the predominant factor causing high toughness and ductibility is the high level of chemical cross-linking in the block copolymer. This is expected to provide a more highly cross-linked butadiene phase, assuming that the block-styrene segments are much less cross-linkable. This may restrict to some extent the final domain size of the IPN. Previous micrographs (17) show for the 0.2% Dicup cross-linked 1205 semi IPN substantial PS rich domains with a sharp butadiene-rich interface where presumably low levels of polystyrene reside. In the more highly cross-linked copolymer semi IPN, the domain size was smaller and the continuous butadiene-rich phase between PS domains showed less contrast, suggesting a higher level of either PS homopolymer grafted PS on the polybutadiene segments.

Morphology

<u>Microstructural changes during IPN formation</u>: The block copolymers are first mixed with dicumyl peroxide in an internal mixer before compression moulding. Samples sectioned from

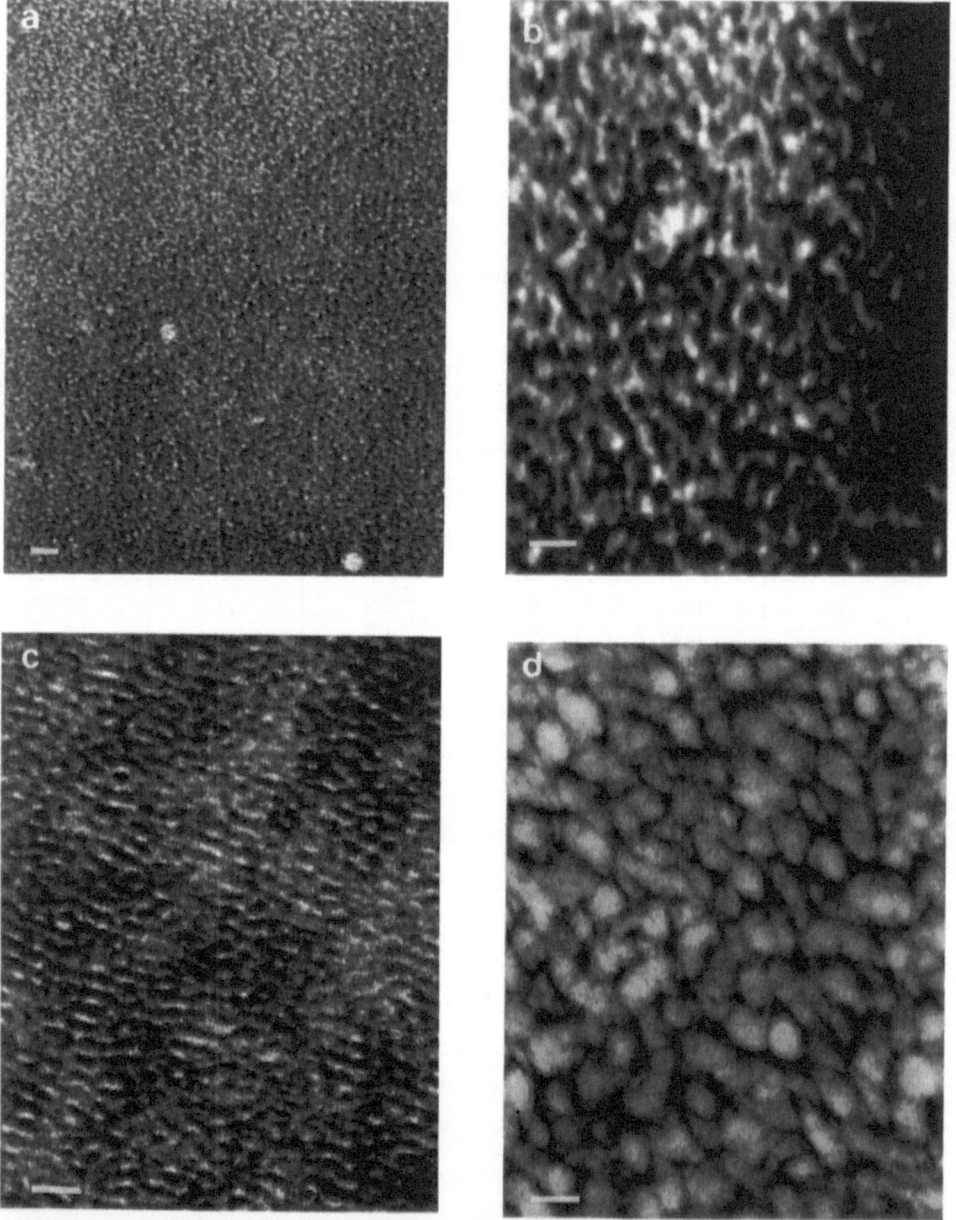

Figure 1. a,b Quenched uncross-linked Solprene 416 (Bar = 100 nm),
 low and high magnification (Bar = 50 nm)
 c: cross-linked, annealed Solprene 416 (Bar = 50 nm)
 d: full IPN based on lightly cross-linked 416 (Bar = 50 nm)

the solidified melt have poorly defined structure, as indicated by Figure 1a. However, at higher magnifications butadiene rich and poor phases can be resolved (Figure 1b). Subsequent curing of the block copolymer (here Solprene 416) yields a product with somewhat better phase definition (Figure 1c) although the dimensions of the domains are still rather small. Studies reported elsewhere (20) confirm by sectioning in various directions that this polymer is made up of distorted rods or cylinders and so the micrograph represents cryogenic sectioning along the cylinders.

We have endeavoured to section styrene-swollen block copolymers, but whilst the monomer makes the sample rigid and easy to cut at -100°C, attempts to preserve the specimens in the TEM have been hampered by the lack of a cold stage. Subsequent IPN products have significantly enlarged styrene-rich, unstained domains, consistent with 70% styrene uptake. Thus in the case of Solprene 416 cut normal to cylinders, domain diameters increase from about 10 nm to 30 nm (Figure 1d) consistent with Yeo's (16) estimates.

Variations in IPN morphology: The following micrographs (Figures 2a-d) show full and semi IPNs, made from lightly and heavily cross-linked Solprene 416.

The main trend seen in this series (and other sets of IPNs including those based on Solprene 1205) is a reduction in domain size as SBS cross-link density increases (eg. from Fig. 2a to 2b) and a decrease with addition of DVB (ie. Fig. 2a to 2c and Fig. 2b to 2d). This change in domain dimension is accompanied by an increase in compatibilization with cross-linking, as recognized with other polymer systems, including the DMA data given below. The lighly cross-linked full IPN (Fig. 2c) appears on the available data to have domains no larger than the corresponding highly cross-linked network (2d) and this aspect is being further investigated.

Dynamic Mechanical Analysis

It is well-known (23) that SBS block copolymers show well-separated transitions corresponding to the constituent soft and hard segments, although the expected large E'' peaks for polystyrene can diminish in unsupported test-pieces when tested in resonant frequency

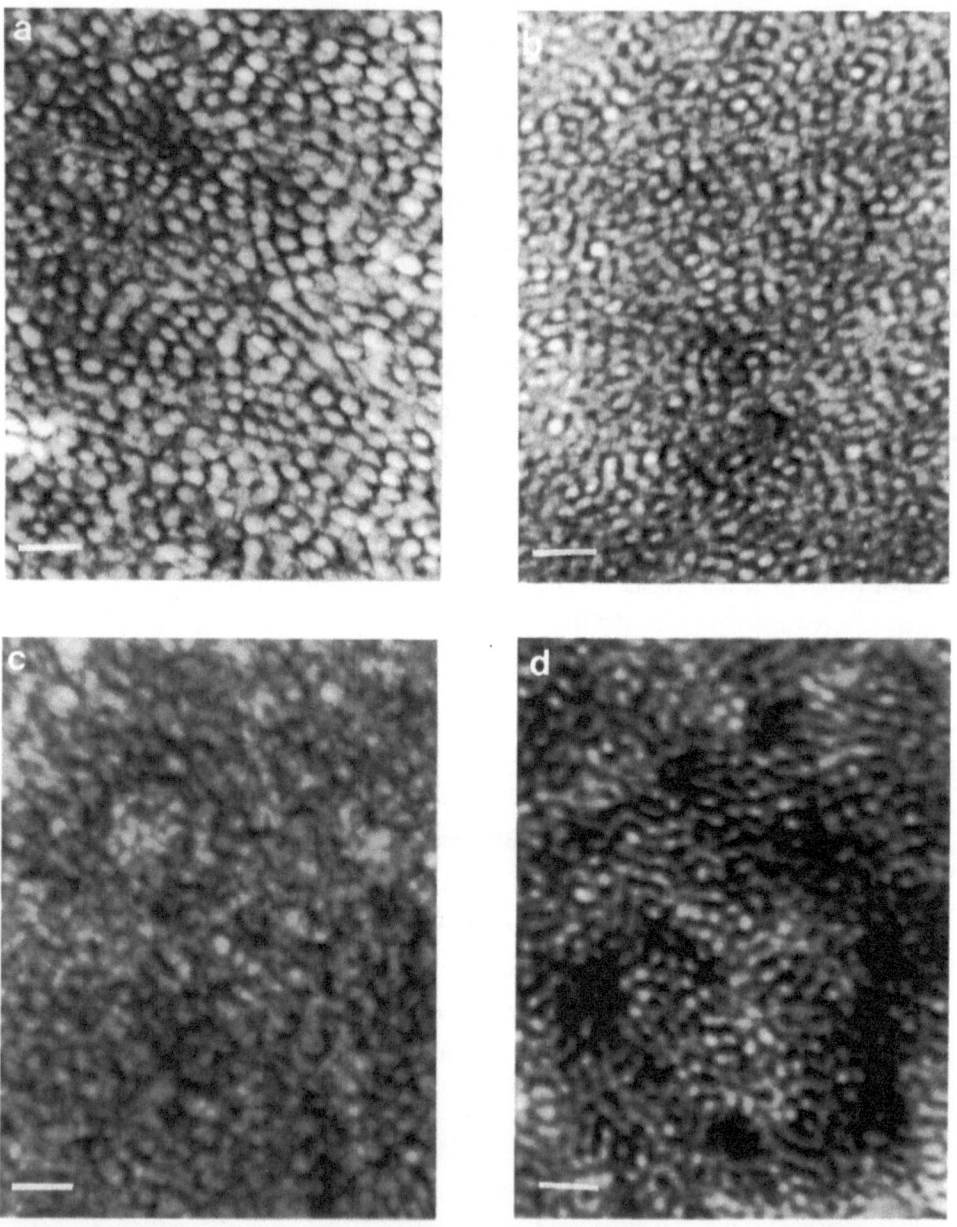

Figure 2.   Morphologies of stained sections of Solprene 416 IPNs (Bar = 100 nm)
             a: low cross-link, semi   b: high cross-link semi
             c: low cross-link, full   d: high cross-link, full

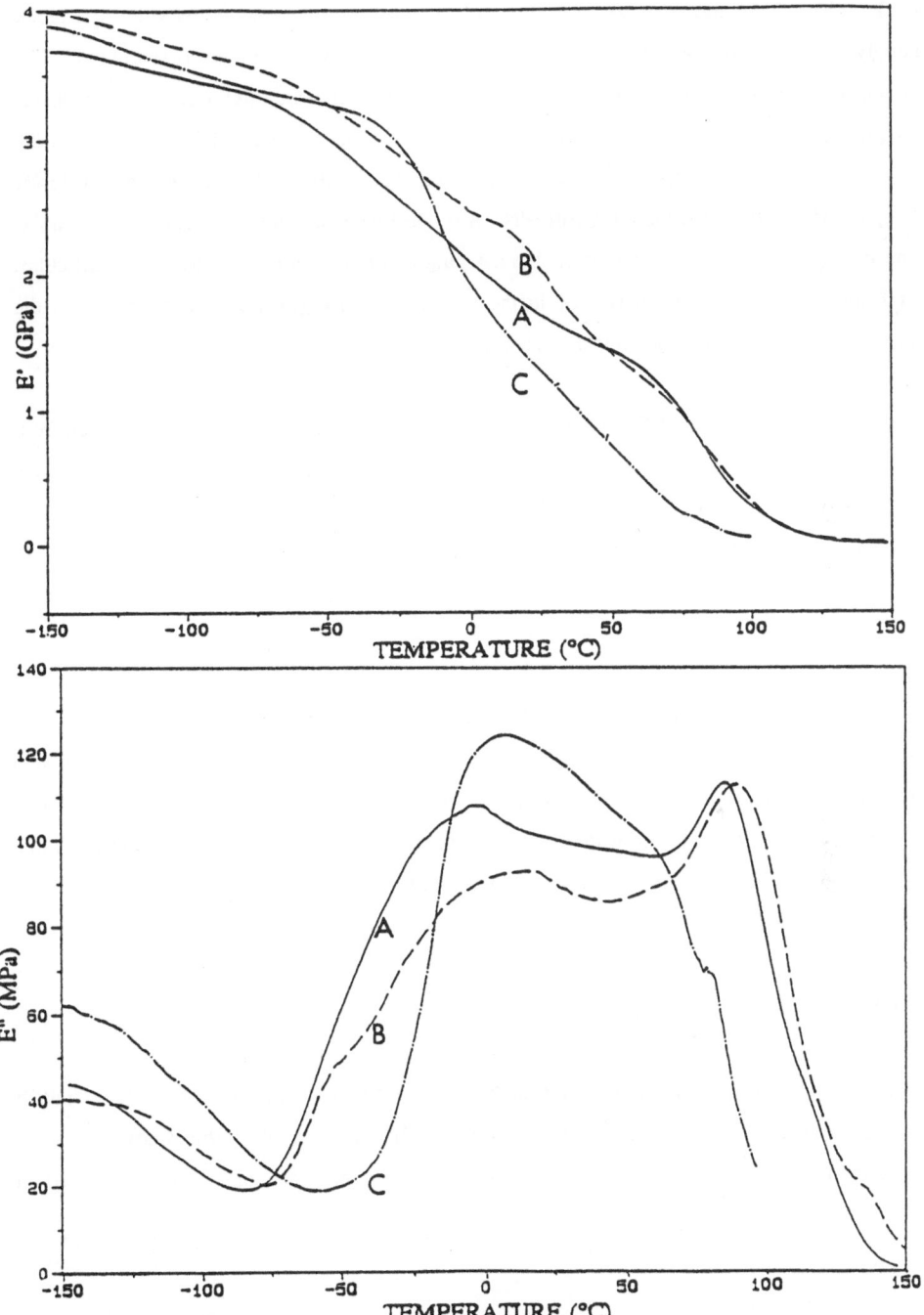

Figure 3a.   Storage Modulus vs Temperature.   3b. Loss Modulus vs Temperature.
Sample Codes A = low cross-linked 1205 Semi IPN. B = low cross-linked 1205 Full IPN
                      C = high cross-linked 1205 Full IPN.

mode. The DMA traces shown in Figures 3a,b below compare a semi-IPN with light block copolymer cross-linking with two full IPNs, at high and low elastomer cross-linking. The storage modulus traces at elevated temperatures are comparable although well below the lower Tg (ie. less than -100°C) there is surprisingly significant difference in stiffness. The E" data, however, reveal substantial differences in thermal transitions, with a shifting of the lower temperatures in the corresponding full IPN. A modest decrease in the PS transition at 105°C confirms greater miscibility and probable merging of the component phrases. In particular, grafting and cross-linking of styrene in the butadiene-rich regions serves to strengthen the interface and reduce compositional differences.

In the highly cross-linked SBS IPN this trend is further accentuated with a broad plateau from 0 to 70°C being observed. This confirms higher miscibility and greater interlocking of the phases. We observe that the full IPNs have greater transparency than the semi-IPNs, also consistent with a more dispersed mosaic structure.

CONCLUSIONS

On the basis of mechanical, thermal and morphological evidence, the formation of full IPNs from highly cross-linked block copolymer precursors entails grafting, cross-linking and segmental movements leading to higher phase miscibility and adhesion. A suitable synthetic mechanism remains to be identified, requiring further labelling of polymer segments so that their migration during IPN formation can be monitored, together with a rigorous structural analysis, possibly entailing solid state NMR and neutron scattering techniques.

ACKNOWLEDGMENTS

The support by the Australian Research Council for part of this study is gratefully acknowledged. Fracture toughness and micromechanics aspects of this study have been assisted by Y-W Mai and part of the electron microscopy has been undertaken by H.S. Byun.

References

1 Brown RA, Masters AJ, Price C, Yuan XF (1988) In: Booth C and Price C (eds) Comprehensive polymer science, vol 2, Pergamon, Oxford, Chapter 6

2 Meier DJ (1969) J Polymer Sci, 26C:81

3 Meier DJ (1970) Polymer Preprints, 11:400

4 Meier DJ (1985) NATO Adv Study Inst Ser E, 89:173

5 Spontak RJ, Williams MC, Agard DA (1988) Polymer 29:387

6 Aggarwal SL, (1985) In Folkes MJ ed, Processing, structure and properties of block copolymers, Elsevier Applied Science, England, Chapt 1

7 Pedemonte E, Turturro A, Bianchi U, Devetta P (1973) Polymer 14:145

8 Koizumi S, Hasegawa H, Hashimoto T (1989) Mutual diffusion of diblock polymer ...., paper 77, Macromolecular Chemistry Section, 1989 Int Chem Cong Pacific Basin Soc, Hawaii, Dec 17-22

9 Cho K, Brown HR, Miller DC (1990) J Polymer Sci B28:1699

10 Echte A, (1987) In Riew C (ed) Rubber-Toughened Plastics, ACS Adv Chem 222:15

11 Sardelis K, Michels HJ, Allen G (1987) Polymer, 28:244

12 Hikmet RM, Callister S, Keller A (1988) In Lemstra PJ and Kleintjens LA (eds) Integration of fundamental polymer science and technology, Elsevier Applied Science, 2:306

13 Nemoto, N (1989) Tracer diffusion of linear polystyrenes in entanglement networks, paper 18, Macromolecular Chem Section, 1989 Int Chem Cong Pacific Basin Soc, Hawaii Dec 17-22

14 Siegfried DL, Thomas DA, Sperling LH (1981) J Appl Polymer Sci, 26:177

15 An JH, Fernandez AM, Sperling LH (1987) Macromolecules 20:191

16 Yeo JK, Sperling LH, Thomas DA (1983) Polymer 24:307

17 Burford RP, Mai Y-W (1991) Advances in IPNs III, Chemtec, Toronto, 75

18 Burford RP, Mai Y-W (1989) Contemporary topics in polymer science, Plenum, NY 6:699

19 Burford RP, Mai Y-W (1989) In Lemstra PJ and Kleintjens LA (eds) Integration of fundamental polymer science and technology, Elsevier Applied Science 3:136

20 Jones JJ, Burford RP (1991) Polymer International 26:000

21 Plati E, Williams JG (1975) Polymer Eng Sci 15:470

22 Byun HS, Burford RP, Mai Y-W (1989) Materials Forum 13:26

23 Matsuo M, Ueno T, Horino H, Chujya S, Asai H (1968) Polymer 9:425

A New Trend of Polymer Science and Technology for Humanity

Seizo Okamura

Faculty of Liberal Arts, Kyoto Sangyo University,
Motoyama, Kamigamo, Kita-ku, Kyoto 603, Japan

Abstract: A consideration is described on the "Properties" of materials as the basis of science and technology, in connection with my early research on textile. The "Functionality" is then discussed in some details and the stream from property to "Sensibility" is viewed through "Functionality" in my own style. Biocompatibility has been regarded as one of the functionalities of polymers and studied in our laboratory a decade ago, as one of the prestages of the sensibility or humanity concepts. Finally, the arts in science are briefly imagined in two ways; one, comprehensive approach downward (anti-clockwise, in the top part of Figure 2) from the arts to science, and the other, individual one upward (in the bottom part of Figure 2) from science to the arts.

PROPERTY OF MATERIALS (THE PRIMARY PERFORMANCE)[1]

My early research works in the period from 1937 to 1940 were commenced with the measurements of physical properties of textile such as breaking strength and elongation of artificial wool-like fibers which were made from soybean proteins.

Then I thought that experimental values might be different according to measuring conditions. For materials in practical use, it is well known that the property is based primarily upon the method of measurement. Property (one of the answers: A) depends on measurement (one of the questions: Q). Essentially, properties (A) are not only stored in materials, but also newly elicited from the materials by new kind of experiment (Q).

In this manifestation, materials are considered to be a part of the "Nature" and then the measurement turns to be a part of normal experiments in natural science. Those processes between Q and A are very significant as the bridge between the "existence" (materials) and the "consciousness" (thought) by which scientific researches are eventually realized.

My early researches aimed for getting a new type of man-made fiber from various kinds of proteins, which are easily available on those days, and for measuring various kinds of properties in the light of so-called textile chemistry. This was done about a half century ago and now become macromolecular chemistry.

FUNCTIONALITY (THE SECONDARY PERFORMANCE)[2]

In practical use of materials, many kinds of properties have been involved and several complicated problems have been raised. For example, a single Q deduces several As, or several Qs induce a single A. If the material possesses time-dependent properties, the properties of this material might change from simple A to time-dependent A and then appear being "active" which contrasts to being "passive" as in usual materials. Recently those complex properties are fashionably termed functionalities, which are the secondary performance.

Generally speaking, the property of materials is non-living (governed only by space-axis), but the functionality of materials spans to a living field (the passive component, governed by time-axis beyond space). Recently, the stream from property to functionality has extended deeply into so-called sensibility(-ble) or sensitivity (-tive), which are the tertiary performance. Sensibility is accessible by living things (the active component, governed by memory-axis beyond time and space). The primary, secondary and tertiary performances, in other words, property, functionality and sensibility, correspond to non-livings, plants and animals including human beings, or in other words, axes of space, time and memory (brain in the case of human beings), respectively.

SENSIBILITY (THE TERTIARY PERFORMANCE)[3]

My recent research interest (1970-1975) has been concentrated into the biomedical applications of polymers; for instance, artificial trachea made of collagen-coated polymer membrane. To improve biocompatibility of the material surface, soluble collagen was immobilized on the surface of artificial organ materials. Investigations were carried out to find suitable conditions for a ballance between biodegradation of immobilized collagen and in vivo tissue regeneration. The rate of biodegradation of coated collagen was controlled by the radiation-induced crosslinking reaction.

My research interest has been turned from usual natural science with non-living things to biological polymer science with living things. In parallel with this, my mental stream ran from property to functionality, and ultimately to sensibility. On these days I have been personally interested in general relationship between piece and total, part and whole, individual and assemble (similar to "Teil und Ganz" in Heisenberg), in which so-called "piling-up image" is spontaneously set up in my mind as sketched in the following figure.

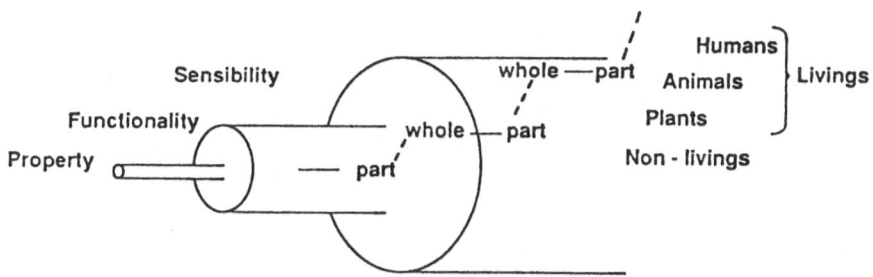

Figure 1   Piling-up Images of Three Stages

In the field of textile chemistry, a similar trend is recognized, too. The functionality of fiber or the sensibility of textile such as resilience of fabrics has become the focus of interest.

NEW TREND[4]

The Society of Polymer Science, Japan has publised several articles under the title "KANSEI TO KOBUNSHI, Sensibility and Polymers" in the November issue of Volume 39, 1990 of a journal "Kobunshi". In this volume, Dr. M. Okamoto (Toray Co.) gives an openning remark on "KANSEI, sensibility". "Kansei" in polymer

materials are discussed by some other authors: Dr. S. Yamaguchi
(Kuray Co.) on textiles, Dr. Y. Murase (Yamaha Co.) on music
instruments, Drs. M. Tanaka and S. Kumagai (Shiseido Co.) on
cosmetics, Drs. T. Ichikawa and W. Saito (Dainihon Printing Co.)
on hollography, and Prof. Y. Komata (Odawara Women's College) on
the taste perception.

Recently, especially since November in 1990, Japanese polymer
scientists have been strongly interested in this new aspect of
polymer science and technology. The interest in "Kansei of
polymer" will push scientific approach to humanity through the
consideration of sensibility, and make the science approach the
arts.

FROM THE ARTS TO SCIENCE[5]

For a possible way for the arts approaching to science, we will
consider pushing the arts (humanity) to science (logics). On
this matter a famous classical work has been published in 1873
by Dr. Gustav Theodor Fechner, which was entitled "Experimentelle Aesthetik". He proposed that painting products should be
subjected to rating by art critics in six orders from 0
(worthless) to 5 (very excellent). Then he compared the rating
by experts with his own. His procedure is based on an empirical
equation consisting of the terms of senses on the form, color or
harmony of the art products. It was done 120 years ago.

Later in this century, the market of textile products followed
principally the same way to grasp sensibility in fabrics,
especially in costful suits. I would like to mention briefly
the work of Prof. S. Kwabata (Kyoto University), which was done
in the same line as the sinsibility research and presented at
the International Conference on the Textile Science held at
Liberee, Czechoslovakia, on September 16-18, 1991.

First, the human sensibility for stuits was rated by the experts
on suits as follows.

Table 1 Grade and Rating

Grade	Not used	poor	fair	average	good	excellent
Rating	0	1	2	3	4	5

Then, the rating values were correlated with statistical values, so-called fabric hands, (secondary performance) such as stiffness, fullness or flexibility of clothes. Finally, the values of fabric hands were compared with other empirical values, e.g., mechanical properties and surface characters of fibers (primary performance). He succeeded in obtaining excellent suits fabricated with polyester fibers, instead of wool fibers, by using the "Experimentelle Aesthetic"-like methods (but with more delicate processes).

These ways of investigations are regarded as possible ways of the arts (sensibility) approaching to science via functionality. It is regretted that the judgements by experts are the black box for science. However, these processes are useful in practice and helpful for further logical improvement.

FROM SCIENCE TO THE ARTS[6]

We are now in the era in which the concept of molecular design dominates in all disciplines of chemistry, including polymer science and technology. A new way of science approaching to the arts from an opposite site is now opening logical processes without supposition of black box. The new approach proceeds from the property research to the functionality research and hopefully to the sensibility research, which corresponds to the arts.

I would like to mention some examples of researches in this direction, which were kindly informed by Dr. Y. Tsunoda [Kao Co.].

First, Professor Jacquelline Belloni-Cofler and coworkers reviewed photographic development reactions, and clarified them in terms of physicochemical aspects (Endevour, New Series, vol.15, No.1, pp.2-9, 1991). They have clearly shown the importance of formation of silver nuclei by incident light, as a potential catalyst for further dark reactions. Also they stated in the article that "ever since the first results, achieved by Louis Daguerre (1839), the key step of the development reactions has benefitted from innumerable empirical improvements" and then emphasized that "recent physicochemical basis has been fully understood" and thus "this enormous gain of research results in the sensitivity (fortunately the same as sensibility) has led to the manufacture of the emulsion of much finer particles and of much shorter exposure time for photographic film (by several orders of magnitude)". The modern photography seems to be a typical example of using highly "functional" materials which were produced on the basis of molecular design.

Secondly, I would like to refer to the work of R. D. Michell, W. J. Nebe and W. H. Hardam (J. Imaging Sci., 30, 215, 1986). They obtained an active photosensitizer by irradiating transparent precursor molecules (inactive in photoreactions), which are mixed in polymer matrix. Further irradiation stimulated it for photoreactions, by which patterns were formed. These processes might be represented as two step photochemical amplification. These treatments shift "property" toward "sensibility" via "functionality".

The third example is the work of H. Ito and C. E. Willson (Polym. Eng. Sci., 23, 1019, 1983). They made an immortal acidic catalyst by photoirradiation of a sensitizer (onium salt). In this reaction, removal of protective groups analogous to enzyme reactions took place. Incorporation of different kinds of reactive species in polymer matrix can yield diverse answers in response to different kinds of stimulations or questions. Approach from science to the arts seems to be possible in the near future.

TENTATIVE CONCLUSIONS[7]
The relationship between polymer science and the arts was discussed on the basis of the consideration of property,

functionality, and sensibility. My own research on physical properties of fiber and assay of biocompatibility of artificial organ materials was referenced to. The relationship is schematically shown in Figure 2.

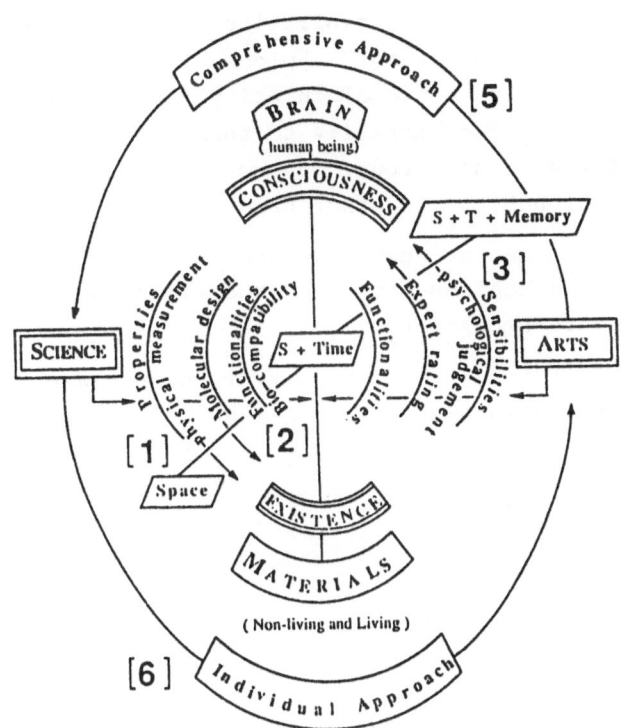

Figure 2   Relationship between Science and Arts
            through Brain and Materials

As shown in Figure 2, the human brain and materials (polymers) seem to be homogeneously fused into one concept through considerations on the relationship between polymer science and the arts in the individual or comprehensive approach.

Finally, I would like to put emphasis on the importance of macromolecular concepts. The characteristics of macromolecules are usually summarized in two points; one is a linearity and length of the molecule having many sites of property, functionality, and sensibility, and the other is flexible interactions among these groups induced by free rotation and translation of the long molecule. Also, the human brain is similarly characterized as "property", "fuctionality" and "sensibility" of sensory organs and their flexible interactions. Exchange of human sensibilities sometimes produce new expressions such as "beautiful", "pleasant", "lovely" and so on. In a similar way, the macromolecularity frequently appears and incorporates with functionality to turn out to be a hardware (the kind of functional group) in one hand and a software (the degree of functionality) in other hand.

Investigations on macromolecular concepts are very analogous to those on brain mechanisms, in which the relation between polymer and brain is more important.

The importance of researches on polymer science and the arts are getting recognized in science as well as in technology.

Liquid Barrier and Thermal Comfort Properties of Surgical Gowns

Elizabeth A. McCullough

Institute for Environmental Research, Kansas State University
Seaton Hall, Manhattan, Kansas 66506 U.S.A.

For many years, protective garments have been worn by surgical personnel to prevent the contamination of the patient from microorganisms. More recently, medical personnel have been concerned about being exposed to diseases from the body fluids of the patient [e.g., hepatitis B (HBV), human immunodeficiency virus (HIV), and acquired immunodeficiency syndrome (AIDS)]. The Association of Operating Room Nurses and testimony at the preliminary hearings of the Occupational Safety and Health Administration have recommended that surgical gowns repel all of the body fluids typically found in the surgical setting (1,2). Consequently, the liquid barrier properties of fabrics used to produce these protective garments need to be measured so that these products can be compared by hospital personnel.

A variety of test methods have been developed that measure a fabric's resistance to liquid penetration. However, manufacturers, trade associations, and government agencies do not agree as to which test method is the best predictor of liquid barrier performance, particularly in the surgical environment.

Manufacturers have developed new nonwoven fabrics for surgical gowns that provide improved liquid barrier protection. However, the technology necessary to make a fabric impervious to liquid penetration may also increase the fabric's resistance to evaporative heat transfer. Consequently, there may be a trade-off between adequate liquid barrier protection (i.e., safety) and thermal comfort in the operating room.

Therefore, the purpose of this study was:

1) to evaluate the liquid barrier (safety) properties of representative fabrics used to make surgical gowns--using different test methods and challenge liquids,

2) to measure the insulation value and evaporative resistance provided by surgical gown ensembles measured with a thermal manikin, and

Y. Imanishi (Ed.)
Progress in Pacific Polymer Science 2
© Springer-Verlag Berlin Heidelberg 1992

3) to measure the thermal comfort and clothing comfort sensations perceived by medical personnel while wearing surgical gown ensembles in a simulated surgical environment.

PROCEDURES

Six surgical gown fabrics were evaluated for **liquid barrier properties**. The fabrics ranged from regular to impervious and included the following: a traditional woven 50% cotton/50% polyester muslin, reusable; Kimberly-Clark Evolution[R], regular, disposable; American Converters, Baxter Optima[R], regular, disposable; Johnson & Johnson, Surgikos Barrier 450[TM], regular, disposable; W.L. Gore prototype 10120, impervious, disposable; and Kimberly-Clark Evolution[R] Specialty, impervious, disposable. Randomly selected samples of fabric from the gown front and sleeve were subjected to **seven different liquid barrier test methods**. The methods were compared with respect to their ability to differentiate between regular (liquid resistant) and impervious (liquid proof) fabrics (3,4,5).

1. IST 80.7-70, Saline Repellency of Nonwovens (INDA Mason Jar Test): The fabric is used to seal an inverted Mason jar containing a liquid, and the time of penetration of the liquid through the fabric is noted. (In this study, the test was arbitrarily stopped at one hour.)

2. AATCC 42-1985, Water Resistance: Impact Penetration Test: Five hundred ml of liquid is allowed to spray from a height of 2 ft against the taut surface of a test specimen backed by a preweighed blotter. The blotter is then reweighed, and the difference in weight used to determine the amount of liquid that penetrated.

3. AATCC 127-1985, Water Resistance: Hydrostatic Pressure Test: A test specimen, mounted under the orifice of a conical well, is exposed to 10 ml of a liquid and subjected to water pressure increasing at a constant rate until three points of leakage appear on its surface. The pressure level at which penetration occurred is given. (In this study, the test was arbitrarily stopped at 100 cm of water pressure.)

4. Kimberly-Clark Blood Strike Through Test: Drops of liquid are placed on top of a fabric over a preweighed blotter. Pressure of 1 psi is applied quickly, and the

blotter is then reweighed. Liquid strike through is measured in grams and converted to a percentage of the volume that is used to challenge the fabric.

5. ASTM FXXXX (draft) Standard Test Method for the Resistance of Protective Clothing Materials to Biological Fluids conducted at 1 psi (based on ASTM F 903): The fabric is subjected to a liquid at 0 psi for 5 min, at 1 psi of pressure for 1 min, and at 0 psi for 54 min. Time of penetration is recorded.

6. ASTM FXXXX (draft) Standard Test Method for the Resistance of Protective Clothing Materials to Biological Fluids conducted at 2 psi: Procedure is the same as that given above only 2 psi is used instead of 1 psi.

7. Gore Elbow Lean Test: The fabric is placed over an ink pad saturated with synthetic blood and covered with a preweighed blotter. Pressure is applied quickly with an elbow, and the blotter is then reweighed, and the difference in weight is used to determine the amount of liquid that penetrated.

A variety of liquids are present in the surgical environment. Therefore, each gown was tested with four challenge liquids: distilled water, bovine blood, synthetic blood, and 70% isopropyl alcohol.

For the physical thermal tests, the surgical gowns were tested over a standard surgical ensemble on a thermal manikin in an environment chamber (ASTM F 1291) (4). The resistance to dry and evaporative heat transfer and the permeability index values were measured according to procedures given by McCullough et al. (6).

Eighteen human subjects with surgical experience evaluated the thermal comfort and clothing comfort characteristics of the six surgical gowns. Each gown was worn with a standard surgical ensemble. Each subject wore the traditional woven gown first, followed by the regular nonwovens gowns, and then the impervious gowns (in random order within groups).

The nude subjects and their surgical garments were weighed separately before and after the test sessions to determine the amount of sweat generated while wearing the surgical ensemble. The subjects were tested in groups of four of the same sex with two subjects working on each side of a 1 m (3.3 ft) table. During the two hour test period, the subjects completed a series of eight different tasks (12 minutes each) that were designed to stress them mentally and physically.

The experiment was conducted in an environmental chamber at 24°C (76°F), 60% relative humidity, air velocity less than 0.15 m/s (< 30 ft/min), and 15 air changes per hour. These conditions represent the "worst case" scenario based on the acceptable operating room conditions recommended by ASHRAE (7). The subjects were acclimated to the environmental conditions for 45 minutes while dressing and being instrumented with skin sensors. The skin temperature sensors were placed on the inner lower arm, chest, and calf. The subjects' skin temperatures and subjective responses were recorded every 30 minutes during the two hour session.

On the Thermal Sensation Scale, subjects were asked to indicate how they felt using a nine-point scale ranging from very cold to neutral to very hot. On the Clothing Comfort Sensation Scale, subjects were asked to indicate the degree to which they sensed the following characteristics while wearing the surgical gown: comfortable, binding, lightweight, flexible, noisy, soft, scratchy, stiff, rough, papery, staticy, slippery, clingy, clammy, dry, and sticky. Subjects were also asked to give their overall preference for the best regular gown and the best impervious gown after all gowns were worn.

RESULTS
Liquid Barrier Tests

For the statistical analysis, the results for each test specimen were coded either one for a pass (i.e., no liquid penetration) or zero for a failure. The effects of fabric type and challenge liquid on the liquid penetration of the fabrics were analyzed using separate two-way analyses of variance, Fisher's LSD post hoc comparison tests, and Spearman's rank order correlations. The level of statistical significance was set at 0.05. The statistical tables and raw data are available in the technical report (8).

The impervious Gore prototype gown provided the highest amount of protection from liquid penetration on all of the tests performed in this study. The other impervious gown, the Evolution[R] Specialty, provided the same degree of protection as the Gore gown on all tests except the Mason Jar Test and the Biological Fluid Resistance Test (2 psi). The regular gowns exhibited the same liquid barrier properties (i.e., failed most of the tests), except that the Evolution[R] gown performed significantly better than these gowns on the Impact Penetration Test and the KC Blood Strike Through Test. The woven cotton/polyester gown failed every test and consistently provided the least liquid barrier protection of all the gowns.

All of the surgical gown fabrics failed more often with 70% isopropyl alcohol than with any other challenge liquid, regardless of test method (probably because of its low surface tension). Except for the KC Blood Strike Through Test, the bovine blood produced the highest passing rate (i.e., least challenging) for all gowns and for all other test methods. Synthetic blood yielded results similar to bovine blood on the Impact Penetration, Hydrostatic Pressure, Biological Fluid Resistance (1 and 2 psi), and the KC Blood Strike Through Tests and results similar to distilled water (or saline solution) on the Mason Jar, Hydrostatic Pressure, Biological Fluid Resistance (2 psi), and the KC Blood Strike Through Tests.

Results of the KC Blood Strike Through Test, the Impact Penetration Test, the Hydrostatic Pressure Test, the Biological Fluid Resistance Test (2 psi) were highly correlated (i.e., they differentiated between different types of gowns in a similar manner).

Thermal Resistance and Evaporative Resistance Tests

According to the manikin tests, the impervious Evolution[R] Specialty had a higher resistance to evaporative heat transfer and a lower permeability index (i.e., a lower percent of evaporative cooling possible) than the other gowns. However, the impervious Gore prototype had thermal characteristics similar to those of the regular gowns (i.e., resistance values were lower).

Thermal Comfort Evaluation with Human Subjects

The effect of gown type on the subjects' thermal sensations and clothing comfort sensations were analyzed using separate analyses of variance and Fisher's LSD tests. The thermal resistance values of the gowns and fabrics, the amount of moisture retained in the surgical clothing, and the mean skin temperature of the subjects were also analyzed using the same statistical procedures.

Approximately five times more moisture (i.e., unevaporated sweat) was found in the Evolution[R] Specialty gown and surgical ensemble than in the other gowns and their auxiliary garments after the subjects wore them for two hours. Subjects had a significantly higher mean skin temperature when wearing the Evolution[R] Specialty gown than when wearing the others, although the magnitude of the difference was small. We failed to find differences in the thermal sensations of subjects due to the type of surgical gown they were wearing--probably because all gown ensembles had a total insulation value of 1.6 clo.

The results of the Clothing Comfort Sensation Scale indicated that the impervious Evolution[R] gown was rated significantly less comfortable than the other gowns; the impervious Gore prototype was rated as comfortable as the regular gowns. The subjects perceived the Kimberly-Clark Evolution[R] and Evolution[R] Specialty gowns as less flexible and less lightweight than the other gowns; they also rated the Evolution[R] more noisy than the others. Subjects felt that the Evolution[R] Specialty gown was less soft than all of the other except for the regular evolution[R]. Subjects rated the woven cotton/polyester gown significantly less papery than the Evolution[R], Gore prototype, and Evolution[R] Specialty, but not the Optima[TM] or Barrier 450[TM]. Thus, subjects were not able to distinguish between nonwovens and wovens with respect to their papery feel. Subjects perceived the Evolution[R] Specialty to be significantly more clammy than the cotton/polyester gown, the Barrier 450[TM], and the Optima[TM], and less dry than the other five gowns.

When subjects compared the regular gowns, more preferred the Baxter Optima[TM] than the others; when the impervious gowns were compared, more subjects preferred the Gore prototype than the Evolution[R] Specialty.

IMPLICATIONS

Impervious gowns should be worn by surgical personnel in the operating room rather than regular gowns for optimum liquid barrier protection. Although the nonwoven regular gowns provided some barrier protection (according to some tests with some liquid challenges), they cannot be considered barriers to bloodborne pathogens. However, when comparing the regular gowns tested in this study, the Kimberly-Clark Evolution[R] is recommended because it provided more protection than the others. The woven cotton/polyester gown is not recommended for use because it had the poorest liquid barrier protection, and it was not more comfortable to wear than the other regular gowns tested in this study because it exhibited superior liquid barrier properties while its thermal comfort properties were similar to those of regular surgical gowns.

When evaluating the liquid barrier properties of surgical gowns, the Kimberly-Clark Blood Strike Through Test, the AATCC Impact Penetration Test, the AATCC Hydrostatic Pressure Test, and/or the draft ASTM Biological Fluid Resistance Test (2 psi) should be used. The Mason Jar Test and the Biological Fluid Test at 1 psi did not clearly differentiate the regular gowns from the impervious gowns and therefore, are not recommended. Synthetic blood should be used as the primary challenge liquid in penetration testing because it is easy and inexpensive to prepare, it generates results similar to real bovine blood, and its penetration through a fabric is easy to detect visually.

Surgical personnel would like to be assured that the fabric in their surgical gowns will pass all of the liquid challenges typically found in the surgical environment. However, it is very costly for industry to discard an entire lot of fabric because one specimen fails a liquid barrier test. To date, there is no agreement concerning the number of samples that need to be tested and the percentage of failures that are acceptable for fabric used in surgical textiles. Ideally, for the protection of health care workers, the failure rate should be 0%, but this may be an unreasonable expectation due to quality control limitations in manufacturing and handling, and due to experimental variance in testing. Producers and users of surgical gowns need to address this issue.

RECOMMENDATIONS FOR FURTHER STUDY

Methods for sealing the metal surfaces of the test devices against nonwoven fabrics should be explored because leaks may be confused with penetration and recorded as a test failure. Test methods which quantify the liquid penetration rather than detect it with visual inspection should be developed and compared with others.

The scope of this study did not include bacteriological challenges or the use of multiple challenges (e.g., prewetting the fabric with sweat and then exposing it to a liquid challenge or a series of liquid challenges). However, these parameters are important for the safety of medical personnel and should be investigated.

The gown fabrics in this study need to be compared to other nonwoven gowns available on the market and with prototypes that will soon be available.

Besides the traditional cotton/polyester muslin gown, other reusable gowns should be tested because the demand for reusable gowns is predicted to increase as employers are required to provide protective clothing for their workers (2). Residues and abrasion from the laundering process are reported to compromise the liquid barrier properties of the gown fabric over time and should be investigated.

REFERENCES

1 Association of Operating Room Nurses (1988) AORN Journal 47(2):572-576

2 Sharbaugh RJ (1990) Idea '90 Proceedings of the International Nonwovens Conference and Exposition Conference, Washington DC, 233-237

3 American Association of Textile Chemists and Colorists (1988) AATCC technical manual, Research Triangle Park, NC: AATCC

4 American Society for Testing and Materials (1990) Annual book of ASTM standards, Philadelphia, PA: ASTM

5 Association of the Nonwovens Fabric Industry (INDA) (1982) INDA standards, New York: INDA

6 McCullough EA, Jones BW, and Tamura T (1989) ASHRAE Transactions 95(2):316-328

7 American Society of Heating, Refrigerating, and Air-Conditioning Engineers (1987) HVAC systems and applications 23.1-23.12, Atlanta: ASHRAE

8 Schoenberger LK and McCullough EA (1990) Liquid barrier and thermal comfort properties of surgical gowns, IER Report #90-07, Manhattan, KS: Kansas State University

Fluctuations and Beauty

Toshimitsu MUSHA

Department of Applied Electronics, Tokyo Institute of Technology
Nagatsuta, Midoriku, Yokohama, JAPAN 227

Abstract: Acoustic frequency fluctuations of sound of music in general have the so-called $1/f$ spectrum in which power spectrum density is inversely proportional to Fourier frequency f. It was for a long time a mystery, but recently we found that this nature was closely related to physiological phenomena. Rhythmical phenomena in biology such as the heartbeat, spontaneous discharges of a neuron, clapping, the alpha rhythm observed on scalp potentials, and intervals of action potential impulses propagating on a nerve axon, have also $1/f$ spectrum. It is conjectured that music is unconsciously composed such that it simulates basic biological rhythm fluctuations. This coincidence suggests that external stimulations which obey $1/f$ fluctuations in time or in space evoke comfortable sensations.

Many people share common standards of evaluations of fine art and music regardless of the fact that they live(d) in different eras, have different cultural backgrounds and religions, etc. The only possible properties they can share in common would be physiological functions. Therefore, it is likely that the sensation caused when one looks at beautiful things must be deeply related to some physiological phenomena. Based on this principle, we have generated nice musical sounds, a beautiful 3-dimensional surface, color patterns and textures from mathematically generated $1/f$ number sequences and 2-dimensional $1/f$ number arrays.

1. Introduction

When one hears a certain piece of music for the first time, one understands in most cases that it is music and not noise. Why is it possible? Both musical sounds and noisy sounds are nothing but acoustic vibrations, so that there must be some common nature in various acoustic vibrations of musical sounds which characterizes musical acoustic vibrations as it is.

On playing music one reads a score, in which musical notes indicate acoustic frequencies. Therefore, characteristics of musical sounds are considered to be in the time evolution of acoustic frequency. Voss presented a paper[1] at the First International Symposium on $1/f$ Fluctuations, which was organized by the present author in Tokyo in 1977, that acoustic frequency fluctuations of musical sounds have in general a power spectral density which is approximately in inverse pro-

Y. Imanishi (Ed.)
Progress in Pacific Polymer Science 2
© Springer-Verlag Berlin Heidelberg 1992

portion to Fourier frequency f, the so-called $1/f$ spectrum. We also tried the same analysis about a variety of music. Except for certain modern musical pieces, his finding was correct. This nature of musical sounds had been a mystery for a long time until we found the reason for it through a series of research on biological rhythm fluctuations.

The results of our analysis of musical sounds are shown in Figs.1a and 1b. Both show basically $1/f$ spectrum with some spikes at particular Fourier frequencies corresponding to *vibrato* of violins and *tempo* of the music. The present author defines musical sounds in the following way.

"Acoustic vibrations should be said to be musical sounds if frequency fluctuations have 1/f spectrum."

The acoustic vibrations of musical sound do not in general have a sine waveform because various musical instruments are involved and higher harmonics of the fundamental tone are included, and hence the instantaneous frequency is defined as half the number of zero-crossing of the sound waveform per second.

When walking in the forest we sometimes hear a murmuring of a water stream, and it relaxes our mental stress. It is sometimes said that a murmuring of a water stream in the forest is music played by *Nature*. Is it really music in an exact sense of meaning? A murmuring was recorded on magnetic tape and was spectrum analyzed, and it was found that it really belonged to musical sounds according to our definition.

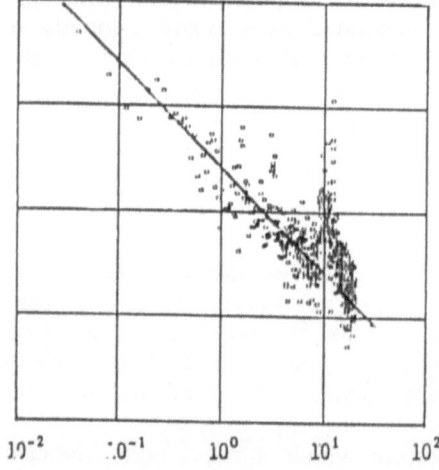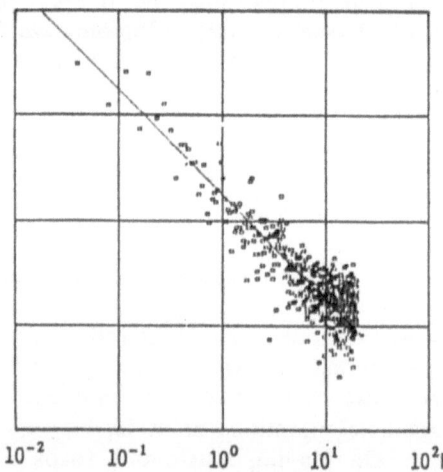

Fig.1a Power spectral density of frequency fluctuations in Vivaldi's "Four Seasons" *Spring*.

Fig.1b Power spectral density of frequency fluctuations in Bethoven's Symphony "Pastral" the 1st *movement*.

There are an infinitely large variety of acoustic frequency sequences or melodies which satisfy the definition of musical sound. The all possible melodies make up a statistical set or *ensemble*. Existing pieces of music are members of this statistical ensemble, but still there are many, many other possible acoustic waveforms (or melodies) which can be musical sounds. Some of them are waiting for human composers and others will never been composed by human composers because they cannot be played by musical instruments or they are not so attractive. We tried to pick up these members mathematically and played them with musical synthesizers after proper arrangements made by a musician. Some of them became very nice background music. When this music was played with a synthesizer in such a way that the tempo was slightly fluctuated as 1/f, it sounded milder than when it was played in a constant tempo. The 1/f-fluctuated tempo simulates the music played by human players as will be described in §2.

2. Biological Rhythm Fluctuations

For the biological body fluctuations play important roles in maintaining its life, and a variety of biological fluctuations are found to have 1/f fluctuations. These findings will be described below.

2.1 Heart rate

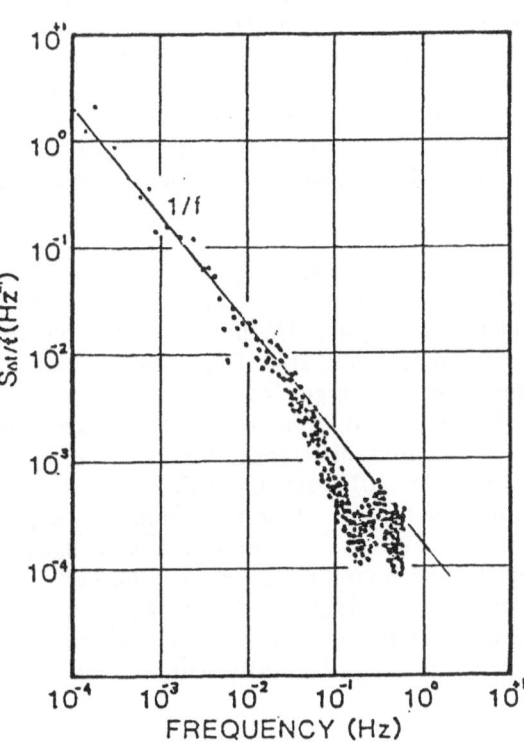

Fig.2 Power spectral density of heart rate fluctuations of a normal subject. An enhancement of the spectral level at 0.3 Hz is caused by respiration.

In a normal quiet subject, the heartbeat looks very regular. When it is measured precisely, however, a slight fluctuation of the period will be found from beat to beat. The fluctuation of the heartbeat period was found[2] to have $1/f$ spectrum for frequency below 0.01 Hz, above which the spectrum was approximately in proportion to $1/f^2$ as is shown in Fig.2. This result was my surprize because $1/f$ fluctuations are ubiquitously observed in electric conductivity in any type of electric conductors, and the physical mechanism is not clear as yet regardless of its history as long as more than 60 years. Moreover, it was also observed that heartbeat period fluctuations of an *embryo* became small, in other words, more regular when its mother got ill, and activity of the heartbeat fluctuation recovered its activity when the mother became healthy. The heartbeat fluctuation is a measure of the vitality of life.

2.2 Spontaneous discharges of a neuron

The heartbeat is triggered by an action potential impulse emitted by a neuron which is located in the *sinus node* on the right atrium of the heart. Therefore, it is likely that the heartbeat fluctuation is attributed to fluctuations of neuronal discharge intervals. This conjecture has been confirmed by the experiment with a giant neuron of the African snail which is as large as 0.2 mm and can be recognized by bare eyes. Snail neurons will have the same basic properties as those in human neurons. A neuron was separated from the snail body and a glass microelectrode was inserted into it to monitor the membrane potential relative to the outside potential. Immediately after the electrode insertion very irregular spontaneous discharge intervals were observed, but about after 20 minutes they became very regular as is shown in Fig.3. A small fluctuation is observed in the interval and its power spectral density is shown[3] in Fig.4.

As is clear in Fig.4, the power spectral density is of $1/f$ type, and the spectral levels of fractional period fluctuations for the human heartbeat and neuronal discharges are in the same order of magnitude. it then follows that $1/f$ heartbeat period fluctuations are mainly attributed to neuronal discharge fluctuations. A new problem arises: "Why is the neuronal discharge interval subject to $1/f$ fluctuations?" "Is it related to a living state of a neuron?" Of course a dead biological cell will not show any potential difference between the inside and outside of the biological cell. Recently similar spiky behavior of the membrane potential was found with an artificial membrane separating two ionic solutions with different ionic concentrations, and discharge interval fluctuations of this artificial membrane also showed $1/f$ power spectrum density.[4] Furthermore, it was shown by computer simulation that if the membrane conductance for ionic current passage has $1/f$ fluctuations, the discharge interval will subject to $1/f$ fluctuations.[5] Therefore, $1/f$

heartbeat fluctuation may be a manifestation of general $1/f$ conductance fluctuations of electric conductors and has nothing to do with the living state.

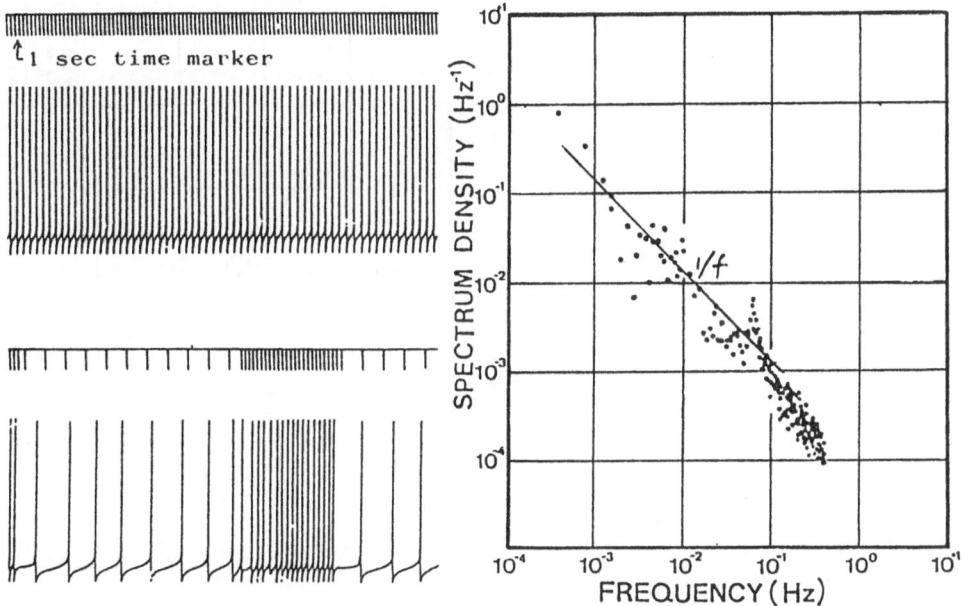

Fig.3 Membrane potential of a giant neuron of African snail.

Fig.4 Power spectral density of discharge intervals as shown in Fig.3.

2.3 Clapping

Typical rhythmical body movement is clapping. Clapping can be made surprisingly regular at a rate from 2 to 3 claps per second. When one tries it no body could notice irregularity of clapping intervals. Why can one make such a regular rhythmical motion? Where in the body do we have a biological clock? The clapping, however, cannot be so regular as a quartz oscillator, and a small amount of fluctuations were detected by putting a microswitch on a palm which was tapped by the other hand. The interval between on's of the switch was measured successively by a counter. This experiment was performed in two different modes. In one mode, a subject tried to synchronize his clapping with an electric metronome, and in the other mode he listened to metronome ticking for several seconds and the metronome was switched off while the subject continued tapping. The records in these two modes are plotted in Fig.5. It is noted that fluctuation amplitudes were almost the same for these two modes and no drift was observed in the free tapping over 700 taps. Our body has a time keeping function. Taps as many as 3000 times were performed and recorded ten times with breaks, and the power spectral densities were evaluated in these two modes, the result being shown in Figs.5 and 6; in the latter five subjects were investigated.

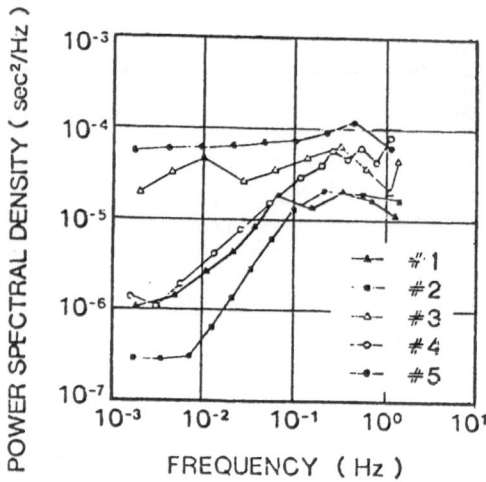

Fig.5 Power spectral densities of metronome clapping intervals in five subjects.

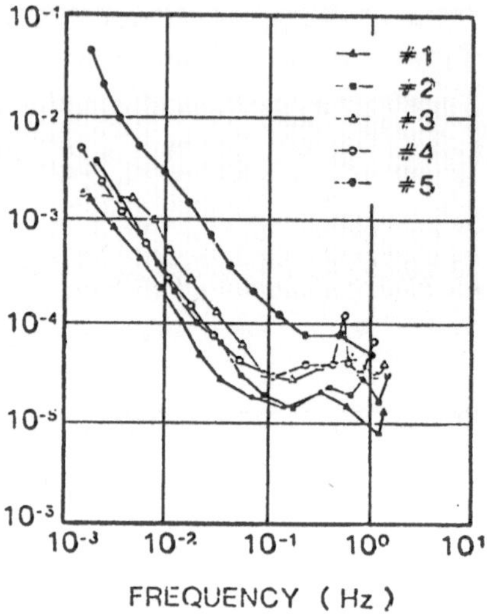

Fig.6 Power spectral densities of free clapping intervals in five subjects.

Judging from these two different power spectrum densities metronome tapping interval fluctuations occurred randomly and there was no correlation from tap to tap. The biological clock did not work in this case. However, when a subject made free tapping without a metronome, the basic biological rhythm fluctuations are manifested in the clapping interval. Although the spectral level differs from subject to subject, the spectral shapes are all the same.[6] The spectral shape refers to the control mechanism and the spectral level refers probably to sensitivity of each subject.

2.4 Others

When eyes are closed the sine-wave-shaped large-amplitude alpha rhythm appears in the scalp potentials, of which frequency fluctuates within a frequency range 8-13 Hz, and this almost periodic waveform is clearly seen in the spontaneous electroencephalogram. The instantaneous frequency is defined from the interval between consecutive zero-crossing points. Its power spectral density is found to be of $1/f$ in a certain range of the Fourier frequency. It is another example of $1/f$ fluctuations of biological rhythm phenomena. When a subject is in a noisy environment this Fourier frequency range is reduced. Therefore, it seems that mental tension reduces $1/f$ fluctuations.

Action potential impulses propagate down a nerve axon. The axon shows a strong nonlinearity as a transmission line, and hence the conduction speed of an action potential impulse depends on time distances from the foregoing impulses. Therefore, the time relation of impulses in an input impulse train in general is modified during the propagation. We have investigated this property with giant axons of squid. The giant axon is as thick as 0.8 mm or so, and it was laid in a chamber filled with sea water. Impulses were launched randomly at one end of the axon. The propagation of impulses was monitored with electrodes laid underneath the axon all in parallel to one another. It was found that the time relation of action potential impulses was modified after launching and immediately reached a stable state. The modulation of impulse intervals in a steady state showed $1/f$ spectrum.[7]

It is concluded that $1/f$ fluctuations are very well adapted to our body, namely in rhythm fluctuations as well as in transmission property of the biological signal. It is probably not limited to human body but also in biological systems in general.

3. Patterns of Beauty

We now know that frequency fluctuations of musical sounds in general have $1/f$ spectrum. On the other hand, it is found that biological rhythm is in general subject to $1/f$ fluctuations. This coincidence will not be accidental. Music is composed such that its sound (fluctuations in the acoustic frequency and audio power) makes people feel comfortable and relaxed, and as a result musical sounds fluctuations automatically have the same property as biological rhythm fluctuations. Therefore, it would be natural to assume that music is composed such that it simulates the biological rhythm fluctuations, probably because external stimulations with $1/f$ fluctuations evoke nice feeling within our body. As $1/f$ biological fluctuations are not limited to human bodies but observed in biological bodies, musical sounds will give positive effects to animals. We often hear that musical sounds accelerate production of milk by cows, growth of vegetable, and so on. These effects are understandable from the present findings.

3.1 Grain of woods

After finding that the heartbeat period is subject to $1/f$ fluctuations, the present author drew parallel lines on a sheet of paper such that their space intervals are equal to heartbeat time intervals, and realized that it looked very much like a wood-grain pattern. Therefore, the spectrum of wood-grain pattern was investigated in terms of spatial frequency. The result is shown in Figs.7a and b corresponding to

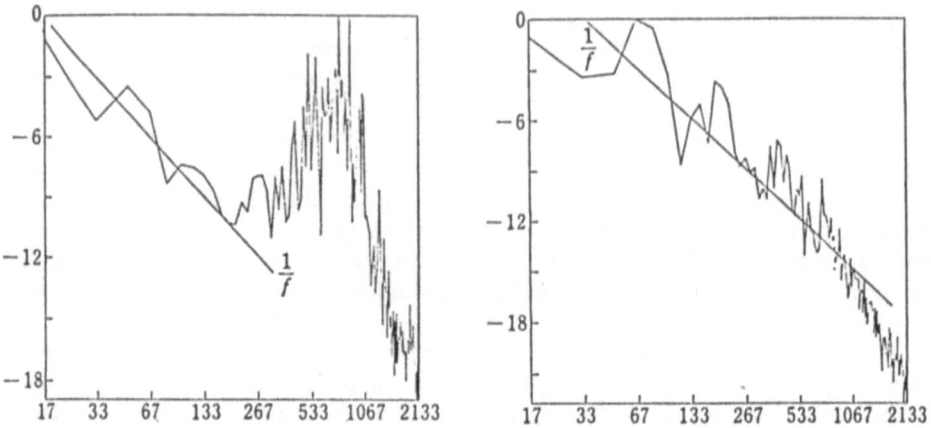

Fig.7 Power spectral densities of straight (left) and cross (right) grain of redwood. The vertical and horizontal axess indicate spectral density in dB and spatial frequency in lines per meter.

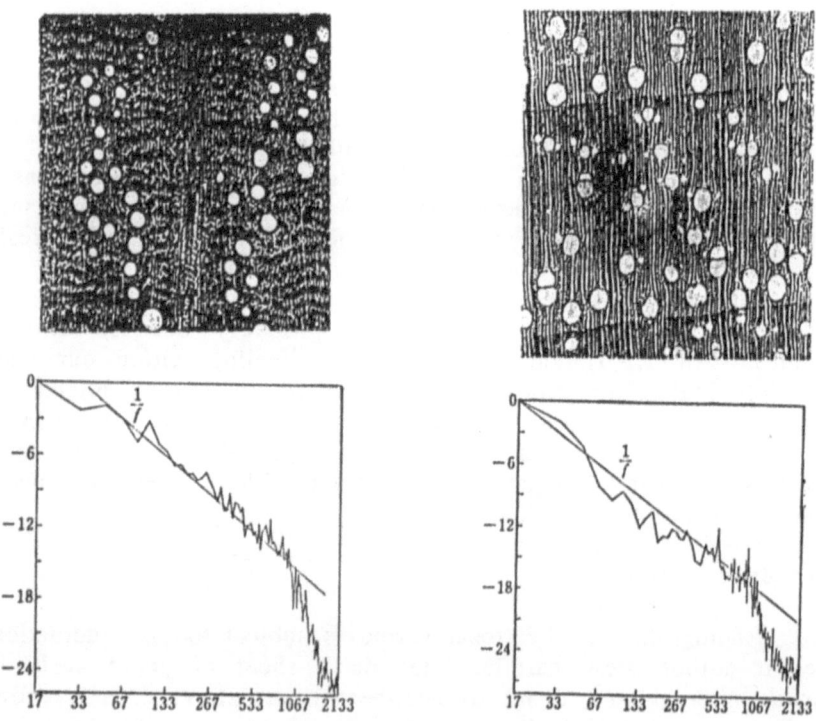

Fig.8 Microscopic texture of oak (left) and camphor tree (right). The corresponding power spectral densities of these spatial pattern along the horizontal axis are show below.

straight and cross grains of woods. An enhancement of the spectral level at spatial frequency near 500-1000 lines/m in Fig.7*a* refers to the mean interval of grain stripes, which is 1/500 m (= 2 mm) to 1/1000 m (= 1 mm). As was expected the spectrum in either case is very close to 1/*f* type. Moreover, Fig.8 shows the spectrum of microscopic texture pattern which is observed on a cross section of wood, and again it shows 1/*f* spectrum.

Some people like to see wood-grain patterns on interior surface of their houses, and they love unpainted wood surfaces which exhibit beautiful wood-grain patterns with spatial 1/*f* spectrum. The Japanese traditional tea house which is a manifestation of beauty of simplicity makes use of natural curved, distorted wood as pillars and beams which no doubt give us mental tranquility.

3.2　Wall paper

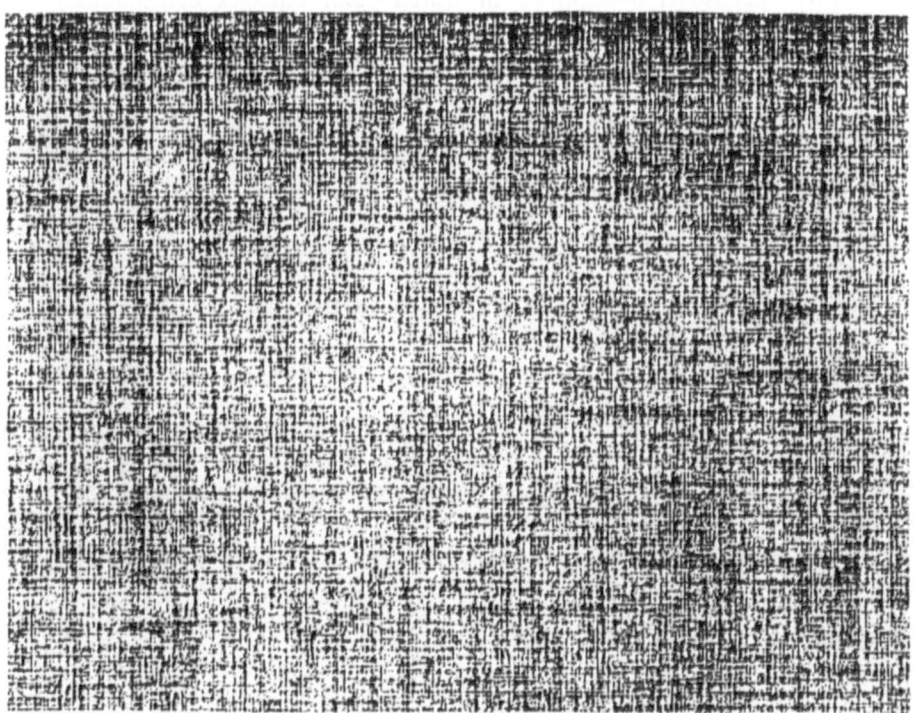

Fig.9　2-dimensional dot pattern where the dot density modulation in the horizontal and vertical directions has 1/*f* spectrum in the spatial frequency.

Special type of the wall paper is sold very well and a salesman brought this pattern to me. This is a very simple pattern. Threads running lengthwise and crosswise have non-uniform thickness, giving some taste to a pattern on the paper. The way of non-uniformity generates comfortable sensations in people who see it. It seemed to me that the modulation is very similar to $1/f$ fluctuations. Based on this idea a similar pattern was generated by means of 2-dimensional $1/f$ pattern in the following way.

1)A rectangular plane is divided into small square pixels.

2)Two-dimensional $1/f$ array of positive integers n_{jk} is generated mathematically, whose variation along the x and y axes have $1/f$ spectrum.

3)Dots as many as n_{jk} are printed randomly in pixel (j,k).

Now a 2-dimensional dot density pattern has $1/f$ spectrum in the horizontal as well as vertical direction. The pattern generated in this way is shown in Fig.9.

In a similar fashion, $1/f$ colored patterns can be generated in the following way.

1) A color palette is determined in which color changes continuously in any sense.

2) The palette is divided into, for instance, 16 sections or 16 discrete colors.

3) These discretized colors are numbered from 0 to 15.

4) A 2-dimensional matrix of positive integers, which has $1/f$ spectrum in the x and y directions, are reformed into matrix of integers between 0 and 15.

5) Numbered colors are allotted to the corresponding pixels to finish a color pattern.

(As the color pattern cannot be reproduced in the paper is it not shown.)

4. Corrugated surface

Instead of colors, height can also be given to the pixels to generate a forest of vertical bars. Now let these bars of different lengths be covered with a continuous surface by means of a proper interpolation. An example of such $1/f$ surfaces is shown in Fig. 10. This 3-dimensional shape was made with plaster and it gives nice feeling when touched with hand.

5 Beauty is in Ordered Disorder

Straight lines, circles, rectangles, and such geometrical patterns do not exist in *Nature*. In designing houses people use geometrical lines such as rectangles, straight lines, circles, and so on, mainly because materials in such shapes are easily mass produced. Such geometrical patterns have a certain kind of beauty, in other words, beauty of artificial kind. Regularity which is very often the case in planning new cities seems to generate this kind of beauty. Now that we know what kind of disorder leads to beauty, we have to re-examined our basic attitude to designing our environments. What we call 'ordered disorder' shape does not necessary require more cost than mass produced design.

Fig.10   A corrugated surface in which cross-sectional curves in the lateral and longitudinal directions have $1/f$ fluctuations.

[1] R. F. Voss (1977) Proc. Symp. on $1/f$ Fluctuations 199.

[2] M. Kobayashi and T. Musha (1982) IEEE Trans. on Biomed, Eng. BME-29: 456.

[3] T. Musha, K. Inoue and H. Takeuchi (1983) IEEE Trans. on Biomed. Eng. BME-30: 194.

[4] H. Akabane and T. Musha (1990) Jpn J. Appl. Phys. 29: L1866.

[5] M. Sano, H. Nakauchi, H. Akabane and T. Musha Jpn. J. Appl. Phys. 29: 2186.

1. R. Voss

[6] T. Musha, K. Katsurai and Y. Teramachi (1985) IEEE Trans. on Biomed. Eng. BME-32: 578.

[7] T. Musha, Y. Kosugi, G. Matsumoto and M. Suzuki (1981) IEEE Trans. Biomed. Eng. BME-28: 616.

Polymer Materials Which Appeal to Kansei

Miyoshi Okamoto

The Okamoto Research Laboratory, TORAY Industry Inc.,
3-2-2 Sonoyama, Ohtsu, 520 Japan

Out of the total sum of 15 billion dollars of annual textile imports, approximately 7 billion dollars are imported from Italy and France. Most of those textiles are of a good sense, and thus called as *KANSEI Sho-hin*, the sensible goods. Sensibility -*KANSEI*- can be defined in various ways as, for example:

(1) Emotional capacity to respond to sensory stimuli.
(2) The biggest part of human nature yet unpolished even in this day.
(3) An expression to summarize the incomprehensibility of consumers' taste.
(4) A key concept to understand a future trend of commercial products.
(5) An antonym to *RISEI* (reason, i.e., an intellectual power to evaluate physical factors).
(6) A sense common to the sense of the times and life feeling.
(7) Sensitivity to the significance of information.
(8) An attractive condition in contrast to a necessary condition.
(9) **An aesthetic sense to external appearance** and comfortableness (in the case of clothes).
(10) Characteristics representing a symbolic value rather than a functional value.
(11) A natural aptitude or a good taste for finding an aesthetic value.
(12) Mental and physical agreeableness.
(13) A word which is employed to distinguish one's perspective view.
(14) A capability to perceive the emotional change caused interactively.
(15) An emotional wear in a sense that reason is a logical wear.
(16) A movement of human mind (the sense of memory).

It may be critical in the following discussion to define explicitly what sensibility does imply. Sensibility is inherent to a human being which is charaterized by its diversity resulted from environments, cultures, religions, laws, customs, etc., and thus is concerned with a symbolic value of individuals. However, it may include many other functions behind our consciousness.

We evaluate textiles in terms not only of the physical factors (*RISEI*) such as tensile strength and durability, but also of the mental factors (*KANSEI*) whether we like or not without reason. More recently, there appeared other factors to lead the fashion, which cannot be classified in two categories mentioned above. Those factors, which include exclusiveness, specialization, distinctiveness, sophistication, etc., represent the change of mentality by human intelligence. Those factors can be summarized by the term of *GOSEI*, which literally means wisdom but may be defined as the intellectual capacity to create a unified concept upon the sensory experience received through *KANSEI*. *GOSEI* thus binds *KANSEI* with intelligence, and acts as a communicator between a consumer who buys a product and a particular man or group who designs or shares it. *GOSEI* is able to create a new sense of value, so that a product should attract *GOSEI* as well as *KANSEI* (sensibility) and *RISEI* (reason) to have a good selling prospect.

Y. Imanishi (Ed.)
Progress in Pacific Polymer Science 2
© Springer-Verlag Berlin Heidelberg 1992

336

Let review some products developed in Toray from the view point of *KANSEI* and *GOSEI*.

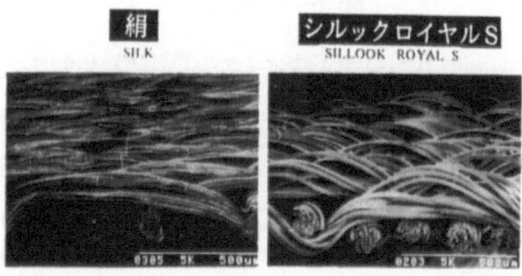

Fig. 1: Comparison of silk and Sillook Royal S™ texture

Fig. 1 compares the texture of natural silk and Sillook Royal S™ made of polyester. Silk is distinguished by its characteristic luster, vivid colouring, puffiness, draping and silk-scrooping. Sillook Royal S is silk-like not only from its hand, but also from a visual and auditory standpoint as shown in Fig. 2. It has a three-petal shaped cross-section with micro-slits of the order of a few μ's at each pedal top. Fig. 3 is a SEM photograph of the cross-sectional view of Sillook Royal S filaments, which are characterized by the three-petal shape and micro-slits. These micro-slits absorb the reflected light, and provide the vivid deep colour and elegant anisotropic luster simultaneously. The micro-slits effectively prevent the worn-out shaping of the cloth, and introduce the elegant silk-scrooping and the pleasant cloth-rustling sound when two edges of a micro-slit are touched and rubbed.

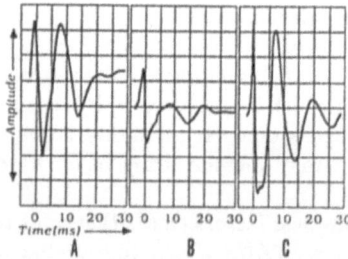

Fig. 2: Wave forms of scrooping sound of *DONSU*. A; Fabric made of side slit fibers, B; Fabric made of conventional silk-like fibers, and C; Fabric made of real silk.

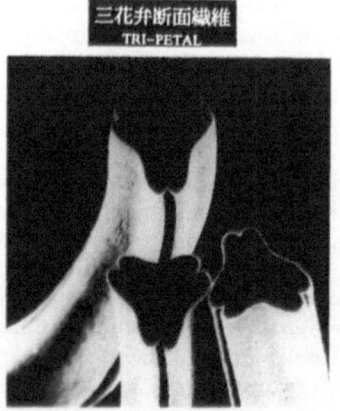

Fig. 3: Three-petal shaped cross-section with micro-slits

Fig. 4: New designs by Sillook Royal S™

Fig. 5: New designs by Sillook Royal S™

Fig. 6: New designs by Sillook Royal S™

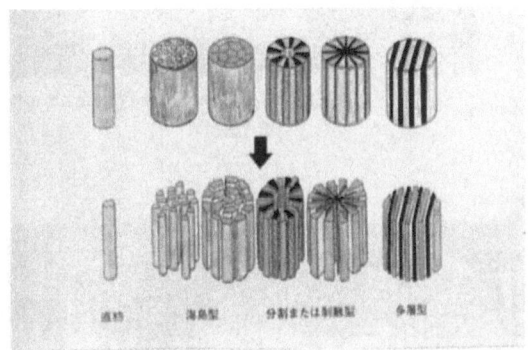

Fig. 7: Ultra-fine fiber technology by composite spinning

Silk-like polyester was first developed by mimicing natural silk. Silk-like triangular cross-section affords the characteristic luster and crispness to polyester. Silk fineness and puffiness was achieved by developing the ultra-fine fiber spinning technology and blending fibers of different shrinkage, respectively. The change of fiber surface structure yields silk-scrooping. Now silk-like polyester surpasses natural silk in some extent as demonstrated by recent fashions (see Figs. 4, 5 & 6).

The development went further to produce much finer fibers composed of monofilaments less than 0.7 denier. Here ultra-fiber technology was developed on the basis of various composite spinning methods as illustrated in Figs. 7 & 8.

The Shingosen (ultra-fine polyester fiber) project was first started by mimicing silk, but went further to surpass silk by innovating extremely fine filament (see Fig. 9), resulting a completely new material which opened a possibility of creating a new genre of expression in the fashion world. Shingosen provides a new environment of materials where no natural fiber can satisfy fashion designers. At first, the R&D was directed to improve the physical properties of polyester fiber. Silk was always an ultimate goal for synthetic fibers, and in the next step polyester fiber with a triangular cross-section was produced by mimicing silk. Since the modified cross-section yarn alone was found not sufficient to mimic silk, the fabric structure was modified to resemble silk fabric by caustisizing. The biomimetics came to the end to suggest what should be done next, and thus a special emphasis was examined on a particular characteristic to surpass silk in at least one respect.

Fig. 8: New designs by using a peach-skin type fabric

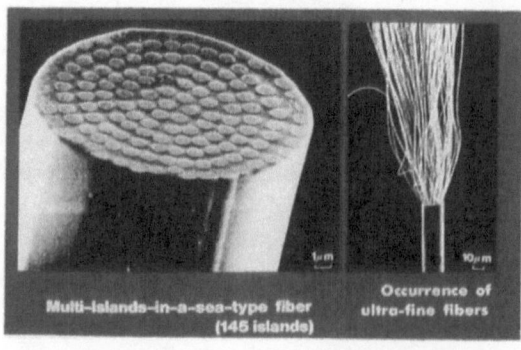

Fig. 9: Ultra-fine fiber technology of multi-islands-in-sea type

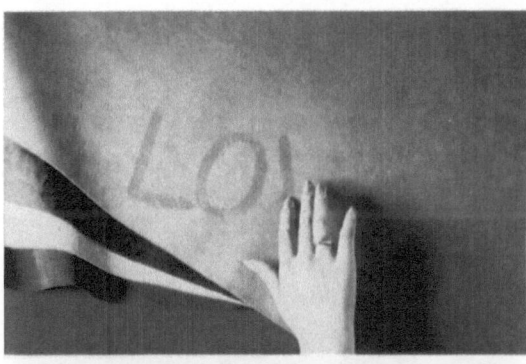

Fig. 10: Ecsaine,™ a non-woven suede-
type artificial leather

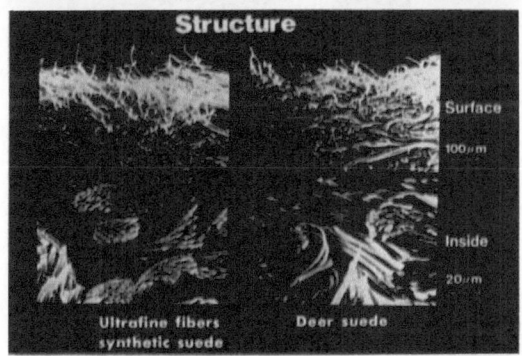

Fig. 11: Structure of ultra-fine fibers
synthetic suede and dear suede

Fig. 12: Ecsaine™ samples with various
colors

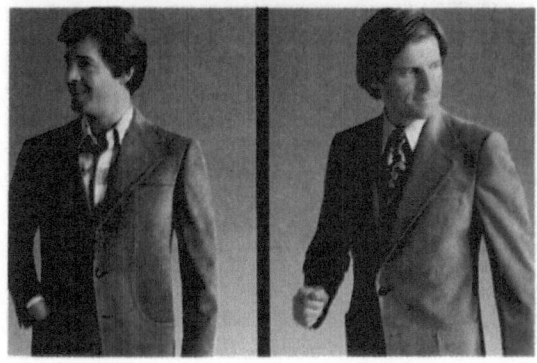

Fig. 13: New designs by Ecsaine™

340

Fig. 14: New designs by Ecsaine™

Fig. 15: New designs by Ecsaine™

Fig. 16: All weather golf gloves by Ecsaine™

Fig. 17: Interiors by Ecsaine™

The micro-denier project resulted in a non-woven suede-type artificial leather; Ecsaine™, (Alcantara™ in Europe or Ultrasuede™ in the U.S.A.), and a wiping cloth; Traysee™. Natural leather is composed of high-density fabric made of extremely fine fibers. Ecsaine™ (see Fig. 10) is a raised suede type of artificial leather, resembling natural leather from its structural standpoint. Fig. 11 compares the structure of Ecsaine™ and dear suede, where Exsaine™ exhibits much finer structure to yield more flexibility. Thus Ecsaine™ possesses the characteristics of the moisture vapour transmission and hand feel of natural leather. Furthermore, Ecsaine™ finds much wider application in the clothing and interior field, distinguished by a variety of colours (Fig. 12), washability and flexibility. The examples of its application are shown in Figs. 13, 14, 15, 16, 17 & 18.

Fig. 18: New goods by Ecsaine™

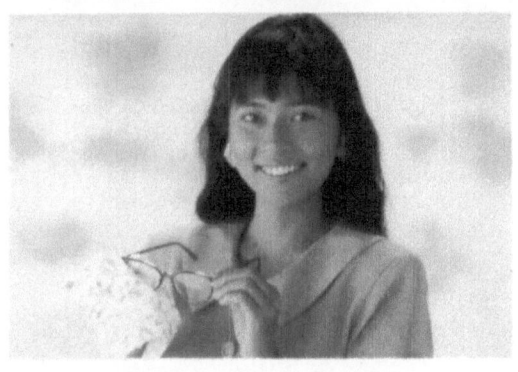

Fig. 19: Toraysee™, a high-functional
wiping cloth

Toraysee™ (Fig. 19) is on the other hand a high density woven fabric with ultra-fine fibers, which possesses pores containing micro-pockets (see Fig. 20). These micro-pockets absorb dust or oil, and functions as a highly-efficient wiping cloth. Its high-density surface is also suited for printing. This particular example of Fig. 21 shows the Tokyo map printed on Toraysee, and you will see how fine it is. This printing capability may add an intellectual pleasure to the physical function.

Another example shows the application of thermochroism to enjoy the change of clothes colour tone according to environmental conditions as demonstrated by swimming costumes and ski wears made of Sway™ (Figs. 22 & 23). Sway™ is the thermochromic clothing material. Here the microcapsules containing the dyestuff (acting as a donor), the chromophore agent (acting as an electron accepter) and the achromatizing agent, are dispersed homogeneously over the basic fabric, and coated with polyurethane resin (see Fig. 24). The colour changes thermoreversibly, and gives a pleasant surprise (Fig. 25).

Fig. 20: Comparison of the fabric surfaces of Toraysee™ and nylon taffeta

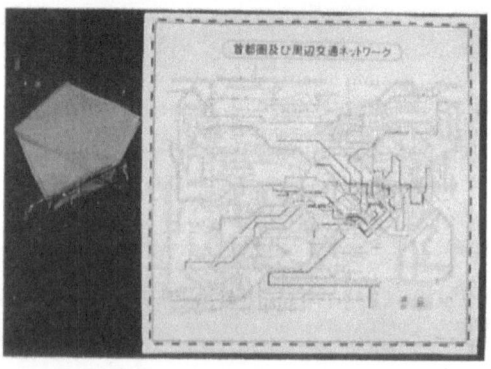

Fig. 21: Tokyo map printed on wiping cloth made of Toraysee™

Fig. 22: Swimming costumes of Sway,™ changing color tones as temperature decreases from left to right.

Fig. 23: Ski wears by Sway,™ changing color tones as temperature decreases from left to right.

As shown in these examples, people requires not only physical functions, but also some other factors which are not necessarily important from the practical point of view. Today, flexibility, imagination, comfortableness and quick response are four major factors important in everyday life, and those factors may be incorporated in *KANSEI* or *GOSEI*. *KANSEI* and *GOSEI* are difficult subjects to deal with as science. Here we are forced to modify our view on science. For example, *fuzzy* and *neuro* are two words often found in the advertisement of commercial goods. Both *fuzzy* and *neuro* belonged to the *KANSEI* field, but now have an explicit definition in the field of science. Although they do not understand the science of *fuzzy* or *neuro*, these words attract consumers since the words fill the gap between *RISEI* and *KANSEI*. The concept of *fractals*, *chaos* and *1/f fluctuation* can be applied to the polymer technology including fiber, paint, plastics, etc. to open a new prospect. The plastic surface with a fractal structure affords more naturalness. Printed wooden or marble pattern is an application of fractals. The 1/f fluctuation theory has been applied to translate the paintings into music.

In conclusion, an individual *KANSEI* can be developed and transformed in a suitable form of *GOSEI* to be applied in the field of technology. *RISEI* is a pre-requisite in the development of polymer materials. However, the *KANSEI* and *GOSEI* factors will play a vital role in technology to create a real richness as indicated in Fig. 26.

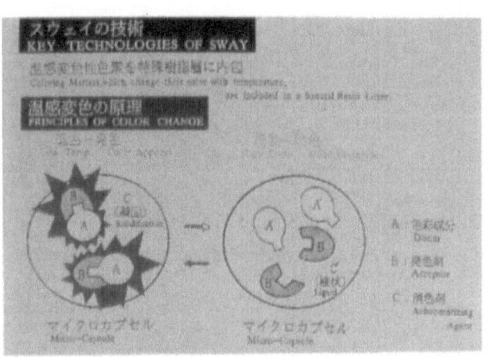

Fig. 24: Thermochromic mechanism of Sway™

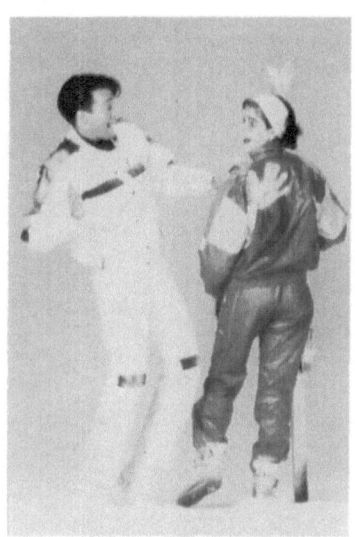

Fig. 25: Hand print on the Sway™ ski wear. Palm warmness causes a thermochromic change.

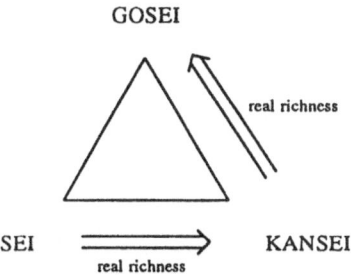

Fig. 26: Schematic relation between *KAN-SEI* and *GOSEI*

Learning Through Nature for the Creation of New Textile Fabrics

Shinji Yamaguchi

Fiber Research & Development Group 1,
Technical Research Center,
Kuraray Co., Ltd.
Sakazu, Kurashiki, 710, Japan

1. Preface

In 1950's and 60's when natural fiber materials were exhausted and synthetic polymers ,invented to provide the fibers as substitute materials, had been adapted to clothings, durable and easy-care materials were essential factors for that purpose.

It was presumed that there lay tremendous distance to reach our targets by simply studying primary structure of polymer chains in order to imitate hand and appearances of natural fibers and textiles selected through our lengthy history.

In the 1970's in this context,the extensive researches on natural fibers, in particular, silk, wool,its yarns, and textiles, have made remarkable progresses: further reseaches on properties of fiber assemblies and analyses of their hands have made a considerable progress,[1] and then these seriese of researches have led to silk-like and wool-like modifications.

Particularly polyester easily given modifications attracted wide attentions as a foundamental material to imitate natural fibers. Refferring to respective starting points of imitations, a silk-like fabric was derived from the filaments of trilobal cross sections molded after those of silk yarns and an alkaline weight reduction treatment after the process to avoid sericins of silk yarn; and wool-like fabric also from the crimps given through a texturizing process modeled after spiral crimps of wool fibers. Furthermore,it was witnessed that some of shape factors be under controls during polymer component or in process of spining and drawing, it was made posible to obtain the shapes which any conventional natural fibers could have hardly offered. The fibers of which shape factors are controlled on a micron order exceeds a level of imitation, and havebeen recognized as innovative materials from a view point of a function and sensitiveness. This is why it is today called Shin-Gosen.

2. Beauty of The Silhouette

The beauty of a silk Kimono among Japanese traditional wears is not limited to its colors and patterns. Taking an example of a silhouette of a retreating figure of a walking women in Kimono,her delicate movements of body lines could be obviously senseed. One of these factors stems from an easiness to move when the fabric transform in bias (shearing deformation), and from the nature of easily retaining its born shape because the silk fabric has little frictional resistance (a small histeresis loss). As to the born properties of silk, fibroinis covered with sericin in a stage of raw silk, and then sericin is removed through a scouring process and fibroin fiber in a shape of trilobal cross section remains as it is.

In the above process, fiber-to-fiber spaces grew among fibers,and the silk fabric turned into the fabric which moved easily and was rich in bounce.

As a result of born nature of silk , silky synthetic fabrics have been developed on its full scale. As shown in Table-1, continuous pursuits of silky synthetic fibers have been carried out without satisfactions with past achievements. Polyester fibers are featured with trilobal cross sections,and their surfaces were peeled off with an alkaline weight reduction by which its mutual frictional resistance amang fibers were minimized just like silk yarns.

Y. Imanishi (Ed.)
Progress in Pacific Polymer Science 2
© Springer-Verlag Berlin Heidelberg 1992

Table-1: The History of Development of Silky Synthetic Fibers

```
      1965              1975              1985
      |                 |                 |
   d  | 1st Generation
   e  |   A single yarn
   v  |   Fine deniers
   e  |   Different shape of yarn cross sections
   l  |   Up-grade by weight reduction treatment
   o  | ----------------  ----------------------------- → S T R A I G H T
   p  |
   m  | 2nd Generation
   e  |   Fibers of mixture of different shrinkages
   n  |   Mixtur of differnt shape of yarns
   t  |   Up-grade by high weight reduction treatment
   s  | --------------------  -------------------- → D I F F E R E N C E
      |                                           O F   S H R I N K A G E
      |
      |    3rd Generation
      |      Mixture of different deniers
      |      Looos by the air mingle and interlaces
      |      Fiber-minuteness
      |      ------------------------  ------------- → M I N U T E N E S S
      |
      |       The New Generation
      |         Silk waves
      |         Multi-ply of yarn structures
      |         Spontaneous extention by crystallization
      |         Natural disorders
      |         -------------  ------ → I R R E G U L A R
      ↓                              S T R U C T U R E
```

To cover an insufficient bulkiness, taking place together with a decrease of yarn stiffness in an alkaline weight reduction process, intermingled yarns with different shrinkage ratios among yarns to which are given the bulkiness by giving the difference of shrinkage ratio among yarns, and the yarns spliced with loops and slacks by an air-jet method have been developed: the second and third generation have been developed with these types of yarns.

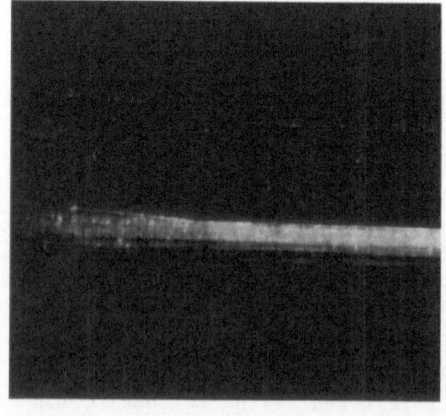 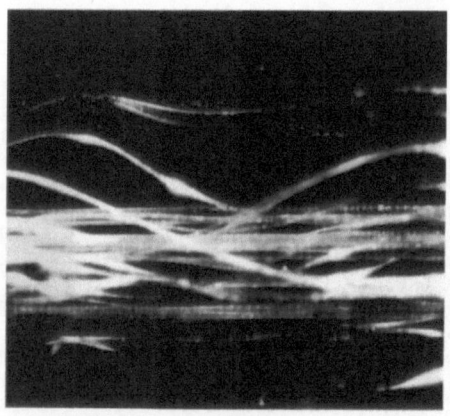

before shrinkage   after shrinkage
Fig.1 Fibers of Mixture Different Shrinkage

The silk, however, provided numbers of proprietary characteristics worthy
of learning, namely, softness, suppleness and the like which the silk
exclusively offers. For more profound studies, it was required to go back
to the studies on the silk worm. When the silk worm forms a cocoon, it
gives waves to the yarns as it fixes the cocoon by spining yarns out,swing-
ing its mouth like in a shape of figure 8.It was presumed that silk waves
, when to weave into the fabrics, turned resilient element, and did a
significant job to bring forth softness and bulkiness to the fabrics.
Bisides, thin and thick yarns are intermingled in a reeling process to
unwind the cocoon, and consequently it was acknowledged in the course of
time that it was not proved thorough imitation to simply double uniform
and straight regular polyester yarns.
It has been further evident that during a process of scouring and dyeing
silk fabrics, silk yarns swell through absorbing water, and lose their
volume through drying and dehydrating; as a result,the spaces among fibers
grew wider and the fabrics changed to the supple ones and showed up beautiful silhouettes.Adapted these analyses to polyester fabrics, "Shin-gosen"
, as a type of new generation, have been developed with various kinds of
development methods. These models are shown in Fig.2 .

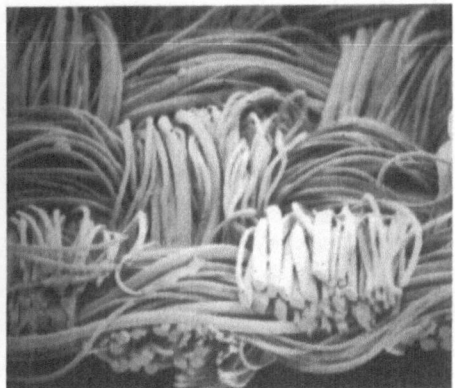

Regular Polyester   "Shin-Gosen" Kuraray Nymhas
Fig.2 Silk Like Fabrics Structure

In the course of pursuing the hands of the silk, the combinations and
fusions of multiplied technologies have been advanced in various fashions.
For instance, one of the technical skills to minimize an internal stress
between warp and weft caused by an effect of a spontaneous extention through crystallization process, the skill to mix micro denier yarns, and the
skill to give a concave and convex in a sub-micron order for a control on
surface touch is turned in. Regarding the cross section of fibers,a lot of
kinds of configurations have been developed other than a trilobal cross
section.
The technologies to imitate the silk woven and knitted fabrics have not
been limited not only to merely match the hands, but to draw out broader
characteristics, and then independently stepped forward. As a result of it
, the textile materials able to express beauties of silhouettes for
feminines, have been created, and they are presumably well known as the
fabrics beyond genuine natural materials.

3. From Touch to Function
There is a word "Numeri-Ramy touch" to represent the smoothness of
woolen fabrics. The touch varies from a condition of fluffs on a fabric
surface to another. The method to measure a feeling of a ramy touch is

carried out by rubbing a fabric with a sensor modeled after finger prints, and is specified with a friction coeffecient and its variation ranges;the more ramy touch the fabric is,the smaller a numerical value is. A mink fur , cashimere fabric, calf suede and such are their most representing examples. The ramy touch of a calf suede and suede touch , representing "a writing effect" brought forth by variation of fluff directions on a leather surface, depend chiefly upon conditions of thier micro fibers.
From a view point of imitating a form, a man-made leather, which has ever been drastically dedicated to that purpose, can be identified as one of master pieces in terms of a success in a conscientious reproduction of fluff conditions of micro fibers.
 Modeled after a structure of collagen fiber bundles in a cow leather, a sea -island is made, and its micro fibers are bundled through extracting its sea component, and then they are made into non-woven fabrics, which ultim- ately turn man-made leathers through following processes.

Fig. 3   Structure of Collagen
           Micro-Fiber Bundles

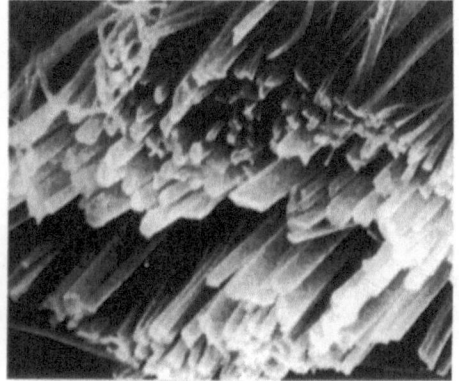

Fig.4 Micro-Fiber Made from Sea-
       Island Fiber for Man-Made Leather

Among the technogies of making micro fibers, there has been developed a couple of methods to directly spin through minute spinnerets, or a sea- island fiber and a partition fiber etc.. The technology of spining micro fibers,initiated from the development of man-made leathers,has been applied to that of woven and knitted fabics in general.
   The touch of fluffs of micro fibers just like downy hairs of the human babies and surface of peaches offers the unprecedented touch,and constitu- ted one of the new fields of Shin- Gosen.
Besides, the high density fabrics of micro fibers through which vapour gets but not does liquid like a water-drop, excellently· works for a demonstration

Fig.5 Partition Type Micro Fiber

of a vapour-permeable and water

-proof effect, and plays significant rolls for uses in a coat,wind breaker ,ski wear requiring specific performances. When micro fibers are used in wiping cloths, they fill their functions to remove dusts as well as oily layers from the surface.

4. Learn through Nature

The most serious subject to the synthetic fiber makers in a highly matured society and keen competitions against natural fibers in pursuit of satisf- actions of their customers,is how sophisticatedly they are able to manifes t 'the sensitiveness' appealing to human senses through colors of merchan- dises(colordevelopment and luster), patterns and designs.   There are a lot of things to learn through natural fibers, namely, some of them are the fields where they belong to as well as ingenious structures of nature; this is quite identical to the world of colors.

  The Morpho Butterfly, the Sun Butterfly in a Japanese name, living in a basin of the River of Amazon enchants people all over the world with its transparent cobalt blue color and beatiful metalic luster.

On the wing surfaces of the butterfly, the slit fins are orderly in row, and their pitches are about 0.7μm long . Their cross sections have stairs like on a ladder and a structure of 9 to 10 stairs resembling a lattice whose pitch is about 0.2μm each. The light reflects refactively through these sections, and colors develop by an occurance of interference just when the light goes through; this color development is further emphasized between slit openings,which charms us with the mysteriously vivied beauty in cobalt blue. The reflection rate performs an important roll to make use of a color development of an interference color. As the reflection rate of regular fibers is low, it is almost impossible to develop deep and vivid colors.When the structure of deep slit opening like wings of the butterfly is formed on the fabric surfaces,it has become possible to produce beauti- ful fabrics similar to velvets, contrasting well with a deep shade and luster by which regular plain fabrics hardly expressed.

Polyester multi-spiral yarn fabrics of specific collocation structures is shown Fig.6 in contrast to color developed sections of the butterfly's scales.   The fabrics with contrasts of deep color shades and lusters emphasize silhouettes of clothings, and are appreciated as the fabric materials of higher grade.

Wing Scale Surface of Morpho Butterfly   Polyester Multi-Spiral Yarn's
                                             Fabrics of Specific Collocation

Fig.6

Tones of dyed fabrics, composed of a color development,color brightness, and color depth are related to optical properties of fibers. The larger

a refraction rate is, the larger reflection rate on the surface is, and therefore dyed fabrics appear light brownish, and the color depth is hard to get.The refraction rate of the fibers is 1.48 to 1.62 . Polyester fiber belongs to the highest level among the fiber materials, and there existed a problem in developing and depending colors,which brought forth some bad influence to dyed fabrics in black and deep colors of a low reflection rate, and accordingly chemical modifications ran up against a brick wall.
 However, the arrangements of the natural world have been utilizing its optical behaviours with skills far beyond human intelligence. One of them is known as "Moth Eye Principle" [2] and the eyes of a moth, flying in the air toward evening, have characteristic concave and convex structure.
Same the the principle, it was discovered that dyed fabrics appear deeper in shade and increase its clarity by coarsening fabric fibers,and increasing its fineness of a concave and convex up to 0.2 to 0.7μm in light wave length order. As a mean of making a fine concave and convex structure of a specific size, firstly ultra-fine particles, which have a similar refraction rate to fibers and an average diameter below 100nm, are uniformly dispersed in polyester fiber without cohesion ; then the fiber surfaces are treated with an etching method using regular alkaline solution.
 The fine concave and convex structure is formed by a difference of solubility against alkali between potyester fibers and the particles uniformly dispersed in polyester fibers. The potyester fiber of fine concave and convex structure, "Micro crater fiber-SN2000" , provides not only a deeper color shade,but an improvement of luster,a dissolution of glittering under direct rays(for a particular use in the artificial hair), a color brightness of printed fabrics,modifications of hand and handle of fabrics.

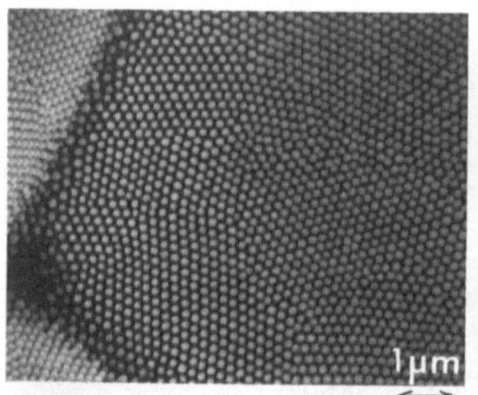

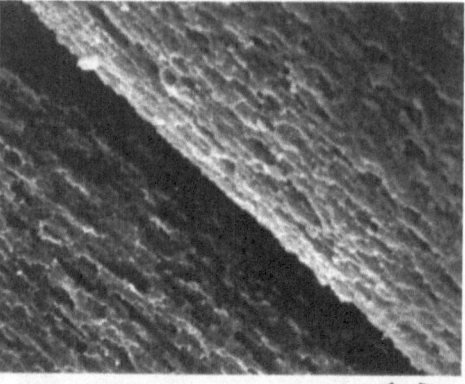

Fig.7 Characteristic Concave and
 Convex Structure of Night-Moth Eye

Fig.8 Micro Crater Polyester Fiber

 As far as black-dyed fabrics are concerned, the blackness of SN-2000 has been manifested its excellent characteristics for a use in the black formal wears, and been modified with coating a resin of low refraction rate ; besides, this blackness has been by far upgraded by a thin film on the surface, nearly 100nm thick, which the plasma polymerization method could just feature;by this method, SN-2000 successfully provided the coal-blackness which any natural fiber has not ever been able to offer, and has outdone a level of natural fibers.

5.Conclusion
 Though we have been continuously learning through Nature and its models, and repeating numbers of imitations,we have not utilized yet its ingeneous and minute structures.

Our sincere hope is that the clothings of higher comfort and sensitiveness to pleasantly excite the human beings will be born from one to another, through analyses of proprietary functions of structures and developments of new innovative materials, based on bio-mimetic accesses to Nature.

6. References
1)S.Kawabata(1980)The Standardization and Analysis of Hand Evaluation
            The textile Machinery Society Japan
2)C.G.Bernhard(1967)Enddeavor, $\underline{26}$, 79

Author Index

Subject Index